CORPORATE SAFETY COMPLIANCE

OSHA, ETHICS, AND THE LAW

Occupational Safety and Health Guide Series

Series Editor

Thomas D. Schneid
Eastern Kentucky University
Richmond, Kentucky

Published Titles

Corporate Safety Compliance: OSHA, Ethics, and the Law
by Thomas D. Schneid

Disaster Management and Preparedness
by Thomas D. Schneid and Larry R. Collins

**Managing Workers' Compensation: A Guide to Injury Reduction
and Effective Claim Management**
by Keith R. Wertz and James J. Bryant

Motor Carrier Safety: A Guide to Regulatory Compliance
by E. Scott Dunlap

Occupational Health Guide to Violence in the Workplace
by Thomas D. Schneid

Physical Hazards of the Workplace
by Larry R. Collins and Thomas D. Schneid

CORPORATE
SAFETY
COMPLIANCE

OSHA, ETHICS, AND THE LAW

THOMAS D. SCHNEID

CRC Press
Taylor & Francis Group
Boca Raton   London   New York

CRC Press is an imprint of the
Taylor & Francis Group, an **informa** business

CRC Press
Taylor & Francis Group
6000 Broken Sound Parkway NW, Suite 300
Boca Raton, FL 33487-2742

© 2008 by Taylor & Francis Group, LLC
CRC Press is an imprint of Taylor & Francis Group, an Informa business

No claim to original U.S. Government works
Printed in the United States of America on acid-free paper
10 9 8 7 6 5 4 3 2

International Standard Book Number-13: 978-1-4200-6647-0 (Hardcover)

Library of Congress Cataloging-in-Publication Data

Schneid, Thomas D.
 Corporate compliance: OSHA, ethics, and the law/ Thomas D. Schneid.
  p. cm. -- (Occupational safety and health guide series)
 Includes bibliographical references and index.
 ISBN 978-1-4200-6647-0 (alk. paper)
 1. Industrial safety--Standards--United States. 2. Industrial safety--Moral and ethical aspects. 3. United States. Occupational Safety and Health Administration.
I. Title. II. Series.

 T55.S3282 2008
 363.110973--dc22   2008011620

Visit the Taylor & Francis Web site at
http://www.taylorandfrancis.com

and the CRC Press Web site at
http://www.crcpress.com

Disclaimer

The author and his students have attempted in every way to ensure that the information contained in this text is accurate and up-to-date; however, laws, regulations, and case decisions are being modified and changing virtually on a daily basis. To this end, it is imperative that safety professionals research and acquire the assistance of local counsel prior to depending solely on the information in this text.

This text was specifically designed to serve as a "starting point" for students and safety professionals to identify the potential sources of the law or regulation that may be applicable to the specific circumstances. Safety professionals and students should be aware that not all areas of potential legal liability have been covered in this text. It is imperative that safety professionals appropriately research the issue at hand and review all matters with local counsel.

The author has attempted to identify the issues that have the greatest frequency within the confines of this text. The author provides no warranty, either expressed or implied, as to the accuracy of the law, standards, regulations, or other information contained in this text. Although suggestions and recommendations are offered, the author does not intend this text to provide specific legal counsel with regard to individual circumstances. Competent legal counsel should always be sought to assist with specific circumstances and situations.

Contents

Foreword

The mere mention of the word OSHA (Occupational Safety and Health Administration) scares many people. OSHA has often been the much-maligned federal agency that takes the brunt of attacks from both management and labor organizations.

Take the time to carefully study this text, an authoritative manual on OSHA, and the specific procedures and standards. One of the best ways to have to never utilize the knowledge in this text is to never have an OSHA case. Realistically, the very best way to avoid the administrative issues here is to operate your facility, organization, and site in the safest possible way.

OSHA standards are *minimums* in the workplace, *not* maximums.

Control your hazards by working to create a safe work environment. Involve your employees, management, and OSHA to create and maintain a safe and healthy working environment.

Do not let safety be forgotten once you get the "new insurance policy" in effect.

Make safety a way of life, from the top down, and from the bottom up.

Make sure that the newest employee has been trained and educated in safety and health procedures.

Merely training employees to, for example, always wear their safety glasses is not nearly as effective as educating them as to what your program means and how it affects them. After making the program meaningful to them, then provide the appropriate training to do the job correctly.

Provide the tools and resources to do the job safely, and you may just use this book as a reference. But, if things do not go that way, follow the expert suggestions as outlined here.

I have worked with Tom Schneid for over 20 years, and I can tell you from first-hand experience that he will help guide you through this maze.

While I served as a disaster safety officer for FEMA (Federal Emergency Management Agency), I often called upon Tom for advice — when I was in the field at the Oklahoma City Bombing in 1995, to the World Trade Center attacks in 2001. This book provides some sage advice.

The best advice is to operate safely, be prepared, and make safety your way of life, whether at home or on the job.

Michael J. Fagel, Ph.D.
Certified Emergency Manager
Aurora, Illinois

Acknowledgments

The author would like to thank his wife, Jani, and his girls, Shelby Lyn, Madison Renee, and Kasi Nicole, for their support and for giving up their time in order that this text could be written during every available evening and almost every weekend. A lot of softball games were missed during the writing of this text.

Special thanks to Sheila Patterson and Tanya Ritter for their assistance in assembling this text and for editing my constant misspellings.

And particular thanks to my graduate students in my Ethics in Loss Prevention and Safety class during the summer session, 2007. These graduate students assisted in researching many of the ethical issues identified in this text.

And last, thanks to my parents, Bob and Rosie, for their efforts and sacrifices to ensure that my brothers and sister, as well as myself, learned the value of education and service to others.

About the Author

Thomas D. Schneid is a tenured professor in the Department of Safety, Security, and Emergency Management (formerly Loss Prevention and Safety) at Eastern Kentucky University and serves as the Graduate Program Director for the online and on-campus Master of Science program in Safety, Security, and Emergency Management.

Schneid earned a B.S. in education, M.S. and CAS in safety, an M.S. in international business, and a Ph.D. in environmental engineering as well as his Juris Doctor (J.D. in law) from West Virginia University and LL.M. (Graduate Law) from the University of San Diego. He is a member of the bar for the U.S. Supreme Court, Sixth Circuit Court of Appeals, and a member of the bars of a number of federal districts as well as the Kentucky and West Virginia state bar associations.

Schneid has authored and coauthored 15 texts and more than 100 articles on various safety, fire, emergency medical service (EMS), disaster management, and legal topics.

1 Introduction

The whole of science is nothing more than a refinement of everyday thinking.

Albert Einstein

Safety, in my biased opinion, is the noblest of professions in this modern industrial world. I base this opinion on the fact that safety professionals are often the lone voice of reason within the management hierarchy, championing the safety of the worker in the arena of productivity that pushes for ever-increasing profits. Safety professionals work long hours, will never achieve the very top of the corporate heap and make middle of the road salaries, but own a compassion and drive to work each day to ensure that each and every worker goes home in the same condition he or she arrived in at the beginning of the day. Although many workers do not realize it, the safety professional is in the trenches each and every day fighting to improve their workplace and the quality of their work environment. This noblest of corporate warriors will lose as many of the hard-fought battles as he or she wins, but each win steadily moves the safety program one step closer to creating or maintaining the safest workplace possible for his or her employees.

Safety professionals work alone, as part of the team and as a leader depending on the circumstances. Safety professionals must often "make do" with what is available and utilize their imagination to create new and innovative things from what is available. In this world where there is substantial pressure to survive and make a buck, safety professionals are often low on the corporate priority list for funding but are fully expected to achieve the results often holding the program together through sheer will and elbow grease. The achievement of these results is often not attainable given any number of changes in the workplace which are beyond the control of the safety professional.

The safety profession is often a "high-wire act." The myriad of risks, some of which can be controlled and some that are beyond our scope, can create a precarious environment where risks and dangers, internal and external, controllable and uncontrollable, are part and parcel of the daily work environment. Management support is essential in order to control the controllable and the uncontrollable, ranging from risks of nature to internal strife, which must be planned for and contained as best as possible.

The safety professional can wear many hats during the average day. On various educational levels and often in multiple languages, the safety professional must educate all facets of the organization regarding safety components and programs. The safety professional may be required to switch on a moment's notice to other hats ranging from disciplinarian to counselor, from coach to student, from executive to trainer. The hats will vary depending on the day, and the safety professional is expected to be an expert in all areas.

The safety profession is demanding, exciting, and very seldom the same from day to day. But above all, the safety professional is making an impact on the lives of others, whether they know it or not. Often silently, frequently unheralded, and always unsung, the safety professional works daily to make the work environment safer for all employees. When employees return home at the end of their shift in the same manner as when they arrived, success has been achieved, if only for the day. The safety profession is the noblest of professions because lives are saved, risks of injuries and illnesses are eliminated, and employees can go on with their productive lives because of the unsung and unheralded professional work of the men and women in this service.

Pat yourself on the back. You are making a difference!

2 History of OSHA

The world hates change, yet it is the only thing that has brought progress.

Charles F. Kettering

To appreciate the history of the Occupational Safety and Health Act (hereinafter referred to as the OSH Act),[1] we need to place the OSH Act in context with the social changes that were going on in the United States during the 1960s. The decade of the 1960s was the era of the Vietnam War; the assassinations of John F. Kennedy, Robert Kennedy, and Martin Luther King, Jr.; the hippie movement; protests in the streets; Kent State; burning draft cards and bras; marijuana and LSD among other drugs, and, above all, a youthful call for social change in America. During this decade and the decade that followed, Congress enacted numerous laws, such as the Civil Rights Act,[2] which changed the landscape not only within the law but also within our workplace and our society.

Initially started in the late 1960s and early 1970s, the ninety-first Congress enacted the OSH Act, and President Nixon signed the OSH Act into law on April 28, 1971.[3] The OSH Act was the first comprehensive federal law that governed the health and safety of workers in primarily the private sector. Although there were already existing laws that governed certain industries, such as mining, and laws covering federal contractors, the OSH Act applied to literally all private-sector industries and companies except those in which other federal agencies exercised statutory authority that encompassed safety and health within their jurisdiction.

The OSH Act generally applies to every employer engaged in business that affects interstate commerce and which has at least one employee.[4] The OSH Act requires every employer engaged in interstate commerce to furnish employees "a place of employment … free from recognized hazards that are causing, or are likely to cause, death or serious harm."[5] As permitted under the OSH Act, approximately twenty states petitioned and were granted individualized safety and health programs for their states. These programs, known as "state plan" programs, require that the state's program possess all of the required elements and be as inclusive as or more inclusive than the federal Occupational Safety and Health Administration (OSHA) requirements. Some of the states that have elected to possess their own state plan occupational safety and health program include Kentucky, Iowa, and Washington.

The OSH Act provided for the establishment of three independent branches under which the OSH Act would be managed and regulated. OSHA, located in the

---

[1] 29 U.S.C.A. Sections 653 and 654.
[2] 42 U.S.C.A. Sections 2000E through 2000E-17.
[3] 29 U.S.C.A. Section 653.
[4] 29 CFR Section 1975. Note that the terms "interstate commerce" and "employee" have specific meanings under the OSH Act.
[5] 29 U.S.C.A. Section 654(a) (1).

Department of Labor, receives the most notoriety and is the enforcement branch under the OSH Act. The National Institute for Occupational Safety and Health (NIOSH) provides research and testing on matters of safety and health and is located in the Department of Health and Human Services. The Occupational Safety and Health Review Commission (OSHRC) is the independent judicial branch and provides one of the administrative appeal processes under the OSH Act.

OSHA is housed within the U.S. Department of Labor and has the authority under the OSH Act to investigate, inspect, issue citations, and propose penalties for safety violations.[1] NIOSH, the education and research arm of the OSH Act, is housed but not combined with the Centers for Disease Control and Prevention and is housed in the Department of Health and Human Services. The Occupational Safety and Health Review Commission (OSHRC), the court system established under the OSH Act, is an independent federal agency.

Often called "state plan" programs, the OSH Act permits states to adopt their own safety and health programs. At this point in time, more than half of the states have adopted their own safety and health programs with the approval and oversight of OSHA and the U.S. Department of Labor. In general, the state plan safety and health program standards are required to be as stringent as or more stringent than the federal standards.

Among the many requirements provided under the OSH Act and the standards, two baseline standards are of utmost importance. Under Section 5(a)(1), also known as the "general duty clause," every employer is required to maintain its place of employment free from recognized hazards that are causing or are likely to cause death or serious physical harm to employees. The second baseline standard is Section 5(a)(2) that requires the employer to comply with all promulgated OSHA standards.

The OSH Act established three basic ways in which an OSHA standard could be promulgated. Under Section 6(a), the Secretary of Labor was authorized to adopt national consensus standards and establish federal standards without the rule-making procedures normally required under the OSH Act. This authority granted to OSHA ended on April 27, 1973. The second method by which OSHA may promulgate a standard is under Section 6(b) that establishes the procedures to be followed in modifying, revoking, or issuing new standards. This is the normal method by which OSHA establishes new standards today. The third method provided under the OSH Act but seldom utilized is the emergency temporary standard under Section 6(c). Emergency temporary standards may be issued by the Secretary of Labor if employees are subject to grave danger from exposure to substances or agents known to be toxic or physically harmful and a standard is necessary to protect employees from harm. These standards are effective immediately upon publication in the *Federal Register* and are in effect for a period not to exceed 6 months.

As noted above, the enforcement function under the OSH Act is provided to OSHA. OSHA compliance officers are empowered by Section 8(a) to inspect any workplace covered by the OSH Act, subject to the limitations set forth in the case law. An OSHA compliance officer is required to present his or her credentials to the owner or manager prior to proceeding on an inspection tour, and the owner or

---

[1] 29 U.S.C.A. Section 656-658.

manager possesses the right to accompany the compliance officer during the inspection tour. Employees or union representatives also possess the right to accompany the compliance officer on his or her inspection tour. After the inspection, the compliance officer will hold a "closing conference" with the employer and employee representatives at which time the safety and health conditions and potential citations or violations will be discussed. Most compliance officers do not possess the authority to issue "on-the-spot" citations but must confer with the regional or area director before issuing citations. Compliance officers possess the authority to shut down an operation that is life threatening or that places employees in a position of imminent danger. As discussed later in this text, safety professionals are encouraged to prepare their companies and organizations for an OSHA inspection and develop programs and strategies to insure all rights and responsibilities provided under the OSH Act are being adhered to by all affected parties.

Following the closing conference, the compliance officer is required to issue a report to the area or regional director. The area or regional director usually decides whether to issue a citation and assesses any penalty for the alleged violation. Additionally, the area or regional director sets the date for compliance or abatement for each of the alleged violations. If a citation is issued by the area or regional director, this notice is mailed to the employer as soon as possible after the inspection, but in no event can this notification take place more than 6 months after the alleged violation occurred. Citations must be in writing and must describe with particularity the alleged violation, the relevant standard and regulation, and the date of the alleged violation. The appeals process is discussed in detail later in this text.

Of particular interest to safety professionals working in the public sector is the fact that state and local governments are not covered by the OSH Act due to the fact that state and local governments are not considered employers.[1] Although some state and local governments have voluntarily requested full coverage or partial coverage under the OSH Act or by state plans, most public employees are not covered under the OSH Act. As of this writing, "twenty-six states have exercised their option to operate their own state OSHA programs. Of those, 22 state plans cover both private sector employees and state and local government employees, while four state plans cover public employees only."[2] However, legislation has been offered in the past, and proposed legislation is currently pending before Congress to expand OSHA's jurisdiction and provide coverage to all federal, state, and local public employees as well as some private-sector employees who are not currently covered.[3]

The above synopsis of the OSH Act is in no way all inclusive but is provided to the reader to establish the framework of reference for the chapters to follow in this text. For further information, visit the exceptional Web sites provided by these agencies, including OSHA at www.osha.gov, NIOSH at www.cdc.gov/niosh, and the Occupational Safety and Health Review Commission at www.oshrc.gov.

---

[1] 29 U.S.C.A. Section 652(5). Also note that the OSH Act does not cover business on Indian reservations due to the fact that it would violate Indian treaty rights. See *Navajo Forest Products Industries,* OSHRC Docket No. 76-5013.

[2] *Congress Pushes OSHA Coverage for Public Workers,* Occupational Hazards, July 2007, at p. 10.

[3] Id. The proposed bill is called "Protecting America's Workers Act."

3 OSHA Inspection Procedures

He who has learned to disagree without being disagreeable has discovered the most valuable secret of a diplomat.

Robert Estabrook

When should safety professionals prepare for an OSHA inspection — how about yesterday! It is imperative that a safety professional begin to prepare for an OSHA inspection the day he or she starts the job and work with the management team and employees to achieve and maintain compliance with the OSHA standards. Additionally, the safety professionals should discuss the company philosophy and strategies for handling an OSHA inspection with the management team and legal counsel in order that the company or organization's rights are protected before, during, and after the inspection.

For most safety professionals, being within OSHA's jurisdiction is a clear issue. However, if there is any doubt, the first issue a safety professional needs to clarify is whether federal OSHA has jurisdiction over the company or organization. If not, does a state plan have jurisdiction? Is your company outside of both state and federal jurisdictional lines?

For most companies, you will be either a "fed plan" or a "state plan" state for OSHA jurisdictional purposes. For virtually every private-sector employer and some public-sector employers, jurisdiction lies with either federal OSHA or one of the state plan states. Literally every private-sector and some public-sector employers in the United States that employ one or more persons and is engaged in a business that in any way affects interstate commerce is within the scope of the federal OSH Act. The phrase "interstate commerce" has been broadly construed by the U.S. Supreme Court, which has stated that interstate commerce "goes well beyond persons who are themselves engaged in interstate or foreign commerce."[1] In essence, anything that crosses state lines, whether a person or goods and services, places the employer in interstate commerce. Although there are exceptions to this general statement, interstate commerce generally has been "liberally construed to effectuate the congressional purpose" of the OSH Act.[2]

Upon finding that you are covered by the OSH Act, a safety professional must distinguish between a state plan jurisdiction and federal OSH Act jurisdiction. If its facilities or operations are located within a state plan state, an employer must comply with the regulations of its state. Safety professionals should contact their state department

---

[1] See, for example, *NLRB v. Fainblatt*, 306 U.S. 601, 604-05 (1939). See also *U.S. v. Ricciardi*, 357 F.2d 91 (2nd Cir), *cert. denied*, 384 U.S. 942, 385 U.S. 814 (1966).
[2] *Whirlpool Corp. v. Marshall*, 455 U.S. 1, 8 O.S.H. Cases 1001 (1980).

of labor to acquire the pertinent regulations and standards. If facilities or operations are located in a federal OSHA state, the applicable standards and regulations can be acquired from any area OSHA office, in the Code of Federal Regulations in book 29, Section 1910 and 1926, or on the OSHA Web site (www.osha.gov).

A common jurisdictional mistake occurs when an employer operates multiple facilities in different locations. Safety professionals should ascertain which state or federal agency has jurisdiction over the particular facility or operations and which specific regulations and standards apply. Although most standards within the state plans mirror the federal OSHA standards, there may be specific state requirements that exceed the federal OSHA standards.

It is important that safety professionals fully understand how OSHA and the state plan states acquire the standards and "make" or promulgate new standards. First, and most important to safety professionals, is the fact that the OSH Act requires that a covered employer comply with specific occupational safety and health standards and all rules, regulations, and orders issued pursuant to the OSH Act that apply to the workplace.[1] The OSH Act also requires that all standards be based on research, demonstration, experimentation, or other appropriate information that is often a battleground in the promulgation of new standards.[2] The Secretary of Labor is authorized under the Act to "promulgate, modify, or revoke any occupational safety and health standard," and the OSH Act describes the procedures that the Secretary must follow when establishing new occupational safety and health standards.[3]

The OSH Act authorizes three specific ways to promulgate new standards. From 1970 to 1973, the Secretary of Labor was authorized in Section 6(a) of the Act to adopt national consensus standards and establish federal safety and health standards without following lengthy rule-making procedures.[4] Many of the early OSHA standards were adapted mainly from other areas of regulation, such as the National Electric Code and American National Standards Institute (ANSI) guidelines. However, safety professionals should note that this promulgation method is no longer in effect.

The usual method of issuing, modifying, or revoking a new or existing OSHA standard is set out in Section 6(b) of the OSH Act and is known as the informal rule-making process. The informal rule-making process requires notice to interested parties, through subscription in the *Federal Register* of the proposed regulation and standard, and provides an opportunity for comment in a non-adversarial administrative hearing.[5] The proposed standard can also be advertised through magazine articles and other publications, thus informing interested parties of the proposed standard and regulation. This method differs from the requirements of most other administrative agencies that follow the Administrative Procedure Act in that the OSH Act provides interested persons an opportunity to request a public hearing with oral testimony. It also requires the Secretary of Labor to publish in the *Federal Register* a notice of the time and place of such hearings. Safety professionals are permitted to testify at these hearings.

---

[1] 29 U.S.C. Section 655 (b).
[2] Id.
[3] 29 C.F.R. Section 1911.15 and 29 C.F.R. 1910.
[4] 29 U.S.C. Section 1910.
[5] 29 U.S.C. Section 655(b).

Although not required under the OSH Act, the Secretary of Labor has directed, by regulation, that OSHA follow a more rigorous procedure for comment and hearing than other administrative agencies. Upon notice and request for a hearing, OSHA must provide a hearing examiner in order to listen to any oral testimony offered. All oral testimony is preserved in a verbatim transcript. Interested persons are provided an opportunity to cross-examine OSHA representatives or others on critical issues. The Secretary must state the reasons for the action to be taken on the proposed standard, and the statement must be supported by substantial evidence in the record as a whole.

The Secretary of Labor has the authority not to permit oral hearings and to call for written comment only. Within 60 days after the period for written comment or oral hearings has expired, the Secretary must decide whether to adopt, modify, or revoke the standard in question. The Secretary can also decide not to adopt a new standard. The Secretary must then publish a statement of the reasons for any decision in the *Federal Register*. OSHA regulations further mandate that the Secretary provide a supplemental statement of significant issues in the decision. Safety and health professionals should be aware that the standard as adopted and published in the *Federal Register* may be different from the proposed standard. The Secretary is not required to reopen hearings when the adopted standard is a "logical outgrowth" of the proposed standard.[1]

The third method for promulgating new standards, and the one most infrequently used, is the emergency temporary standard permitted under Section 6(c).[2] The Secretary of Labor may establish a standard immediately if it is determined that employees are subject to grave danger from exposure to substances or agents known to be toxic or physically harmful and if an emergency standard would protect the employees from the danger. An emergency temporary standard becomes effective on publication in the *Federal Register* and may remain in effect for 6 months. During this 6-month period, the Secretary must adopt a new permanent standard or abandon the emergency standard.

Safety professionals should be aware that only the Secretary of Labor can establish new OSHA standards. Recommendations or requests for an OSHA standard can come from any interested person or organization, including employees, employers, labor unions, environmental groups, and others. When the Secretary receives a petition to adopt a new standard or to modify or revoke an existing standard, the normal course is for the Secretary to forward the request to NIOSH and the National Advisory Committee on Occupational Safety and Health (NACOSH) or the Secretary may use a private organization such as ANSI for advice and review.

As safety professionals are aware, the OSH Act requires that an employer maintain a place of employment free from recognized hazards that are causing or are likely to cause death or serious physical harm, even if there is no specific OSHA standard addressing the circumstances. Under Section 5(a)(1), known as the "general duty clause," an employer may be cited for a violation of the OSH Act if the condition causes harm or is likely to cause harm to employees, even if OSHA has not

---

[1] See *Taylor Diving & Salvage Co. v. DOL*, 599 F.2d 622, 7 OSH Cases 1507 (5th Cir., 1979).
[2] 29 U.S.C. Section 655(c).

promulgated a standard specifically addressing the particular hazard. The general duty clause is a catch-all standard encompassing all potential hazards that have not been specifically addressed in the OSHA standards.

Safety professionals are encouraged to take a proactive approach in maintaining their competency in this expanding area of OSHA regulations. As noted previously, the first notice of any new OSHA standard, modification of an existing standard, revocation of a standard, or emergency standard must be published in the *Federal Register*. Safety professionals can use the *Federal Register*, or professional publications that monitor OSHA standards, to track the progress of proposed standards throughout the approval process. With this information, safety professionals can provide testimony to OSHA when necessary, prepare their organizations for acquiring resources and personnel necessary to achieve compliance, and get a head start on developing compliance programs to meet requirements in a timely manner.

OSHA performs all enforcement functions under the OSH Act. Under Section 8(a) of the Act, OSHA compliance officers have the right to enter any workplace of a covered employer without delay, inspect and investigate a workplace during regular hours and at other reasonable times, and obtain an inspection warrant if access to a facility or operation is denied. Upon arrival at an inspection site, the compliance officer is required to present his or her credentials to the owner or designated representative of the employer before starting the inspection. The employer representative and an employee, union representative, or both may accompany the compliance officer on the inspection. Compliance officers can question the employer and employees and inspect required records, such as the OSHA Form 300, which records injuries and illnesses. Most compliance officers cannot issue on-the-spot citations except in specific situations, such as those involving imminent danger. Compliance officers only have authority to document potential hazards and must report or confer with the OSHA area director before issuing a citation.

A compliance officer or any other employee of OSHA may not provide advance notice of the inspection under penalty of law.[1] The OSHA area director is, however, permitted to provide notice under the following circumstances:

1. In cases of apparent imminent danger, to enable the employer to correct the danger as quickly as possible.
2. When the inspection can most effectively be conducted after regular business hours or where special preparations are necessary.
3. To ensure the presence of employee and employer representatives or appropriate personnel needed to aid in inspections.
4. When the area director determines that advance notice would enhance the probability of an effective and thorough inspection.[2]

It should also be noted that the advance notice requirement applies only to the compliance division. OSHA and state plan consultation services, such as the Kentucky Education and Training Division, are usually requested services, and advance notice

---

[1] 29 C.F.R. Section 17(f).
[2] Occupational Safety and Health Law, 2080209 (1988).

is provided. Additionally, activities within specific programs, such as the Voluntary Protection Program, usually provide advance notice for inspection-related activities.

In general, OSHA compliance officers can also take environmental samples and obtain photographs related to the inspection. Additionally, compliance officers can use other "reasonable investigative techniques," including personal sampling equipment, dosimeters, air sampling badges, and other equipment. Compliance officers must, however, take reasonable precautions when using photographic or sampling equipment to avoid creating hazardous conditions (i.e., a spark-producing camera flash in a flammable area) or disclosing a trade secret.[1]

As a general rule, an OSHA inspection has four basic components: (1) the opening conference, (2) the walk-through inspection, (3) the closing conference, and (4) the issuance of citations, if necessary. In the opening conference, the compliance officer may explain the purpose and type of inspection to be conducted, request records to be evaluated, question the employer, ask for appropriate representatives to accompany him or her during the walk-through inspection, and ask additional questions or request more information. The compliance officer may, but is not required to, provide the employer with copies of the applicable laws and regulations governing procedures and health and safety standards. The opening conference is usually brief and informal; its primary purpose is to establish the scope and purpose of the walk-through inspection.

After the opening conference and review of appropriate records, the compliance officer, usually accompanied by a representative of the employer and a representative of the employees, conducts a physical inspection of the facility or worksite. The general purpose of this walk-through inspection is to determine if the facility or worksite complies with OSHA standards. The compliance officer must identify potential safety and health hazards in the workplace, if any, and document them to support issuance of citations.[2]

The OSHA compliance officer can use various forms to document potential safety and health hazards observed during the inspection. The most commonly used form is the OSHA-1 Inspection Report that contains basic information about the company or organization. The OSHA compliance officer usually records this information during the opening conference and walk-through inspection.

Two additional forms are usually attached to the OSHA Inspection Report. The OSHA-1A form, known as the narrative, is used to record information gathered during the walk-through inspection; names and addresses of employees, management officials, and employee representatives accompanying the OSHA compliance officer on the inspection; and other information. A separate worksheet, known as OSHA-1B, is used by the OSHA compliance officer to document each condition that he or she believes could be an OSHA violation. One OSHA-1B worksheet is completed for each potential violation noted by the compliance officer.

When the walk-through inspection is completed, the OSHA compliance officer usually conducts an informal closing conference with the employer or the employer's representative to "informally advise (the employer) of any apparent safety or health

---

[1] 29 C.F.R. Section 1903.9.
[2] *OSHA Manual*, Note 62, at III-D8.

violations disclosed by the inspection."[1] The OSHA compliance officer informs the employer of the potential hazards observed and indicates the applicable section of the standards allegedly violated, advises that citations may be issued, and informs the employer or representative of the appeal process and rights. Additionally, the OSHA compliance officer advises the employer that the OSH Act prohibits discrimination against employees or others for exercising their rights.[2]

In very unusual situations, the OSHA compliance officer may issue a citation on the spot. When this occurs, the OSHA compliance officer informs the employer of the abatement period, in addition to the other information provided at the closing conference. In most circumstances, the compliance officer will leave the workplace and file a report with the area director who has authority, through the Secretary of Labor, to decide whether a citation should be issued; compute any penalties to be assessed; and set the abatement date for each alleged violation. The area director, under authority from the Secretary, must issue the citation with "reasonable promptness."[3] Citations must be issued in writing and must describe with particularity the violation alleged, including the relevant standard and regulation. There is a 6-month statute of limitations, and the citation must be issued or vacated within this time period. OSHA must serve notice of any citation and proposed penalty by certified mail, unless there is personal service, to an agent or officer of the employer.[4]

After the citation and notice of proposed penalty is issued, but before the notice of contest by the employer is filed, the employer may request an informal conference with the OSHA area director. The general purpose of the informal conference is to clarify the basis for the citation, modify abatement dates or proposed penalties, seek withdrawal of a cited item, or otherwise attempt to settle the case. This conference, as its name implies, is an informal meeting between the employer and OSHA. Employee representatives must have an opportunity to participate if they so request. Safety professionals should note that the request for an informal conference does not stay (delay) the 15-working-day period to file a notice of contest to challenge the citation.[5] However, an informal conference can be a good method through which to discuss the citations directly with OSHA and potentially settle the citation prior to contest.

Under the OSH Act, an employer, employee, or authorized employee representative (including, a labor organization) is given 15 working days from when the citation is issued to file a "notice of contest." If a notice of contest is not filed within 15 working days, the citation and proposed penalty become a final order of the Occupational Safety and Health Review Commission (OSHRC) and are not subject to review by any court or agency. If a timely notice of contest is filed in good faith, the abatement requirement is tolled (temporarily suspended or delayed), and a hearing is scheduled. The employer also has the right to file a petition for modification of the abatement period (known as a PMA) if the employer is unable to comply with the abatement

---

[1] 29 C.F.R. Section 1903.7(e).
[2] 29 U.S.C. Section 660(c)(1).
[3] Id. at Section 658.
[4] Fed. R. Civ. P. 4(d)(3).
[5] 29 U.S.C. Section 659(a).

period provided in the citation. If OSHA contests the PMA, a hearing is scheduled to determine whether the abatement requirements should be modified.

When the notice of contest by the employer is filed, the Secretary must immediately forward the notice to the OSHRC, which then schedules a hearing before its administrative law judge (ALJ). The Secretary of Labor is labeled the "complainant," and the employer the "respondent." The ALJ may affirm, modify, or vacate the citation, any penalties, or the abatement date. Either party can appeal the ALJ's decision by filing a petition for discretionary review (PDR). Additionally, any member of the OSHRC may "direct review" any decision by an ALJ, in whole or in part, without a PDR. If a PDR is not filed and no member of the OSHRC directs a review, the decision of the ALJ becomes final in 30 days. Any party may appeal a final order of the OSHRC by filing a petition for review in the U.S. Court of Appeals for the circuit in which the violation is alleged to have occurred or in the U.S. Court of Appeals for the District of Columbia Circuit. This petition for review must be filed within 60 days from the date of the OSHRC's final order. This is discussed in greater length later in this text.

Safety professionals should be keenly aware of not only their responsibilities under the OSH Act but also their rights and the rights of their company or organization.

First, the safety professional is entitled to know the purpose of the inspection, for example, whether it is based on an employee complaint or is a routine inspection. Second, the safety professional or employer representative has the right to accompany the OSHA compliance officer during the inspection. This can be helpful or harmful — helpful in the sense that the safety professional can avoid certain areas during a complaint inspection, but harmful if a major violation is found and the safety professional, in trying to explain, talks him- or herself into more trouble and ends up with a higher fine and a more serious violation.

Safety professionals should know their rights under the OSH Act as well as the U.S. Constitution. As a result of the landmark case of *Marshall v. Barlow, Inc.*,[1] three distinct avenues for addressing OSHA enforcement efforts have developed and can be efficiently used depending on the circumstances. Safety professionals should be familiar with the strategies, strengths, and weaknesses of each strategy.

In *Barlow,* the Supreme Court held that Section 8(a) of the OSH Act, which empowered OSHA compliance officers to search the work areas of any employment facility within the OSH Act's jurisdiction without a search warrant or other process, was unconstitutional[2] (Fourth Amendment of the U.S. Constitution). The Court concluded that "the concerns expressed by the Secretary (of Labor) do not suffice to justify warrantless inspections under OSHA or vitiate the general constitutional requirement that for a search to be reasonable a warrant must be obtained."[2] This decision opened the door to the first strategy. Under this strategy, the safety professional, as representative of the company or organization, would meet the OSHA compliance officer at the door and require OSHA to acquire a search warrant before entering a facility to conduct an inspection. This would require the OSHA compliance officer to acquire an administrative search warrant and provide the employer's

---

[1] 436 U.S. 307, 6 O.S.H. Cases 1571 (1978).
[2] Id.

counsel an opportunity to attempt to quash the warrant. Safety professionals should be aware that the burden of proof for OSHA to acquire an administrative search is relatively low, and thus OSHA is successful in acquiring the administrative search warrant the vast majority of time. This strategy possesses many risks. This approach should be carefully evaluated with the help of legal counsel given the potential pitfalls and the possibility of sanctions against the employer for bad faith.

The second strategy that can be employed by safety professionals is attempting to limit the scope and inspection techniques used by the OSHA compliance officer based upon the type and breadth of the inspection and the operational structure. This is normally an informal process where the safety professional contacts the regional or area director before a voluntary compliance inspection and an agreement is reached on the specific area to be inspected.[1] This type of limitation is usually only applicable to complaint inspections. Remember, the complaint inspection is usually limited to one area of the operations. However, everything the inspector sees, hears, and smells going to and from the area of the complaint is subject to identification of hazards. Thus, the path of least potential observation of a potential hazard may be selected for this journey.

Another area to consider within this strategy is negotiating limitations on photographing or videotaping to protect trade secrets.[2] If an agreement cannot be reached before the inspection, a court order may be acquired to protect the confidentiality of a trade secret.[3] Safety professionals should be aware that through the Freedom of Information Act and individual state Open Records acts, photographs, documents, and other evidence gathered by OSHA during an inspection may be accessible to the public if appropriate protections are not requested.

The strategy most frequently utilized by safety professionals is to permit the warrantless compliance inspection of the facility. This is, in essence, telling OSHA to "come on in" and inspect the operations. Safety professionals should always be with the OSHA compliance officer and document every item that the OSHA compliance officer notes during the inspection. Additionally, all photographs should be taken by the safety professional. In essence, upon completion of the inspection, the safety professional should have the same correspondence documentation and evidence as the OSHA inspector has. If a citation is issued and ultimately disputed, the safety professional cannot request OSHA's evidence in preparation for the hearing. The safety professionals should gather the evidence with the same intensity and detail as if this matter would end up in the U.S. Supreme Court. If no citations are issued or the matter is settled, the documentation may not be needed. However, if the citations ultimately become disputed and the evidence was not gathered at the time of the inspection, it is virtually impossible to acquire the evidence after the fact.

Safety professionals are encouraged to analyze their situation and facility and develop a policy and plan of action that advises the management team as to how to manage an OSHA inspection. This statement should include the policy's purpose; detailed steps to be followed when OSHA compliance officers attempt entry, during the inspection, at the closing conference, and after the inspection; and forms to assist

---

[1] 137 C.F.R. Section 1908.1-11.
[2] 137 C.F.R. Section 1903.11.
[3] Id.

the subordinate in this procedure. In addition, contingencies should be addressed for situations where a selected member of a team is on vacation or unavailable.

OSHA and most state plan programs offer free consultation services.[1] At the employer's request, OSHA or a state plan program can often assist the safety professional with compliance and other issues. This voluntary program will not cite violations except under very limited and serious circumstances. Safety professionals should also review and consider other programs offered by OSHA, such as the Voluntary Protection Program (VPP) or one of the cooperative or alliance agreement programs.

An OSHA inspection does not have to be a dreaded event in a safety professional's life. Preparation is the key to a successful OSHA inspection. Preparing your safety program, preparing your management team, preparing your employees, and preparing yourself will provide the confidence and skillfulness to weather the inspection.

No matter what the circumstances, safety professionals should remember that the OSHA inspector is simply doing his or her job and the inspection is not personal in nature. The OSHA inspector should be treated in a professional manner at all times. Being prepared and designing a strategy before you receive the knock on the door from OSHA provides not only guidance but also confidence that the safety professional can protect the company's or organization's rights before, during, and after the inspection. And remember to document everything during the inspection — there are no "mulligans" or "do overs" after the inspection. You either have the evidence or you do not!

---

[1] 29 C.F.R. Sections 1908.1-11.

4 Corporate Compliance Management

Leadership involves remembering past mistakes, an analysis of today's achievements, and a well-grounded imagination in visualizing the problem of the future.

Stanley C. Allen

Safety professionals working at the corporate level or with multiple operation locations possess substantially more and different responsibilities within the safety function. The corporate-level safety professional is the leader of all of the safety efforts throughout the organization as well as the monitor, coach, motivator, disciplinarian, counselor, and innovator for the safety efforts throughout the entire company or organization. Above all, the corporate-level safety professional must exhibit leadership skills and abilities through which to captain the safety ship through the constantly turbulent waters that are the safety function.

Although there is much written regarding leadership and the skills and abilities necessary to be a good leader, leadership in the safety function can be a uniquely challenging endeavor due to the management hierarchy, the risks involved, and the changing landscape. Additionally, the corporate safety professional possesses the human element that is varied and volatile, the economic variables of the economy, and the changing legal landscape of OSHA, the Environmental Protection Agency (EPA), and related laws. A corporate safety professional, unlike most leaders, is serving as a functional leader within a larger corporate structure but must possess the same skills and abilities to lead the safety efforts within the leadership scope of the parent entity.

Even though it is not feasible to address every aspect of corporate compliance management within the scope of this text, several important aspects of management at the corporate level are addressed below. Safety professionals serving at the corporate level, and those who aspire to this level, should realize that although many of the functions are the same as those at operational levels, the environment is substantially different and the expectations are elevated. The control aspect of the safety job function which is hands on at the operational level is now virtually remote at the corporate level. Concepts, guidance, motivation, and resources are provided from the corporate level, and implementation is performed at the operational level. Corporate-level safety compliance is a vital component of the safety chain, but the management of the function is substantially different than the operational levels of safety.

The safety function does not "make" anything, and the corporate safety function does not provide direct services at the operational level. At the most basic level, the safety function simply manages risks and preserves assets. Given this inherent difference from production, the safety function is always in a fight for the often sought-after resources of the organization. A vital role that the corporate-level safety professional must play is that of teacher or educator for the management hierarchy at all levels as to the importance and beneficial aspects of safety in the operations. In order to acquire the resources necessary for the safety function to operate, it is important that the management team fully understand the functions, responsibilities, and benefits of the safety function at all levels of the organization. Additionally, it is important that the corporate safety professional be able to justify expenditures and show the management team, in terms they understand, the return on investment the safety function provides to the organization.

In order to be able to acquire the appropriate information through which to justify the safety function, it is important that the corporate-level safety professional design a methodology through which to acquire information regarding the various aspects of the safety functions from each operation within the organization. To be able to properly educate and justify the acquisition of resources from the organization for safety purposes, the information gathered must be timely, accurate, and organized in order to show a true picture of the safety function, the deficiencies needing assistance, and the resources needed.

This sounds rather simplistic in theory but is a major endeavor because of the varying laws, regulations, and operations within the organization. In many organizations, trying to compare the safety efforts of one operation against another operation producing the same product line in the same industry and in the same organization can be like comparing apples to oranges due to different state workers' compensation laws, different state plan requirements, and other laws and regulations. Additionally, there are a myriad of variations within the same types of operations ranging from the make-up of the workforce; location of the facility; speed of the operation; age, type, and design of equipment within the operation; and other variables.

The management team demands accurate and reliable safety information from which to base their decisions. This is especially important within the safety function, because not only are substantial dollar amounts involved, but any decision can and will have an impact on the lives and jobs of numerous employees within the organization. It is essential that the corporate safety professional establish a system through which to gather pertinent, timely, and accurate safety-related data from the operational levels and verify this information prior to providing to the management team.

My father always told me, "Liars can figure … and figures can lie." This is especially pertinent in the safety profession where many aspects are not "black and white" in interpretation or fact. Corporate-level safety professionals may have numerous plant-level or lower-organizational-level safety professionals working directly for them or reporting to plant-level management with indirect lines to the corporate level. Acquisition of accurate information from the lower levels to corporate is vital; however, this information and data can be skewed if pressures are placed upon the use of the data or information. For example, one common method of improving safety performance is to develop safety goals or objectives and hold plants and operations accountable for the achievement of the safety goals or objectives. Often, in conjunction with the safety goal or objective, there is positive reinforcement in terms of money, bonuses, and prizes, as well as negative reinforcement in terms of lower bonuses and reprimands. Depending on the structure of the safety goals or objectives, there can be substantial room for interpretation and skewing of the data in order to achieve the proscribed goal or objective. For example, OSHA utilizes a category within the old OSHA 200 and the new OSHA 300 record keeping system which defines and provides guidance as to what constitutes a "recordable" injury or illness. Although the new OSHA record keeping standard is substantially tighter than the old record keeping system, if a company or organization incorporates the OSHA standard into its goals and objectives and utilizes recordable injuries and illnesses as a category from which to establish positive or negative reinforcement, there is a substantial probability that smart safety professionals at the operational level will find loopholes within the system which may reflect a performance level within the company or organizational safety goals or objectives. Thus, the data received from the operational level may reflect improvement in the safety performance where, in actuality, the improvement identified in the data provided was simply a manipulation of the original data within the broadest interpretation of OSHA standard.

Many organizations utilize the data to establish safety goal and objectives for individual operations and specific units within the operations. This can be a very effective tool. However, to ensure a more accurate view of the overall safety performance, corporate safety professionals should also consider the incorporation of safety audits in general. Safety audits should address every element of each compliance program as well as company policies and procedures. Safety audits should be performed not only by the corporate safety professional in order to acquire a first-hand view of the operation but also by other company officials who are not directly affiliated with the specific plant or operational-level operations. The safety audits should be conducted on a periodic basis and be graded in a fair and consistent manner. Although there are many different ways to grade a safety audit, it is important that the operational-level safety professional receive a performance score and deficiencies be directly identified and a plan of action to correct the deficiencies and improve the performance be developed. Thus, on a periodic basis, the operational-level safety professional will receive a report card, and his or her work can be focused on the areas of identified deficiencies. Additionally, this methodology provides the corporate-level safety professional with a fair and accurate view of each of the safety programs and safety professionals under his or her guidance and permits an appropriate comparison among operations.

Above all, the corporate-level safety professional must be the leader of the safety efforts and champion the operational-level safety professional's needs and efforts within the corporate realm. To this end, the corporate-level safety professional must be knowledgeable in all areas of the operations, be able to communicate effectively at all levels, and be able to visualize and justify the long-range efforts of the company-wide safety program. In theory, this sounds reasonable; however, within the corporate hallways, "back armor" may be essential, and justification of even the most basic of safety expenditures can be a battle. The corporate safety professional must pick his or her battles and lay the groundwork to ensure that the selling of the program or safety activity is a success.

The corporate safety professional must be the visionary for the entire safety effort and must be able to identify risks that are on the horizon in order to properly prepare the company or organization before the risk manifests itself. Unlike other professions, the safety professional must continuously survey the minefield of potential risks to identify the validity and impact of the risk as well as the potential of the risk impacting the operations. The risks in the safety field are constantly changing as new technologies, new standards and regulations, and the world market constantly change the risks in the workplace. To this end, presented below are several categories of potential emerging risks for consideration.

CORPORATE RISK ASSESSMENT CONSIDERATIONS

COMPLIANCE RISKS

Some standards possess a higher risk of serious injury or death than others. For example, there is a greater risk of an employee being killed for a violation of the confined space standard than the OSHA record keeping standard. Although both are very important standards and both should be emphasized within the safety function, the risks are substantially different. To this end, certain standards contain a higher risk than others in terms of probability of serious injury or death. These standards should be identified and special emphasis placed on the higher-risk standards. In general, some of these standards include the following:

- Control of Hazardous Energy (LOTO)
- Fall Protection standard
- Confined Space Entry and Rescue standards
- Hazardous Materials Standard

SAFETY IN THE WORKFORCE

- Safety for aging workforce
- Childcare and eldercare in the facilities
- Drugs and alcohol in the workforce
- Domestic violence in the workplace

TRAVEL RISKS

- Consider providing traveling executives with portable 5-minute air packs
- Consider company policy prohibiting staying in hotels above the seventh floor (Most aerial ladder trucks can reach the sixth floor)
- Consider a company policy prohibiting top-level management from traveling together (either commercial or private jet)
- Consider training in hotel evacuation for traveling management team members

FIRE RISKS

- Structural design and materials
- Sprinkler systems

- Water supply
- Response time and capabilities of local fire department
- Fighting fire versus evacuation

NATURAL DISASTER RISKS

- Consider training on tornado risks in hotels or elsewhere (shelter in place).
- Consider alternative evacuation planning.
- Consider routing from location and alternate transport.

EMERGENCY AND DISASTER PREPAREDNESS (IN FACILITIES)

- Must comply with OSHA standards
- Should provide internal as well as external evacuation in tornadic areas
- Set up a command center
- Designate a triage area
- Arrange transport
- Account for personnel
- Work with public sector
- Coordinate media relations

IN-HOUSE SECURITY

- Perimeter security
- Searches of items being brought in and out of the facility
- In-house searches (lockers, etc.)
- Bomb threat procedures
- Violence in the facility
- Threats to management
- Protection of vehicles in parking areas
- Lighting in parking and external areas
- Truck inspection
- Protection on large volumes of chemicals or flammables
- Use of camera

WORKPLACE VIOLENCE

- Warning signs and training to detect
- Safeguarding managers and executives
- Protecting property and equipment

INDUSTRIAL THEFT AND SABOTAGE

- Safeguarding internal communications and information
- Monitoring e-mail and phone
- Understanding Freedom of Information Act (FOIA) and Open Records protections
- Conducting physical inspection and searches

CYBERTERRORISM

- Hackers, information thefts, internal sabotage
- Software sabotage
- Backup systems and internal systems

TERRORISM AND EXTERNAL GROUPS TARGETING SPECIFIC INDUSTRIES

- 9/11 and World Trade building international terrorism
- Oklahoma City and in-country terrorism
- Industry-specific terrorism
- Strike, picketing, and union-related risks

PANDEMICS, AVIAN FLU, AND OTHER AIRBORNE CONTAMINANTS

- Teraflu, masks, travel restrictions, and related issues
- Operational strategies with loss of large percentage of the workforce
- Internal protections such as airflow in facilities

SUMMARY

The above are just a few of the potential categories of the changing risks in our operations. Depending on the specific responsibilities of the corporate safety professional, the scope of risks can range into the areas of security, environmental, fire, occupational safety, human resources, and related functions. A prudent corporate safety professional must survey the risks and prepare the company to address the risks if the risk transforms into reality.

Corporate safety professionals must be visible in the operations. The corporate safety professional is the one person in every company whose name is known by everyone. Spending time at each operation is essential in order to acquire base-level knowledge of the operations and a working level of communications, not only with the management team but also the employees at the plant or operations. Although every corporate-level safety professional is pressed for time, the time spent at the operational level will pay dividends in the knowledge level and in justifying and championing safety efforts to achieve the vision of an exceptional corporate-wide safety program.

5 Preparing for an Inspection

People never improve unless they look to some standard or example higher and better than themselves.

Tryon Edwards

Preparation for an OSHA inspection begins the first day the safety professional is on the job. Working to achieve and maintain compliance with the OSHA standards and beyond is the primary goal of the safety professional. If the safety professional is working in a proactive manner to achieve and maintain compliance on a daily basis, there should be no worries about an OSHA inspection. However, while these efforts are ongoing and if the workplace has not achieved the full measure of compliance, there may be a risk of violation and thus penalty. In short, there is no substitution for a proactive and compliant safety program.

However, this being said, the safety professional is also an agent of or representative of the company or organization. Part and parcel of a safety professional's position and responsibility is to protect the company or organization from potential risks of which an OSHA inspection may be one. Given this potential risk, safety professionals should properly formulate a strategy beforehand and prepare their management team to protect the company's or organization's rights and fulfill their responsibilities before, during, and after an OSHA inspection.

In addition to the day-to-day activities of the safety professional in achieving and maintaining compliance, a formalized plan as to how to handle an OSHA inspection should be discussed and formulated among all levels of the management team. This formalized plan of action and strategy should carefully analyze all options, rights, and responsibilities with specific job responsibilities in order to ensure that "no stone is left unturned" during an inspection.

From a broader viewpoint, there are three basic strategies that can be employed during an OSHA inspection. The first strategy, and the one most utilized by companies and organizations, is to simply waive the company's rights and voluntarily permit the inspection. The second strategy is where the inspection is generated from a complaint and there are protected processes, equipment, or other items in the workplace. This "limiting" strategy depends on the circumstances. For a complaint inspection, OSHA possesses only the ability to view the area in which the complaint has been filed. However, everything the OSHA compliance officer sees as he or she goes and returns from the area in which the complaint has been generated is subject to inspection. Thus, safety professionals should identify the most efficient route to and from the complaint area wherein the OSHA compliance officer possesses a lower probability of viewing any alleged violation. Additionally, under this strategy, the inspection can be limited through discussions with the OSHA regional director in areas where equipment, processes, or other aspects of the operations are protected and confidential. Remember, most OSHA documents are subject to viewing by the public under the Freedom of Information Act[1] or individual state open record laws unless protected.

The third basic strategy would be for the company or organization to require the OSHA compliance officer to acquire an administrative search warrant prior to gaining entry. This strategy is based upon the U.S. Supreme Court decision in *Marshall v. Barlows* (436 U.S. 307, 6 O.S.H. Cases 1571 [1978]).[2] Utilizing this very technical and legalistic strategy will require the guidance of legal counsel, and every step in this process should be carefully prepared in advance of any inspection.

No matter which strategy your company or organization employs, it is essential for the safety professional to develop a systematic approach through which all future evidence is gathered throughout the inspection. OSHA does not provide the company or organization the information and evidence they have gathered during the inspection. If the evidence is not gathered at the same time as the OSHA inspector is gathering his or her evidence during the inspection, if citations are issued and the company or organization elects to contest the citation, the safety professional may be attempting to build a case with no supporting evidence. Thus, prudent safety professionals should gather and document all potential evidence during the inspection in preparation for a citation being issued and the matter going to contest. If citations are not issued or the matter is not contested, the safety professional can simply file the documentation. However, it is extremely difficult to acquire evidence after the inspection to support a contest down the road. In essence, proper preparation to document every aspect of the inspection should take place on every inspection. It is impossible to know exactly what evidence will be important — everything should be documented and preserved as potential evidence.

As part of the overall strategy, safety professionals should prepare the management team in advance and provide training for management team members as to their specific responsibilities and expectations before, during, and after an OSHA inspection. In addition to this training, specific team members responsible for photographic documentation, sampling, and related activities should be provided the appropriate equipment and be trained in the use of this equipment.

Provided below is a general checklist of some of the responsibilities and activities that may be needed to prepare your management team members. Safety professionals should remember that a management team member should accompany the OSHA inspector or OSHA inspection team members at all times when they are in the operations.

---

[1] 5 U.S.C. Section 552.
[2] Supra. Also see Appendix E of this text.

OSHA INSPECTION CHECKLIST

The following is a recommended checklist for safety professionals preparing for an OSHA inspection:

1. Assemble a team from the management group and identify specific responsibilities in writing for each team member. The team members who should be given appropriate training and education should include, but not be limited to:
   a. an OSHA inspection team coordinator
   b. a document control individual
   c. individuals to accompany the OSHA inspector
   d. a media coordinator
   e. an accident investigation team leader (where applicable)
   f. a notification person
   g. a legal advisor (where applicable)
   h. a law enforcement coordinator (where applicable)
   i. a photographer
   j. an industrial hygienist
2. Decide on and develop a company policy and procedures to provide guidance to the OSHA inspection team.
3. Prepare an OSHA inspection kit, including all equipment necessary to properly document all phases of the inspection. The kit should include equipment such as a camera (with extra film and batteries), a tape player (with extra batteries), a video camera, pads, pens, and other appropriate testing and sampling equipment (that is, a noise level meter, an air sampling kit, etc.).
4. Prepare basic forms to be used by the inspection team members during and following the inspection.
5. When notified that an OSHA inspector has arrived, assemble the team members along with the inspection kit.
6. Identify the inspector. Check his or her credentials and determine the reason for and type of inspection to be conducted.
7. Confirm the reason for the inspection with the inspector (targeted, routine inspection, accident, or in response to a complaint).
   a. For a random or target inspection:
       – Did the inspector check the OSHA 200 Form?
       – Was a warrant required?
   b. For an employee complaint inspection:
       – Did the inspector have a copy of the complaint? If so, obtain a copy.
       – Do allegations in the complaint describe an OSHA violation?
       – Was a warrant required?
       – Was the inspection protested in writing?
   c. For an accident investigation inspection:
       – How was OSHA notified of the accident?

    – Was a warrant required?
    – Was the inspection limited to the accident location?
  d. If a warrant is presented:
    – Were the terms of the warrant reviewed by local counsel?
    – Did the inspector follow the terms of the warrant?
    – Was a copy of the warrant acquired?
    – Was the inspection protested in writing?
 8. Describe the opening conference:
  a. Who was present?
  b. What was said?
  c. Was the conference taped or otherwise documented?
 9. Note what was recorded:
  a. What records were requested by the inspector?
  b. Did the document control coordinator number the photocopies of the documents provided to the inspector?
  c. Did the document control coordinator maintain a list of all photocopies provided to the inspector?
 10. Describe facility inspection:
  a. What areas of the facility were inspected?
  b. What equipment was inspected?
  c. Which employees were interviewed?
  d. Who was the employee or union representative present during the inspection?
  e. Were all the remarks made by the inspector documented?
  f. Did the inspector take photographs?
  g. Did a team member take similar photographs?[1]

Following the closing conference and upon departure of the OSHA inspector or OSHA inspection team, the management team members should assemble and the documentation from the inspection should be collated. If OSHA is to return to the facility for a protracted inspection period, the documentation should be assembled at the end of each day. Careful assembly of this documentation, noting date, time, and other pertinent information on all documents, will permit this documentation to be appropriately utilized by the safety professional and legal counsel when preparing the case for contest or hearing.

There is absolutely no substitution for a well-managed safety and health program. Safety professionals should ensure that their upper-level management team realizes that they cannot get by on a shoestring safety and health program. Each aspect of the safety and health program is important, and this includes preparing for an OSHA inspection. The above checklist was provided as an example of what can be done prior to an OSHA inspection. Although an OSHA inspection can be viewed as adversarial, safety professionals should remember that OSHA inspectors are just doing their jobs to the best of their abilities. Please treat the OSHA compliance officers with respect and courtesy and they will generally be fair and even helpful.

---

[1] "Preparing for an OSHA Inspection," Schneid, T., *Kentucky Manufacturer*, Feb. 1992.

If your program is in good shape, the OSHA inspection is verification of your good work. If violations are noted, this may be assistance to the safety professional to motivate management to provide the necessary resources to achieve compliance. Above all, an OSHA inspection is not a personal attack on your safety and health efforts. Most safety professionals and OSHA have the same objectives — to create and maintain a safe working environment for your employees.

6 OSHA Penalty Structure

Money has little value to its possessor unless it also has value to others.

Leland Stanford

If you want to get a group of safety professionals' attention, simply talk about the huge fine OSHA just levied against a company for a multitude of safety violations. The safety professionals thank the gods it was not their company but then crowd around to find out the details as if they were watching the aftermath of a car accident. OSHA is part and parcel of what a safety professional does in his or her career. OSHA is interwoven into everything a safety professional does in his or her daily activities and job responsibilities. OSHA is, to a greater or lesser extent depending on the company, the essence of safety in the United States.

Penalties within the OSHA scheme are simply the "stick" through which is to make companies who are not in compliance create a workplace for their employees which is in compliance with the OSHA standards. And let us dispel all of the rumors right here at the beginning of the chapter. First, OSHA does not get the money from the penalties — it goes into the general fund. Second, OSHA compliance officers do not have a quota of violations and penalties they are required to write each month. Third, OSHA inspectors do not get a bonus for writing more citations. Fourth, the OSHA compliance officer does not get a cut of the penalties — OSHA's goal, in essence, is the same as all safety professionals — namely, to create and maintain a safe and healthful work environment in all American workplaces. And you know there is a secret to avoiding OSHA penalties — create and maintain a safe and healthful workplace and there will be no penalties!

This being said, OSHA does possess a monetary penalty scheme to motivate those companies to create and maintain a safe and healthful work environment. Safety professionals, as the leader of the safety efforts and agent of the company or organization, should acquire a working knowledge of the various levels of penalties and the requirements for each level in order to properly represent their company during and after an OSHA inspection. The OSH Act provides for a wide range of penalties, from a simple notice with no fine to criminal prosecution. The Omnibus Budget Reconciliation Act of 1990 multiplied maximum penalties sevenfold. Violations are categorized and penalties may be assessed as outlined in Table 6.1.

Safety professionals should be aware that each alleged violation is categorized and the appropriate fine issued by the OSHA area director. Please note that OSHA inspectors do not issue the monetary penalties. It should also be noted that each citation is separate and carries with it an individual monetary fine. The gravity of the violation is the primary factor in determining penalties.[1] In assessing the gravity of a violation, the OSHA compliance officer and OSHA area director must consider the

---

[1] OSHA Compliance Field Operations Manual (*OSHA Manual*) at XI-C3c (April 1977).

TABLE 6.1
Violation and Penalty Schedule

Penalty	Old Penalty Schedule (in dollars)	New Penalty Schedule (1990) (in dollars)
De minimis notice	0	0
Non-serious	0–1000	0–7000
Serious	0–1000	0–7000
Repeat	0–10,000	0–70,000
Willful	0–10,000	25,000 minimum 70,000 maximum
Failure to abate notice	0–1000 per day	0–7000 per day
Failure to posting		0–7000

severity of the injury or illness that *could* result and the probability that an injury or illness *could* occur as a result of the violation.

In assessing monetary fines to correlate with the type and category of violation, OSHA provides specific penalty assessment tables to assist the area director in determining the appropriate fine for the violation. As can be seen, each category possesses a range of monetary fines depending on the circumstances.

After selecting the appropriate penalty table, the OSHA area director determines the degree of probability that the injury or illness will occur by considering:

1. the number of employees exposed;
2. the frequency and duration of the exposure;
3. the proximity of employees to the point of danger;
4. factors such as the speed of the operation that require work under stress; and
5. other factors that might significantly affect the degree of probability of an accident.[1]

OSHA has defined a serious violation as "an infraction in which there is a substantial probability that death or serious harm could result ... unless the employer did not or could not with the exercise of reasonable diligence, know of the presence of the violation."[2] Section 17(b) of the OSH Act requires that a penalty of up to $7000 be assessed for every serious violation cited by the compliance officer. Where a company or organization possesses duplicate operations, if one process is cited as possessing a serious violation, it is possible that each of the duplicate processes or machines may also be cited for the same violation. Thus, if a serious violation is found in one machine and there are many other identical machines in the enterprise, a single serious violation can possibly be multiplied by a number of the same violations on different but identical machines.[3]

---

[1] Id. at (3)(a).
[2] 29 U.S.C. Section 666.
[3] Id.

Safety professionals should never receive a repeat violation. Not only does this reflect poorly on the management of the safety program, a repeat violation carries one of the greatest monetary liabilities. The three major categories carrying the potentially greatest monetary penalties include "repeat violations," "willful violations," and "failure to abate" cited violations. A repeat violation is a second citation for a violation that was cited previously by a compliance officer. OSHA maintains records of all violations and must check for repeat violations after each inspection. A willful violation is the employer's purposeful or negligent failure to correct a known deficiency. This type of violation, in addition to carrying a large monetary fine, exposes the employer to a charge of an egregious violation and the potential for criminal sanctions under the OSH Act or state criminal statutes if an employee is injured or killed as a direct result of the willful violation. Failure to abate a cited violation has the greatest cumulative monetary liability. OSHA may assess a penalty of up to $1000 per day per violation for each day in which a cited violation is not brought into compliance.

Safety professionals should be aware that in assessing monetary penalties, the OSHA area or regional director must consider the good faith of the employer, the gravity of the violation, the employer's past history of compliance, and the size of the employer. Beginning in 1994, former Assistant Secretary of Labor Joseph Dear noted that OSHA will start using its egregious case policy (which was seldom used prior to 1994). Under the egregious violation policy, when violations are determined to be conspicuous, penalties are cited for each violation, rather than combining the violations into a single, smaller penalty.[1]

In addition to the potential civil or monetary penalties that could be assessed, safety professionals should be aware that OSHA regulations may be used as evidence in negligence, product liability, workers' compensation, and other actions involving employee safety and health issues. OSHA standards and regulations are often considered the baseline requirements for safety and health that must be met, not only to achieve compliance with the OSHA regulations, but also to safeguard an organization against other potential civil actions. The use of OSHA standards in private litigation and interaction with other laws are discussed in depth later in this text.

Safety professionals should be aware that the OSHA monetary penalty structure is classified according to the type and gravity of the particular violation. Violations of OSHA standards or the general duty clause are categorized as "de minimis," "other" (non-serious), "serious," "repeat," and "willful." There are also "Failure to Post" and "Failure to Abate" violations. Monetary penalties assessed by the Secretary vary according to the degree of the violation. Penalties range from no monetary penalty to 10 times the imposed penalty for repeat or willful violations.[2] Additionally, the Secretary may refer willful violations to the U.S. Department of Justice for imposition of criminal sanctions.[3]

Starting with the smallest violation penalty category, a *de minimus* violation of an OSHA standard does not immediately or directly relate to safety or health, and the potential of injury is miniscule. With these types of violations, OSHA usually

---

[1] *Occupational Safety and Health Reporter*, V.23, No. 32, Jan. 12, 1994.

[2] 29 C.F.R. Sections 658(a) and 666(c).

[3] Id. at (e).

either does not issue a citation or issues a de minimis citation. Section 9 of the OSH Act provides that "[the] Secretary may prescribe procedures for the issuance of a notice in lieu of a citation with respect to de minimis violations which have no direct or immediate relationship to safety or health."[1] A *de minimis* notice does not constitute a citation and no fine is imposed. Additionally, there usually is no abatement period, and thus there can be no violation for failure to abate.

The *OSHA Compliance Field Operations Manual* (*OSHA Manual*) provides two examples of when *de minimis* notices are generally appropriate:

1. "In situations involving standards containing physical specificity wherein a slight deviation would not have an immediate or direct relationship to safety or health"
2. "Where the height of letters on an exit sign is not in strict conformity with the size requirements of the standard."[2]

Although usually minor in nature, safety professionals should be aware that OSHA has found de minimis violations in cases where employees, as well as the safety records, are persuasive in exemplifying that no injuries or lost time have been incurred. Safety professionals should take these noted violations seriously and correct the violation as quickly as possible.

Some of the most common violations are other or non-serious violations. These types of violations are issued where a violation could lead to an accident or occupational illness, but the probability that it would cause death or serious physical harm is minimal. Such a violation, however, does possess a direct or immediate relationship to the safety and health of workers.[3] These types of violations often carry a minimal monetary penalty and should be addressed immediately by the safety professional.

Safety professionals should be aware that there is often a very thin line between whether a violation is categorized as serious or non-serious. In distinguishing between a serious and a non-serious violation, the Occupational Safety and Health Review Commission (OSHRC) has stated that "a non-serious violation is one in which there is a direct and immediate relationship between the violative condition and occupational safety and health but no such relationship that a resultant injury or illness is death or serious physical harm."[4] The *OSHA Manual* provides guidance and examples for issuing non-serious violations. It states that

an example of non-serious violation is the lack of guardrail at a height from which a fall would more probably result in only a mild sprain or cut or abrasion; i.e., something less than serious harm.[5]

A number of non-serious violations (which) are present in the same piece of equipment which, considered in relation to each other, affect the overall gravity of possible injury

---

[1] Id. at Section 658(a).
[2] *OSHA Manual* at VIII-B2a.
[3] *OSHA Manual* at VIII-B2a.
[4] See, *Crescent Wharf & Warehouse* Co., 1 OSH Cases 1219 (1973).
[5] *OSHA Manual* at VIII-B2a.

resulting from an accident involving the combined violations … may be grouped in a manner similar to that indicated in the preceding paragraph, although the resulting citation will be for a non-serious violation.[1]

Safety professionals should be aware that the difference between a serious and a non-serious violation hinges on subjectively determining the probability of injury or illness that might result from the violation. The decision as to the category of the violation usually turns on the particular facts of the situation. The OSHRC has reduced serious citations to non-serious violations when the employer was able to show that the probability of an accident, and the probability of a serious injury or death, was minimal.[2]

A serious violation is where "there is a substantial probability that death or serious physical harm could result from a condition which exists, or from one or more practices, means, methods, operations or processes which have been adopted or are in use, in such place of employment unless the employer did not, and could not with exercise of reasonable diligence, know of the presence of the violation."[3]

Safety professionals should be aware that Section 17(b) of the Act provides that a civil penalty of up to $7000 *must* be assessed for serious violations, whereas for non-serious violations civil penalties *may* be assessed.[4] Although the violation and severity will depend on the facts of the situation, the amount of the penalty is usually determined by considering the gravity of the violation, the size of the employer, the good faith of the employer, and the employer's history of previous violations.[5]

Safety professionals should be aware that OSHA has the burden of proving the violation as well as the categorization. For OSHA to prove that a violation is within the serious category, OSHA must show only a substantial probability that a foreseeable accident would result in serious physical harm or death. Thus, contrary to common belief, OSHA does not need to show that a violation would create a high probability that an accident would result. Because substantial physical harm is the distinguishing factor between a serious and a non-serious violation, OSHA has defined "serious physical harm" as "permanent, prolonged, or temporary impairment of the body in which part of the body is made functionally useless or is substantially reduced in efficiency on or off the job." Additionally, an occupational illness is defined as "illness that could shorten life or significantly reduce physical or mental efficiency by inhibiting the normal function of a part of the body."[6]

Once OSHA determines that a hazardous condition exists and that employees are exposed or potentially exposed to the hazard, the *OSHA Manual* instructs compliance officers to use a four-step approach to determine whether the violation is serious:

---

[1] Id.

[2] Secretary v. Sky-Hy Erectors & Equip., 4 O.S.H. Cases 1442 (1976). *But see Shaw Constr. v. OSHRC*, 534 F.2d 1183, 4 O.S.H. Cases 1427 (5th Cir. 1976) (holding that serious citation was proper whenever accident was merely possible).

[3] 29 U.S.C. Section 666(j).

[4] Id.

[5] Id.

[6] *OSHA Manual* at VIIIB1b.

1. Determine the type of accident or health hazard exposure that the violated standard is designed to prevent in relation to the hazardous condition identified.
2. Determine the type of injury or illness which it is reasonably predictable could result from the type of accident or health hazard exposure identified in step 1.
3. Determine that the type of injury or illness identified in step 2 includes death or a form of serious physical harm.
4. Determine that the employer knew or with the exercise of reasonable diligence could have known of the presence of the hazardous condition.[1]

Safety professionals should keep in mind that OSHA is required to prove the causal link between the violation of the safety or health standard and the hazard, reasonably predictable injury or illness that could result, potential for serious physical harm or death, and the employer's ability to foresee such harm by using reasonable diligence.[2]

Safety professionals should be aware that OSHA is not required to show that the employer actually knew that the cited condition violated safety or health standards. The employer can be charged with constructive knowledge of the OSHA standards. OSHA also does not have to show that the employer could reasonably foresee that an accident would happen, although it does have the burden of proving that the possibility of an accident was not totally unforeseeable. OSHA does need to prove, however, that the employer knew or should have known of the hazardous condition and that it knew there was a substantial likelihood that serious harm or death would result from an accident. If OSHA cannot prove that the cited violation meets the criteria for a serious violation, the violation may be cited in one of the lesser categories.[3]

Safety professionals should be aware that the most severe monetary penalties under the OSHA penalty structure are for willful violations. A willful violation can result in penalties of up to $70,000 per violation, with a minimum required penalty of $5000. Although the term "willful" is not defined in OSHA regulations, courts generally have defined a willful violation as "an act voluntarily with either an intentional disregard of, or plain indifference to, the Act's requirements."[4] Further, the OSHRC defines a willful violation as "action taken knowledgeably by one subject to the statutory provisions of the OSH Act in disregard of the action's legality. No showing of malicious intent is necessary. A conscious, intentional, deliberate, voluntary decision is properly described as willful."[5]

Safety professionals should be very cautious when violations of this type are identified. There is little distinction between civil and criminal willful violations other than the due process requirements for a criminal violation and the fact that a violation of the general duty clause cannot be used as the basis for a criminal willful violation. The distinction is usually based on the factual circumstances and

---

[1] Id.
[2] Id.
[3] Id at (4).
[4] See *Cedar Construction Co. v. OSHRC*, 587 F.2d 1303, 6 OSH Cases 2010 (D.C. Cir. 1971); *U.S. v. Dye Constr.* 522 F.2d 777, 3 OSH Cases 1337 (4th Cir. 1975); *Empire-Detroit Steel v. OSHRC*, 579 F.2d 378, 6 OSH Cases 1693 (6th Cir. 1978).
[5] See P.A.F. Equipment Co., 7 OSH Cases 1209 (1979).

the fact that a criminal willful violation results from a willful violation that caused an employee death.

According to the *OSHA Manual*, the compliance officer "can assume that an employer has knowledge of any OSHA violation condition of which its supervisor has knowledge; he can also presume that, if the compliance officer was able to discover a violative condition, the employer could have discovered the same condition through the exercise of reasonable diligence."[1]

Courts and the OSHRC have agreed on three basic elements of proof that OSHA must show for a willful violation. OSHA must show that the employer (1) knew or should have known that a violation existed, (2) voluntarily chose not to comply with the OSH Act to remove the violative condition, and (3) made the choice not to comply with intentional disregard of the OSH Act's requirements or plain indifference to them properly characterized as reckless. Courts and the OSHRC have affirmed findings of willful violations in many circumstances, ranging from deliberate disregard of known safety requirements through fall protection equipment not being provided. Other examples of willful violations include cases where safety equipment was ordered but employees were permitted to continue work until the equipment arrived, inexperienced and untrained employees were permitted to perform a hazardous job, and an employer failed to correct a situation that had been previously cited as a violation.[2]

Safety professionals should attempt in every way to avoid the "repeat" and "failure to abate" categories of violations. Repeat and failure to abate violations are often quite similar and confusing to safety professionals. "Repeat" and "Failure to Abate" tend to reflect badly on you and your program. These citations involve a management failure to correct a known violation. In essence, OSHA cited a violation. Upon re-inspection, the violation of a previously cited standard is found, but the violation may not involve the same machinery, equipment, process, or location. This would constitute a repeat violation. If, upon re-inspection by OSHA, a violation of a previously cited standard is found but evidence indicates that the violation continued uncorrected since the original inspection, this would constitute a failure to abate violation.[3]

The potentially largest civil penalty under the OSH Act is for repeat violations. The OSH Act authorizes a penalty of up to $70,000 per violation but permits a maximum penalty of ten times the maximum authorized for the first instance of the violation. Repeat violations can also be grouped within the willful category (i.e., a willful repeat violation) to acquire maximum civil penalties.

Safety professionals should note that when a previous citation has been contested but a final OSHRC order has not yet been received, a second violation is usually cited as a repeat violation. The *OSHA Manual* instructs the compliance officer to notify the assistant regional director and to indicate on the citation that the violation is contested. If the first citation never becomes a final OSHRC order (that is, the citation is vacated or otherwise dismissed), the second citation for the repeat violation will be removed automatically.[4]

---

[1] *OSHA Manual* at VIII-B1c(4).

[2] Id.; Also, see *Universal Auto Radiator Mfg. Co. v. Marshall,* 631 F.2d 20, 8 OSH Cases 2026 (3rd Cir. 1980); *Haven Steel Co. v. OSHRC,* 738 F.2d 397, 11 OSH Cases 2057 (10th Cir. 1984).

[3] *OSHA Manual* at VIII-B5c.

[4] Id.

When a failure to abate violation occurs upon re-inspection, the OSHA compliance officer has found that the company or organization has failed to take necessary corrective action and thus the violation continues uncorrected. The penalty for a failure to abate violation can be up to $7000 per day to a maximum of $70,000. Safety professionals should also be aware that citations for repeat violations, failure to abate violations, or willful repeat violations can be issued for violations of the general duty clause. The *OSHA Manual* instructs compliance officers that citations under the general duty clause are restricted to serious violations or to willful or repeat violations that are of a serious nature.[1]

Safety professionals should never be cited for a failure to post. A relatively new penalty category, the failure to post violation carries a penalty of up to $7000 for each violation. A failure to post violation occurs when an employer fails to post notices required by the OSHA standards, including the OSHA poster, a copy of the year-end summary of the OSHA 300 form, a copy of OSHA citations when received, and copies of other pleadings and notices.

Although the probability of being inspected is relatively low for the average U.S. workplace, OSHA inspections do happen, and the monetary penalties can be substantial for identified violations. Safety professionals should possess a working knowledge of the OSHA monetary penalties as well as the elements required to prove the violation category. A safety professional should educate his or her management team to the potential risks of monetary penalties for non-compliance and take a proactive approach in eliminating or minimizing the underlying potential violations. Additionally, safety professionals, in the event of a citation, should analyze every aspect and element of the alleged violation in order to identify any challenges that may be available regarding the categorization of the alleged violation. Remember, OSHA will attempt to cite the highest category possible which can be supported by the evidence. Safety professionals should ensure that OSHA possesses this proof and that the proper category of monetary penalty is assessed in the event of a citation.

---

[1] Id. at XI-C5c.

7 Defenses to an OSHA Citation

A long dispute means that both parties are wrong.

Voltaire

As part of a clear and careful analysis, a prudent safety professional should identify the possible defenses that may be available as well as the probability of success for each defense. As identified in this chapter, safety professionals should be aware that not all defenses are available in every circumstance, and some defenses may be more viable than others. There are a number of defenses that may be applicable to any given situation. The main defenses tend to fall into four basic categories, namely, (1) constitutional challenges, (2) denial of the violation of the standard, (3) challenging the standard, and (4) challenging the process of the citation. Although there may be other defenses available depending on the circumstances, most defenses fall within one or more of these categories.

Under the category of constitutional challenges, these defenses are the broadest in scope and are framed around one or more of the amendments to the U.S. Constitution. Safety professionals should be aware that the U.S. Supreme Court has yet to find any provision of the OSH Act to be unconstitutional.[1] However, the U.S. Supreme Court has ruled on three cases that "show a chronological development of increasing sophistication in constitutional attacks."[2] In *Atlas Roofing Co. v. OSHRC* (Occupational Safety and Health Review Commission), the Court found that the OSH Act's penalty provisions did not provide a right to a jury trial.[3] The second major decision was in *Marshall v. Barlows* (discussed previously and in Appendix E of this text) requiring OSHA to show administrative probable cause in order to acquire an inspection warrant.[4] In the third major case, the Court found that the benzene standard was improperly promulgated.[5]

In addition to the above cases, there have been constitutional challenges based upon primarily Article III of the Constitution and virtually every Amendment of the Constitution as well as other rights that have been recognized in the Constitution. Before embarking on a constitutional challenge, safety professionals are cautioned to discuss these possible defenses with legal counsel. There is a substantial likelihood

---

[1] Connolly, W.B., and Crowell, D.R., *The Practical Guide to the Occupational Safety and Health Act*, ALM, New York, 1991.
[2] Id at 2.01.
[3] 430 U.S. 442, 97 S.Ct. 1261, 51 L.Ed.2d 464 (1977).
[4] 436 U.S. 307, 98 S.Ct. 1816, 56 L.Ed.2d 35 (1978).
[5] *Ind. Union Dept v. Am. Petroleum Institute*, 448 U.S. 607, 100 S.Ct. 2844, 65 L.Ed. 2d 1010 (1980).

that any constitutional challenges to an OSHA citation will result in protracted litigation and substantial costs.

More often, safety professionals utilize more mundane defenses such as the factual defense to deny the violation of the standard. These fact-based defenses include but are not limited to such defenses as the isolated incident defense, greater hazard defense, performing work impossibility defense, "we didn't know" defense, and "machine wasn't working" defense. These defenses can be utilized at the informal conference stage as well as on appeal to the OSHRC or beyond and are usually based upon the evidence acquired before, during, and after the actual compliance inspection.

In the isolated incident defense, the safety professional must have evidence to prove that "(1) the violation charged resulted exclusively from the employee's conduct; (2) the violation was not participated in, observed by, or performed with the knowledge and/or consent of any supervisory personnel; and (3) that the employee's conduct contravened a well-established company policy or work rule which was in effect at the time, well publicized to the employees, and actively enforced through disciplinary action or other appropriate procedures."[1] Additionally, safety professionals should be aware that the written compliance program that is the basis for the alleged violation must be in place and functional and employees must be trained on the specific safety procedures.[2]

Safety professionals are cautioned that the isolated incident defense is an affirmative defense that must be raised by the company or organization at the beginning of the case.

The greater hazard defense is again based on the facts of the situation and, in essence, argues that compliance with the standard would create a greater hazard and risk of injury to employees than the current procedures or process already in place. Although this affirmative defense carries a heavy evidentiary burden for the safety professional to prove that the standard creates a greater hazard, this defense has been successful in several circumstances, such as fixed guardrails, fall protection, and unique equipment.[3]

The impossible to perform work defense is based on the facts of the situation and involves proving that compliance with the applicable standard is impossible to achieve due to the specific work or equipment requirements. In essence, the safety professional must provide that there is no feasible way in which to achieve compliance with the applicable standard requirements. Safety professionals must show a good faith effort to attempt to comply, and this effort should be supported by feasibility studies, consultant's reports, and related documentation.[4] Of particular importance to safety professionals when utilizing this defense is that no representative of the employer can admit or imply admission that compliance is possible or the defense will fail.[5]

---

[1] *The Practical Guide to the Occupational Safety and Health Act*, supra n. 1. Also see, *Otis Elevator Co.*, 5 OSHC 1514 (1979); *Raymond J. Pitts, Inc.*, 4 OSHC 1058 (1976); *Bill Turpin Painting, Inc.*, 5 OSHC 1576 (1977).

[2] *OSHA Field Manual* at X-(E)(1)(c).

[3] See *Ashland Oil, Inc.*, 1 OSHC 3246 (ALJ, 1973); *Ind. Steel Erectors, Inc.*, 1 OSHC 1497 (1974).

[4] See *Int. Paper Co.*, 2 OSHC 3173 (ALJ, 1974); *U.S. Steel Corp. v. OSHRC*, 537 F.2d 780 (3rd Cir. 1976); *Reynolds Metal Co.*, 3 OSHC 2051 (1976).

[5] *Taylor Bldg Assoc.*, 5 OSHC 1083 (1977); Dic-Underhill, a Joint Venture, 5 OSHC 1271 (1977).

One of the most commonly utilized defenses is the "we didn't know" or lack of employer's knowledge defense. Although the employer's knowledge is part of OSHA's burden to prove when citing serious and willful violations, the lack of knowledge can be reversed and utilized as a defense in certain circumstances. OSHA possesses the duty to prove the employer had actual or constructive knowledge of the alleged violation. If the safety professional can prove that the company or organization did not have knowledge, OSHA is not able to meet its burden in proving the case.[1] This burden is clear for serious violations given the specific language in the OSH Act but is unclear as to non-serious and other categories of violations.[2]

Another commonly utilized affirmative defense, if the facts support the defense, is the "machine is not working" defense. This defense focuses on the primary issue that the machine in which the violation was noted was not operational at the time of the inspection and thus no employees were exposed to the alleged hazard. If the machine is not operating at the time of the inspection, is locked or tagged for repairs, is isolated or stored, or related situations, this factual defense may be available.[3]

In the area of procedural defenses, safety professionals will be focusing on the jurisdictional components as well as specificity of OSHA's allegations and adherence to the requirements of the OSH Act. In essence, the safety professional should look to ensure that the entity is under OSHA's jurisdiction and that OSHA did everything properly in issuing the citation.

One of the first procedural defenses to review is whether the company or operations in which the inspection was conducted and the citation was issued is within the jurisdiction of OSHA. If the operation is outside of OSHA's jurisdiction, such as an agricultural operation or other operation that has been exempted from the OSH Act, this defense may be employed. Additionally, if the company or operation is covered under a state plan program and the inspection was conducted by federal OSHA, this may be outside of the jurisdictional boundaries.

Safety professionals should also look at the time period between when the alleged violation occurred and the issuance of the citation. If the citation is issued more than 6 months following the completion of the inspection, this defense may be viable.[4] Safety professionals should evaluate this possible defense closely especially if OSHA has amended the citation in any manner.[5]

As noted previously in this text, OSHA is required to transmit the Notice of Contest filed in a timely manner (15 working days) by the company or organization to the OSHRC "within 7 days of receipt."[6] If the timely filed Notice of Contest is not filed within the 7 days with the OSHRC, this may be utilized as a defense.

Additionally, OSHA possesses the burden of providing proper service of the citation to the company or organization. OSHA is required to "within a reasonable time after the termination of inspection or investigation to notify the employer by

---

[1] *Brennan v. OSHRC*, 511 F.2d 1139 (9th Cir. 1975).
[2] OSH Act Section 17(k); 29 USC Section 666(j).
[3] See *Horne Plumbing & Heating Co. v. OSHRC*, 5218 F.2d 564 (5th Cir. 1976); *Dunlop v. Rockwell Int. Constr. Co.*, 540 F.2d 1283 (6th Cir. 1976).
[4] See *Brennan v. Chicago Bridge & Iron*, 1 OSHC 1485 (1974).
[5] See *Duane Smelser Roofing Co.*, 4 OSHC 1948 (1976).
[6] OSHRC Rule 32, 29 CFR Section 2200.12.

certified mail of the penalty, if any, proposed to be assessed."[1] This could potentially be an important defense for safety professionals whose mail address is different than the inspection site location given the 15-day Notice of Contest time limitation. However, safety professionals should be aware that the OSHRC tends to look unfavorably upon companies or organizations who utilize this defense when it can be inferred from the evidence that the company or organization possessed actual knowledge of service but the service was to an employee or agent of the company or organization who did not possess authority to accept service.

A procedural defense often used by companies and organizations who utilize temporary workers, loaned employees, and subcontractors is the lack of an employment relationship to the employees involved in the cited violation. This defense closely correlates to the multi-employer worksite rule discussed in depth in Chapter 13. In essence, this defense argues that OSHA cannot cite the company or organization because the hazard was created and controlled by another entity or company. The key issue for safety professionals is that of control of the workplace.[2] Safety professionals should review this possible defense closely in light of the recent decision in *Secretary of Labor v. Summit Contractors, Inc.* (See Appendix E.)[3]

Safety professionals may consider employing the "vagueness" defense where the standard, especially a new standard, "is not construed to give effect to the natural and plain meaning of its words."[4] In essence, OSHA has a duty to ensure that the standard is written in "clear and concise language so that employers will be better able to understand and observe them (standards)."[5] Safety professionals should be aware that industry customs are considered and every word of the standard should be examined. Safety professionals should examine the particular circumstance as applicable to the standard rather than a general claim that the standard is vague.[6]

Safety professionals should be aware that OSHA possesses a duty to state the alleged violation "with particularity."[7] This requirement has created the defense of a lack of particularity, especially in circumstances involving companies or organizations that were not able to correct alleged violations because the citation was not specific and situations where the company or organization was not provided fair notice of the violation. Although there are no specific guidelines in this area, the OSHRC has vacated citations where information is missing and information has not been sufficiently clear.[8] This is an affirmative defense, and the OSHRC has adopted a "totality of circumstances" test in analyzing this defense.[9]

Safety professionals should be aware that there are other procedural defenses that can be utilized depending on the circumstances. Some of these additional procedural defenses include that the standard was improperly promulgated defense, the

---

[1] OSH Act Section 10(a); 29 CFR Section 659(a).

[2] See *Gilles & Cotting, Inc.*, 1 OSHC 1388, rev'd sub nom. *Brennan v. Gilles & Cotting, Inc.*, 504 F.2d 1255 (4th Cir.), on remand, 3 OSHC 2002 (1976).

[3] OSHRC Docket No. 03-1622 (2007).

[4] See *Johnson v. Udall*, 292 F.Supp 738(C.D. Cal 1968).

[5] *Diamond Roofing v. OSHRC*, 528 F.2d 645 (5th Cir. 1976).

[6] See *Santa Fe Trail Transport Co.*, 1 OSHC 1457 (1973), rev'd sub nom.

[7] OSH Act Section 9(a).

[8] See *P&M Sales, Inc.*, 4 OSHC 1158 (1976).

[9] See *Meadows Industries, Inc.*, 7 OSHC 1709 (1979); *Gold Kist, Inc.*, 7 OSHC 1855 (1979).

citation or complaint was improperly amended defense, and the lack of due process defense among other defenses.

Safety directors are advised to seek the assistance of competent legal counsel to guide any attempt to utilize any of the above defenses or other defenses beyond the informal conference stage of any contest. The above defenses are provided to safety professionals not as legal advice but to exemplify the various nature and types of defenses that may be possible depending on the individual circumstances. Careful analysis of the fact and circumstances as well as possible defenses that may be available is essential in the preparation of any contest before the OSHRC or the courts.

8 Criminal Sanctions under the OSH Act and Beyond

If we are to keep our democracy, there must be one commandment: Thou shalt not ration justice.

Learned Hand

The first rule is that safety professionals are responsible for their own criminal acts inside and outside of the workplace. If the act performed by the safety professional is in violation of a criminal law, the individual safety professional may be the responsible party. If the company or organization is responsible for the criminal act, then the company, the officers, or the board of directors may be responsible for the crime. Safety professionals should be aware that outside of their official duties, they are *not* immune from the committing of criminal acts. Safety professionals, working within the scope of their jobs, may be protected by the company or organization.

Additionally, safety professionals should be aware that the criminal law rules and laws are substantially different than the OSH Act and other related civil laws. Depending on the alleged crime, the criminal laws used can be either federal or state criminal laws that can be substantially different in scope and penalty. In most circumstances, the location of the criminal act or the type of act is the deciding factor as to whether the crime would be prosecuted by state or federal authorities. For example, if the safety professional shoots and kills a person on a city street in a manner that constituted the category of murder, the murder would usually be prosecuted by the state authorities. If the murder was committed on federal property, the murder charge would be prosecuted by federal authorities. Certain crimes, such as crimes involving interstate actions (like mail fraud), usually are prosecuted by federal authorities. Each particular crime possesses elements that must be proven by the federal or state prosecutor. The burden of proof is on the federal or state prosecutor to prove each element of the crime. Trial by jury is usually permitted in all criminal cases, and the penalty for the crime is usually proscribed by law.[1]

For safety professionals, the greatest potential of criminal sanctions for work-related activities is under the OSH Act. Although the potential is not great in comparison with the number of workplaces and employees in the United States, OSHA does refer a number of cases to the U.S. Department of Justice for criminal investigation and prosecution. According to Assistant Secretary Edwin G. Foulke, Jr., "since 2001, as part of its strong enforcement program, OSHA proposed more than three-quarters of a billion dollars in penalties for safety and health violations and made

---

[1] See, for example, the Federal Sentencing Guidelines, 18 U.S.C. § 3577.

56 criminal referrals to the Department of Justice, which represents more than 25 percent of all criminal referrals in the history of the Agency."[1]

Safety professionals should be aware that potential criminal liability may exist for individuals as well as companies and organizations through the use of state criminal laws by state and local prosecutors for injuries and fatalities that occur on the job. This utilization of the standard state criminal laws in a workplace setting normally governed by OSHA or state plan programs is controversial as can be seen from the cases below, but it appears to be a viable method through which individual states can penalize safety professionals, companies, and company officers in situations involving fatalities or serious injury. The issue of preemption appears to be settled, thus permitting either OSHA to refer the matter to the U.S. Department of Justice or the individual state prosecuting under their state criminal code, but not both, which would constitute double jeopardy.

The utilization of state criminal laws by prosecutors, at this time, is not preempted by the OSH Act. It should be clarified that the use of criminal sanctions for workplace fatalities and injuries is not a new phenomenon. Under the OSH Act, criminal sanctions have been available since the inception in 1970. In Europe, criminal sanctions for workplace fatalities are frequently used, and in the United States as far back as 1911, the state criminal code was utilized to prosecute the owners in the well-known Triangle Shirt fire in New York which killed over 100 young women.

In recent years, the case that propelled the issue of whether OSHA had jurisdiction over workplace injuries and fatalities and thus "preempts" state prosecution under state criminal statutes for workplace deaths into the limelight was the first-degree murder convictions by an Illinois court of the former president, the plant manager, and the plant foreman of Film Recovery Systems, Inc. for the 1983 work-related death of an employee.[2] In this case, Stephan Golab, a 59-year-old immigrant employee from Poland, died as a direct result of his work at this Elk Grove, Illinois, manufacturer, where he stirred tanks of sodium cyanide used in the recovery of silver from photographic films. In February 1983, Golab "walked into the plant's lunchroom, started violently shaking, collapsed and died from inhaling the plant's cyanide fumes."[3] Following his death, both OSHA and the Cook County prosecutor's office investigated the accident.

OSHA found twenty violations and fined the corporation $4850. This monetary penalty was later reduced by half.[4] The Cook County prosecutor's office, on the other hand, took a different view and filed charges of first-degree murder and 21 counts of reckless conduct against the corporate officers and management personnel and involuntary manslaughter charges against the corporation itself.[5] Under the Illinois

---

[1] Statement of Edwin G. Foulke, Jr., Asst. Secretary, OSHA, U.S. DOL, before the Subcommittee on Workforce Protections, Committee on Ed. and Labor, U.S. House of Representatives, April 24, 2007.

[2] *People v. O'Neil, et al. (Film Recovery Systems)*, Nos. 83 C 11091 & 84 C 5064 (Cir. Ct. of Cook County, Ill. June 14, 1985), rev'd, 194 Ill. App.3d 79, 550 N.E.2d 1090 (1990).

[3] Gibson, "A Worker's Death Spurs Murder Trial," *The National Law Journal*, Jan. 21, 1985, at page 10.

[4] Note: *Getting Away with Murder — Federal OSHA Preemption of State Criminal Prosecutions for Industrial Accidents,* 101 Harv. L.J. 220 (1987).

[5] Id.; Also see Brief for Appellee at 12-20 and Gibson, "A Worker's Death Spurs Murder Trial," *The National Law Journal*, Jan. 21, 1985, at page 10.

law, murder charges can be brought when someone "knowingly creates a strong probability of death or great bodily harm," even if there is no specific intent to kill.[1] The prosecutor's office initially brought charges against five officers and managers of the corporation, but the defendant successfully fought extradition from another state.[2] The prosecutor's intent was to "criminally pierce the corporate veil" to place liability not only upon the corporation but on the responsible individuals.[3]

After a protracted trial, the prosecution was able to obtain conviction of three corporate officials for murder and fourteen counts of reckless conduct; Each was sentenced to 25 years in prison. The corporation was convicted of manslaughter and reckless conduct and fined $24,000.[4] The state court rejected outright the company's defense of preemption of the state prosecution by the Federal OSH Act, and this has sustained appellate challenge.

The decision in *People v. O' Neil* opened many safety professionals' eyes as to the potential criminal liability resulting from their actions or inactions in the workplace.[5] The case was the first time in which a corporate officer had been convicted of murder in a workplace death. As expected, once the door was opened, prosecutors across the country began to initiate similar action utilizing the state criminal codes for workplace injuries and fatalities.

Safety professionals, as well as company officers, are responsible for the safety and health of the employees under their direction. Where a company or organization willfully places an employee or others in harm's way and an injury or fatality occurs, the potential of criminal liability has now become a reality. Safety professionals should be aware that there is a potential for criminal liability not only under the OSH Act but also under individual state criminal codes.

Safety professionals should be aware that fatality and catastrophe investigations conducted by OSHA as well as investigations conducted by state prosecutors are substantially different than the typical OSHA inspection. Safety professionals should seek legal counsel to provide assistance and guidance as soon as feasible following the accident.

In OSHA's Fatality/Catastrophe Investigation procedures available at www.osha.gov, OSHA possesses a specific protocol for these types of investigations.[6] State prosecutors may utilize the state police, criminal detectives, or other related law enforcement agencies to conduct the investigation. This investigation is conducted in the same manner as any other homicide investigation with the same level of detail and thoroughness to withstand challenge in a court of law. Unlike most OSHA inspections, safety professionals should be aware that "everything you say can and will be utilized against you in a court of law." Safety professionals should acquire

---

[1] Id.

[2] Gibson, "A Worker's Death Spurs Murder Trial," *The National Law Journal*, Jan. 21, 1985, at page 10.

[3] Moberg, David, et al., *Employers Who Create Hazardous Workplaces Could Face More Than Just Regulatory Fines, They Could Be Charged with Murder,* supra at page 36.

[4] Sand, Robert, *Murder Convictions for Employee Deaths;General Standards Versus Specific Standards,* 11 Employee Rel. J. 526 (1985-86).

[5] Sand, Robert, *Murder Convictions for Employee Deaths;General Standards Versus Specific Standards*, 11 Employee Rel. J. 526 (1985-86).

[6] OSHA Directive CPL-02-137, 4-14-05.

the guidance of legal counsel before providing any statements to OSHA or state prosecutors.

Safety professionals should be aware that OSHA must be notified within 8 hours for any fatality of multiple injury situation involving multiple employees being hospitalized overnight. Safety professionals should follow the OSHA procedures for notification and ensure that all conversations or messages are properly documented. Although the investigation may be conducted from a criminal perspective, the evidence being gathered can be utilized for the purposes of civil penalties if the criminal action is not pursued. Thus, it is important that the safety professional, as identified in previous chapters, also gather and document every aspect of the investigation.

Safety professionals faced with the potential of a criminal investigation should seek legal counsel as soon as feasible. However, this being said, there is another issue as to whether the safety professional requires legal counsel for him- or herself, as an individual, or can utilize legal counsel representing the company or organization. This is often a very difficult decision given the cost, potential conflicts, and associated matters as they relate to counsel and the company or organization. Additionally, safety professionals should be aware that professional liability insurance and related insurances usually do not provide coverage for criminal matters. A safety professional facing a criminal investigation and possible prosecution under the OSH Act or under the individual state's criminal code should carefully analyze his or her individual situation as well as the company or organization for which he or she is employed to ascertain the proper course. Remember, the rules of a criminal action are substantially different than an OSHA inspection, and the penalties can be far more severe.

9 Occupational Safety and Health Review Commission

If we do not maintain justice, justice will not maintain us.

Francis Bacon

The Occupational Safety and Health Review Commission (OSHRC) is the independent judicial arm created under the OSH Act to adjudicate cases involving OSHA citations. This OSHRC consists of a panel of three judges who are appointed by the President for 6-year staggered terms and confirmed by the Senate. The OSHRC has created a court-like structure with progressive stages for the contest process and possesses specific rules related to this hearing process. (See Appendix B for OSHRC Rules and Procedures.)

Under Section 9(a) of the OSH Act, if the Secretary of Labor believes that an employer "has violated a requirement of Section 5 of the Act, of any standard, rule or order promulgated pursuant to Section 6 of this Act, or of any regulations prescribed pursuant to this Act, he shall within reasonable promptness issue a citation to the employer."[1] "Reasonable promptness" has been found and defined to mean within 6 months from the occurrence of a violation.

Section 9(a) also requires that citations be in writing and "describe with particularity the nature of the violation, including a reference to the provision of the Act, standard, rule, regulation, or order alleged to have been violated."[2] The OSHRC has adopted a fair notice test that is satisfied if the employer is notified of the nature of the violation, the standard allegedly violated, and the location of the alleged violation.[3]

The OSH Act does not specifically provide for a method of service for citations. Section 10(a) authorizes service of notice of proposed penalties by certified mail, and in most instances, the written citations are attached to the penalty notice.[4] Regarding the proper party to be served, the OSHRC has held that service is proper if it "is reasonably calculated to provide an employer with knowledge of the citation and notification of proposed penalty and an opportunity to determine whether to contest or abate."[5]

Under Section 10 of the Act, once a citation is issued, the employer, any employee, or any authorized union representative has 15 working days to file a notice of

---

[1] 29 CFR Section 651 et seq.
[2] Id.
[3] Id.
[4] Id. at Section 10(a).
[5] See, *B.J. Hughes*, 7 OSH Cases 1471 (1979).

contest.[1] If the employer does not contest the violation, abatement date, or proposed penalty, the citation becomes final and not subject to review by any court or agency. If a timely notice of contest is filed in good faith, the abatement requirement is tolled, and a hearing is scheduled. Safety professionals may also contest any part or all of the citation, proposed penalty, or abatement date. Safety professionals should be aware that contests are limited to the reasonableness of the proposed abatement date. Employees also have the right to elect party status after an employer has filed a notice of contest.

A safety professional, as the agent for the company or organization, may also file a petition for modification of the abatement period (PMA) if the company or organization cannot comply with any abatement that has become a final order. If the Secretary of Labor or a safety professional contests the PMA, a hearing is held to determine whether any abatement requirement, even if part of an uncontested citation, should be modified.[2]

Safety professionals should be aware that a notice of contest does not have to be in any particular form and is sent to the area director who issued the citation. The area director must forward the notice to the OSHRC, and the OSHRC must docket the case for hearing.

After pleading, discovery, and other preliminary matters, a hearing is scheduled before an administrative law judge (ALJ). Witnesses testify and are cross-examined under oath, and a verbatim transcript is made. As noted previously, the Federal Rules of Evidence are applicable to this hearing.

Following closure of the hearing, parties may submit briefs to the ALJ. The ALJ's decision contains findings of fact and conclusions of law and affirms, vacates, or modifies the citation, proposed penalty, and abatement requirements. The ALJ's decision is filed with the OSHRC and may be directed for review by any OSHRC member *sua sponte* or in response to a party's petition for discretionary review. Failure to file a petition for discretionary review precludes subsequent judicial review.

The Secretary of Labor has the burden of proving the violation. The hearing is presided over by an ALJ, and he or she renders a decision either affirming, modifying, or vacating the citation, penalty, or abatement date. The ALJ's decision then automatically goes before the OSHRC. The aggrieved party may file a petition requesting that the ALJ's decision be reviewed, but even without this discretionary review, any OSHRC member may direct review of any part or all of the ALJ's decision. If, however, no member of the OSHRC directs a review within 30 days, the ALJ's decision is final. Through either review route, the OSHRC may reconsider the evidence and issue a new decision.

OSHRC review is final, but the factual determinations of the ALJ, especially regarding credibility findings, are often afforded great weight. Briefs may be submitted to the OSHRC, but oral argument, although extremely rare, is also within the OSHRC's discretion.

In this administrative phase of the Act's citation adjudication process, the employer's good faith, the gravity of the violation, the employer's past history of

---

[1] 29 CFR Section 651.

[2] M. Rothstein, *Occupational Safety and Health Law* (2nd ed., 1983), summarized and reprinted in *Employment Law.*

compliance, and the employer's size are all considered in the penalty assessment. The area director can compromise, reduce, or remove a violation. Many citations can be compromised or reduced at this stage.

Although the OSHRC's rules mandate the filing of a complaint by the Secretary and an answer by the employer, pleadings are liberally construed and easily amended. Approximately 90 percent of the cases filed are resolved without a hearing, either through settlement, withdrawal of the citation by the Secretary, or withdrawal of the notice of contest by the employer.[1]

As permitted under Section 11(a) of the Act, any person adversely affected by the OSHRC's final order may file, within 60 days from the decision, a petition for review in the U.S. Court of Appeals for the circuit in which the alleged violation had occurred or in the U.S. Court of Appeals for the District of Columbia Circuit.[2] Under Section 11(b), the Secretary may seek review only in the circuit in which the alleged violation occurred or where the employer has its principal office.[3]

The courts apply the substantial evidence rule to factual determinations made by the OSHRC and its ALJs, but courts vary on the degree of deference afforded the OSHRC's interpretations of the statutes and standards. The burden of proof remains on the Secretary of Labor at this hearing. The Federal Rules of Civil Procedure, Federal Rules of Evidence, and all other legal requirements apply, as with any trial before the federal court. State plan states usually will possess similar requirements for hearings at the state court level.

As discussed in the chapter regarding defenses to an OSHA citation, safety professionals should prepare a strategy beforehand and prepare for the acquisition of potential evidence throughout the inspection. Again, safety professionals should utilize legal counsel for assistance throughout this legalistic process.

Although the warrant process is not initially an OSHRC matter, these types of cases usually result in review by the OSHRC. Safety professionals should be aware that OSHA possesses the authority to conduct inspections and investigations, and this authority is derived from Section 8(a) of the OSH Act, which states:

> In order to carry out the purpose of this Act, the Secretary, upon presenting appropriate credentials to the owner, operator, or agent in charge is authorized to:
>
> 1. enter without delay and at reasonable times any factory, plant, establishment, construction site, or other area, workplace, or environment, where work is performed by an employee of an employer; and
> 2. to inspect and investigate during regular working hours and at other reasonable times, and within reasonable limits and in a reasonable manner, any such place of employment and all pertinent conditions, structures, machines, apparatus, devices, equipment, and materials therein, and to question privately any such employer, owner, operator, agent, or employee.[4]

Given the above, most safety professionals do not require OSHA to acquire an administrative search warrant prior to entry. The decision in *Marshall v. Barlow's,*

---

[1] Rothstein, Knapp, & Liebman, Employment Law.
[2] 29 USC Section 651 et seq.
[3] Id.
[4] 29 U.S.C. Section 657(a).

Inc. (see Appendix E for full case) addressed the issue regarding OSHA's ability to conduct warrantless inspections but opened the door to many related issues. *Barlow's* was not, however, the first case to address the requirement of administrative search warrants. In *Camara v. Municipal Court*[1] and *See v. City of Seattle*[2] (companion cases), the Supreme Court first required a search warrant for nonconsensual administrative inspections. These cases laid the foundation for the *Barlow's* decision and also specified and defined four exceptions to the warrant requirement in administrative inspections. The first three exceptions — namely the consent, plain view, and emergency inspection exceptions — were drawn directly from the law of search and seizure. The fourth, known as the *Colannade-Biswell* or licensing exception, continues to be the most controversial.[3]

Safety professionals should be aware that the usual exception to the search warrant requirement for OSHA used by the vast majority of safety professionals today is simply consent to the safety and health inspection. Valid consent by the safety professional or other agent of the company or organization waives the employer's Fourth Amendment rights and protection. In the administrative setting of an OSHA inspection, consent can be provided by the safety professional if he or she simply fails to object to the inspection.[4] Safety professionals should be aware that this form of consent differs greatly from the criminal investigation requirement that the consent be knowingly and voluntarily given. Additionally, safety professionals should be aware that OSHA compliance officers are not required to inform employers of their right to demand a warrant or even to ask for the employer's consent.[5]

If a safety professional, as an agent for the company or organization, wishes to exercise his or her Fourth Amendment right to require a warrant, this right must be affirmatively exercised and a search warrant required before an inspection. OSHA has instructed compliance officers to answer employers' questions regarding search warrants in a straightforward manner, but they are not required to volunteer information or allowed to mislead, coerce, or threaten an employer.[6]

A major issue for safety professionals regarding the consent exception is whether the individual providing the consent has the authority to do so on behalf of the company or organization. Courts and the OSHRC have provided a broad interpretation of this question, permitting plant managers, forepersons, and even senior employees to provide consent.[7] Safety professionals should also note that OSHA has also found that general contractors may provide consent to inspect a common worksite where other subcontractors are working.[8]

As discussed previously, safety professionals should be aware that a search warrant is not required for equipment, apparatus, and worksites that are in the plain

---

[1] 387 U.S. 523 (1967).

[2] 387 U.S. 541 (1967).

[3] See *Colannade Catering Corp v. U.S.*, 397 U.S. 72 (1970) and *U.S. v. Biswell*, 406 U.S. 311 (1972).

[4] See *Stephens Enterprises v. Marshall*, 578 F.2d 1021 (5th Cir. 1978).

[5] See *Johnson v. Zerbst*, 304 U.S. 458 (1938); *U.S. v. Thriftmart*, 429 F.2d. 1006 (9th Cir.), *cert. denied*, 400 U.S. 926 (1970).

[6] OSHA Memorandum, reported at 8 O.S.H. Re. No. 1, at 3 (June 3, 1978).

[7] See *Dorey Elec. Co. v. OSHRC*, 553 F.2d 357, 5 OSH Cases 1860; Western Waterproofing Co., 5 OSH Cases 1496 (1977); *Stephenson Enterprises, supra* N. 15.

[8] See *Havens Steel Co.*, 6 OSH Cases 1740 (1978).

view of the compliance officer or open to public view. For a compliance officer to issue a citation for a workplace hazard that is within the plain view exception, he or she must (1) be in a place or location where he or she possesses a right to be, and (2) observe (or smell, hear, or acquire through other senses) what is visible or held out to public view.[1] Safety professionals should be aware that the U.S. Supreme Court has set up a significant hurdle for challenging the plain view exception by limiting the right to challenge search and seizure claims to those individuals, companies, or other entities that have an actual and legitimate expectation of intrusion by a governmental action.[2] In the past, simply being the target of the OSHA inspection was usually sufficient for being able to complain of a search or seizure violation by OSHA. This interpretation, as applied to the plain view exception, would permit a compliance officer to observe and issue a citation for workplace violations from a distance off company property (such as from a adjacent hill or a roadway), and if the employer challenged the citation on grounds of search or seizure, the court would not permit the claim due to lack of standing. The plain view exception is often used when a compliance officer observes a violation in a public area, such as a trenching site located on a public street. Safety professionals should be aware of the areas surrounding the facility and operation that are open to public view and should keep the plain view exception in mind.

Safety professionals should be aware that the third exception to the search warrant requirement is an emergency situation. When there exists an urgent threat to human life and a delay in acquiring a search warrant may increase the hazard, or consent cannot be readily obtained from the company or organization because of the emergency situation, and the emergency need outweighs the individual's right to privacy, the OSHA compliance officer may enter the facility without acquiring a search warrant. There are no specific OSHA regulations defining what would constitute an emergency situation, but it is highly likely that a court would find that the emergency situation would require an extreme life-threatening situation likely to cause death or severe injury if the compliance officer did not intervene immediately.

The fourth and most controversial exception to the search warrant requirement is the *Colannade-Biswell* or licensure exception. Before the *Barlow's* decision in 1978, the Supreme Court held in *Colonnade Catering Corp. v. United States* that warrantless non-consensual searches of licensed liquor stores were permitted.[3] Additionally, in *United States v. Biswell*, the Court permitted non-consensual warrantless searches of pawn shops under the Gun Control Act of 1968.[4] The Court found that a business in a regulated industry, such as a liquor store or pawn shop, provided an implied waiver of its Fourth Amendment rights by engaging in these industries.[5]

Safety professionals should be aware that this issue was in flux throughout the 1970s. The *Barlow's* decision settled the issue regarding OSHA's requirement to

---

[1] See *Accu-Namics*, 1 OSH Cases 1751, *aff'd on other grounds*, 515 F.2d 828, 3 OSH Cases 1299 (5th Cir. 1975).

[2] See *Rakas v. Illinois,* 439 U.S. 128, *reh'g denied*, 439 U.S. 1122 (1978).

[3] Supra N. 14.

[4] Id.

[5] Id.

obtain a search warrant for routine compliance inspections, but it also opened the door to peripheral issues involving search warrant requirements.

In basic terms, the *Barlow's* decision required that the probable cause standard be applied to OSHA and other administrative searches. Safety professionals should be aware that issues involving whether the *Barlow's* decision applies to non-routine inspections and whether OSHA may acquire an *ex parte* warrant are still unresolved.[1] Safety professionals should be aware that the probable cause requirement for a search warrant is significantly different from the criminal probable cause standard. In *Barlow's*, the Court defined the administrative probable cause standard:

> A warrant showing that a specific business has been chosen for an OSHA search on the basis of a general administrative plan for the enforcement of the OSH Act derived from neutral sources such as, for example, dispersion of employees in various types of industries across a given area, and the desired frequency of searches in any of the lesser divisions of an area, would protect an employer's Fourth Amendment rights.[2]

Safety professionals should be aware that probable cause for an OSHA or other administrative search can be developed from any of three basic categories: (1) general information about the company's industry, (2) general information about the individual company or organization, and (3) specific information about the company's workplace. The general information regarding the company's industry can be acquired by OSHA from many sources, including industry-by-industry data regarding workplace injuries and illnesses, days lost from work, and other data. General and specific information regarding the company and workplace is usually acquired through employee complaints to OSHA. Other sources of information include that acquired during past fatality or accident investigations, plain view observations by the OSHA compliance officer, and past history of citations on file with OSHA and available on the OSHA Web site (www.osha.gov).

When probable cause is shown by OSHA and an administrative search warrant is issued by a magistrate, the permitted scope of the inspection is usually broad enough to encompass the entire operation. This is often known as a "wall-to-wall" inspection. Courts have generally permitted wall-to-wall inspections so that OSHA compliance officers, who normally are unfamiliar with the operation or facility under inspection, are given great latitude in meeting the intent and purposes of the OSH Act in locating and identifying workplace hazards.

Although safety professionals, as agents of the company or organization, have a constitutional right to require a search warrant before an OSHA inspection, few safety professionals have exercised this right, and OSHA inspections conducted under administrative search warrants continue to be rare. Most safety professionals have found that it is fairly easy for OSHA to acquire an administrative search warrant, and creating an adversarial relationship with the compliance officer might do greater harm than good. Although most employers prefer to maintain good working relationships with OSHA compliance officers, some companies, due to individual

---

[1] An ex parte search warrant can be issued to OSHA without notice to the company or organization. A magistrate can issue the *ex parte* warrant based upon OSHA's presentation of probable cause.

[2] *Marshall v. Barlow's*, 98 S. Ct. 1816 (see full text in Appendix E).

circumstances such as a belief that OSHA is harassing them have exercised their constitutional right to require a search warrant before entry.

In most circumstances, safety professionals should carefully consider their option to require an OSHA compliance officer to obtain a search warrant. Many courts have sanctioned employers seeking a search warrant for contempt of court and other Rule 11 sanctions for frivolous actions, and the likelihood of successfully challenging an OSHA inspection warrant is minimal at best. The decision to require an administrative search warrant of OSHA should be extensively evaluated. Given the potential risks involved in requiring a search warrant, this decision should involve the board of directors, officers, and legal counsel of the organization.

If your employer chooses to require a search warrant, careful planning and preparation should take place before the actual inspection:

1. A policy statement or other directive should be distributed to other management team members informing them of the decision.
2. On-site personnel who will be in charge when the compliance officer arrives at the scene should be trained in all aspects of the pre-inspection and inspection processes and the potential risks involved.
3. All appropriate statements, forms, and other documents should be prepared to be (a) provided to the compliance officer, (b) used for training the team members responsible for documenting the inspection, and (c) used during the actual inspection by team members.
4. On-site team members should be provided with necessary equipment to document all aspects of the inspection (i.e., cameras, noise dosimeters) and must be properly trained to use, maintain, and calibrate the equipment and document test results.

Safety professionals should be aware that there are four basic routes for challenging an administrative search warrant normally available, depending on the court and the circumstances. An employer can (1) seek to enjoin issuance of the administrative warrant in federal district court, (2) refuse to permit the inspection after a warrant is issued and then move to quash the warrant or civil contempt proceedings brought by OSHA, (3) seek to enjoin enforcement of citations in federal district court after the inspection has taken place under protest, and (4) contest the validity of the warrant after the inspection has taken place before the OSHRC.

Safety professionals should be aware that attempting to have the court enjoin OSHA from acquiring an *ex parte* warrant is normally difficult, given the fact that the employer seldom knows that the compliance officer is attempting to acquire an *ex parte* warrant.[1] Safety professionals should be aware that OSHA reintroduced its regulation authorizing *ex parte* warrant applications.[2] Under this regulation, a company's demand for a search warrant for a previous inspection is one of the factors that OSHA will consider in finding that an *ex parte* warrant is "desirable and necessary" for subsequent inspections even before seeking the employer's consent.[3] Refusal to

---

[1] See *Cerro Metal Products v. Marshall*, 467 F. Supp. 869 (E.D. Pa. 1979).
[2] 29 CFR Section 1903.4(d).
[3] Id.

admit an OSHA compliance inspector without a search warrant is only one of the situations where an *ex parte* warrant can be sought, and, conversely, OSHA has not always elected to pursue *ex parte* warrants in circumstances, such as the denial of entry, where grounds existed for pursuing such a warrant.

Safety professionals should be aware that defying the search warrant after the OSHA compliance officer has obtained it from the magistrate or court is potentially the most dangerous route for challenging a search warrant. As was the situation in the *Barlow's case*,[1] the company or organization may move the court to quash the search warrant or wait and defend a refusal to comply with the search warrant in civil contempt proceedings initiated by OSHA. Although the Third Circuit in *Babcock & Wilcox Co. v. Marshall* found that the federal court had authority to test the validity of a search warrant before an OSHA inspection, other courts have found that a motion to quash is not the proper method to challenge a warrant. Safety professionals should be aware that selection of this route carries many other potential dangers. In addition to assessing other penalties, courts have found that some employers lacked good faith in defying the OSHA warrant and placed those employers in civil contempt of court.[2]

The route for judicially challenging OSHA's enforcement authority is usually taken after the OSHA compliance officer completes the inspection under a search warrant. This option has not been successful, as most courts have found that, even if the inspection is completed under a search warrant and under protest, jurisdiction lies with the OSHRC and not the courts.[3] The most frequently used route for challenging a search warrant is challenging the validity of the warrant with the OSHRC after an inspection. The OSHRC does not have authority to question the validity of an administrative search warrant before an inspection, but after the inspection takes place, it does have authority to rule on the constitutionality of an administrative search warrant obtained for the purpose of conducting an OSHA inspection. The safety professional must present any challenges to the administrative search warrant for review by the OSHRC. The OSHRC's authority is consistent with the holding of several courts that the employer must exhaust all administrative remedies before seeking judicial relief.[4]

Safety professionals may want to consider alternatives to requiring an administrative search warrant of OSHA. In many circumstances, understanding the scope of the inspection to be performed and the probability of viewing a potential hazard or a simple telephone call to the regional director can solve most inspection concerns or conflicts. Another alternative is the use of protective orders, which can modify the terms of the inspection, the time of the inspection, and other conditions to make the inspection more reasonable to the employer.

Safety professionals requiring an administrative search warrant should be the last alternative considered when addressing an OSHA inspection situation. In most circumstances, the chances of prevailing or preventing the OSHA compliance officer

---

[1] 436 U.S. 307, 6 OSH Cases 1571 (1978).

[2] See, for example, *Marshall v. Multicast Corp.*, 6 OSH Cases 1486 (D. Ohio 1978).

[3] See *Chromalloy American Corp.*, 7 OSH Cases 1547 (1979).

[4] See Electrocast Steelcast Foundry, 6 OSH Cases 1562 (1978); *Babcock & Wilcox Co. v. Marshall*, 7 OSH Cases 1052; *Marshall v. Chromalloy American Corp.*, Supra n. 32.

from conducting an inspection of the worksite are minimal. In addition, the cost in management time, effort, legal fees, potential loss of goodwill in the community, or the creation of ill will with the local OSHA office generally outweigh the potential benefits of preventing an inevitable OSHA inspection.

Safety professionals should educate their management team as to all aspects of an inspection by OSHA as well as the appeals process through the OSHRC and to the courts. For most safety professionals, OSHA will be voluntarily permitted to inspect the workplace, waiving the Fourth Amendment and other rights, and most contests are after a citation has been issued. However, safety professionals should ensure that their management team is well aware of all rights and responsibilities under the OSH Act and guide the management team in the appropriate direction that is best for the company or organization.

Safety professionals should become familiar with the OSHRC procedures and the OSHRC Web site (at www.oshrc.gov and in Appendix B of this text).

10 Occupational Safety and Health Review Commission Appeal Procedures

Winning isn't everything, but wanting to win is.

Vince Lombardi

When it comes to winning, you need the skill and the will.

Frank Tyger

The primary issue for most safety professionals who have received a citation and have not been able to resolve the issue at informal conference, if an informal conference is even pursued, is whether to appeal the citation to the Occupational Safety and Health Review Commission (OSHRC). With the 15-working-day appeal deadline looming, a careful analysis of the facts, defenses, probabilities of success, impact of acceptance of the citation, and costs should be conducted in order to make this determination.

Safety professionals should examine every aspect of the citation and be prepared to discuss the citation, penalties, defenses, and probabilities with their management team and legal counsel as quickly as feasible in order to make an informed decision within the narrow time constraints. Some of the considerations include the following:

1. What alleged violation(s) were cited?
2. What is the category of the violation(s) (for example, serious, willful, etc.)?
3. What is the monetary penalty?
4. What factual defenses are available?
5. What documentation is available to support the defenses?
6. Are constitutional defenses available?
7. Should the standard itself be challenged?
8. What documentation or evidence does OSHA possess?
9. Who will serve as witnesses?
10. Is there a need for outside experts?
11. What is the status of the case law in this area?
12. Will you use in-house or outside counsel?
13. Will acceptance of the citation impact the operations?
14. What is the cost if we win? If we lose?

After careful analysis and review, if the management team determines that an appeal of one or all of the citations is appropriate, it is imperative that the safety professional or legal counsel file the Notice of Appeal document within the 15-working-day time limitation. Failure to file the Notice of Appeal in a timely manner can result in the loss of all appeal rights.

Safety professionals should be aware that OSHRC hearings are adversarial and on-the-record.[1] To this end, it is highly recommended that legal counsel be acquired to assist in assembling and presenting the case before the OSHRC. In this hearing, the OSHRC administrative law judge (ALJ) presides over and conducts the hearing in accordance with the Federal Rules of Civil Procedure.[2] Additionally, the ALJ will conduct the hearing much like a federal district court trial, and the Federal Rules of Evidence usually apply.[3] However, it should be noted that the ALJ possesses only the power to recommend a decision to the OSHRC, and OSHRC trials tend to reach a decision far more quickly than a civil action in federal court.

Procedurally, upon receipt of the Notice of Contest, OSHA has 30 days to notify and file the complaint with the OSHRC. In the complaint, OSHA is required to identify the alleged violations, proposed penalties, and abatement times for the particular alleged violations that are addressed in the Notice of Contest. Safety professionals must be aware that an Answer must be filed with the OSHRC within 15 days from the receipt of the complaint. The Answer must provide specific responses to the charges. Safety professionals should pay special attention to the fact that any allegations that the safety professional does not respond to are automatically admitted.[4]

Safety professionals should be aware that the OSHRC has the authority to issue subpoenas requiring attendance as well as subpoenas requiring the production of documents. Safety professionals can contact the OSHRC (recommended through their attorney) to acquire the subpoena forms, and service of the subpoena is specified under the Federal Rules of Civil Procedure.

Although it happens infrequently, safety professionals should be aware that employees, unions, and other representatives of the employees are permitted to participate in the hearings before the OSHRC. Usually the safety professional is required to post a notice of hearing, and interested employees, or their representatives, can petition the OSHRC to intervene in the action as interested parties.[5]

Safety professionals should be actively involved in working with their legal counsel in the preparation of the case before the OSHRC. Although legal counsel will be presenting the case, it is important that the safety professional work with and guide counsel in preparing every facet of the case. In general, preparation of the case will include an opening statement, presentation of witnesses on direct and cross-examination, and a closing statement.

The initial burden of proof is on OSHA, or more correctly the Secretary of Labor. OSHA possesses the burden of proving every element of the alleged violation

---

[1] 5 U.S.C. Section 551 *et. seq.*
[2] 29 U.S.C. Section 661(f).
[3] 29 C.F.R. Section 2200.71.
[4] 29 C.F.R. Section 2200.36(a).
[5] 29 C.F.R. Section 2200.20 (a) and (b).

by a preponderance of the evidence.[1] However, once OSHA proves their prima facie case that the citations are supported by the evidence, the burden of rebutting OSHA's evidence shifts to the company to present contrary evidence or provide an affirmative defense to justify the actions or inactions of the company.[2]

As discussed in Chapter 7, there are a number of defenses that may be applicable which fall into four basic categories: (1) constitutional challenges, (2) denial of the violation of the standard, (3) challenging the standard, and (4) challenging the process of the citation. Although there may be other defenses available depending on the circumstances, most defenses fall within one or more of these categories.

Safety professionals should be aware that OSHA offers at least two simplified processes to expedite the appeals process. First, in certain circumstances, OSHA can provide a letter to the company or organization, identifying and offering a reduction in the penalty and monetary fine if the company agrees to the settlement and abates the identified hazard. With this method, the company would simply agree to the reduction in category and monetary penalty, agree to abate the hazard, and sign the agreement. The agreement as well as a check for the penalty amount are forwarded to the OSHA office to resolve the citation. Additionally, again in limited citations, the OSHRC can provide a simplified proceeding in substitution for a full OSHA hearing. This simplified proceeding was designed to save money and time for the company or organization as well as for OSHA. In the simplified proceedings, the Federal Rules of Evidence are not applicable; there is no complaint, answer, or formal pleadings. There would be no discovery and no interlocutory appeals from the decision.[3] The simplified hearing would be heard before an administrative law judge, and a written decision is usually rendered in the case.[4]

Whether a simplified proceeding or a full OSHA hearing, safety professionals should be aware that there are very strict perimeters for appeal for discretionary review to the full OSHRC after the administrative law judge renders his or her opinion. In short, the safety professional and legal counsel will have 20 days from the date the administrative law judge mails a copy of the decision to file a petition to the OSHRC for discretionary review. If this petition is not filed within this time period, the administrative law judge's decision becomes a final order unless a commission member, on his or her own motion and within 13 days following the docketing of the administrative law judge's report, directs review of the administrative law judge's decision.[5]

The safety professional or legal counsel should prepare an original and three copies of the petition for review and file with the executive secretary for the OSHRC within the specified time period. The other party, whether OSHA or the company or the organization, is permitted to file a statement in opposition to the petition for review. Briefs are also permitted in support of or in opposition of the parties' contentions along with the petition for review.[6]

---

[1] See *Bechtel Corp.* 2 OSHC 1336 (1974); *Armour Elevators Co.*, 1 OSHC 1409 (1973).
[2] 29 C.F.R. Section 2200.90.
[3] 29 CFR Sections 2200.200–2200.204.
[4] 29 CFR Section 200.208.
[5] 29 CFR Section 2200.91(b)(2).
[6] 29 CFR Sections 2200.90(b)(2) and 90 (b)(3).

Safety professionals should be aware that the OSHRC is not bound by the administrative law judge's decision or finding of facts. The OSHRC must substantiate its position if the administrative law judge's findings and decision are rejected.[1] Additionally, the OSHRC possesses the authority to issue an order accepting, rejecting, or modifying the citation, the proposed penalty, or any aspects of the decision. However, safety professionals should be aware that unlike some other federal court proceedings, the OSHRC possesses no authority to award attorney fees or assess costs against either party.[2] Thus, safety professionals should ensure that their management team is aware that they will bear all legal costs related to this appeal process.

In certain circumstances, the company or organization or OSHA may wish to appeal an adverse decision beyond the OSHRC to the U.S. Court of Appeals.

The OSH Act provides, "upon the filing of the petition for review, the Court shall have jurisdiction of the proceeding and of the question determined therein, and shall have power to grant temporary relief or restraining order as it deems just and proper, and to make and enter upon the pleadings, testimony, and proceedings set forth in such record, a decree affirming, modifying, or setting aside in whole or in part, the order of the Commission and enforcing the same to the extent it is affirmed or modified."[3]

Procedurally, the safety professional or legal counsel is provided 60 days to file a petition for review from the issuance of the final order by the OSHRC to the U.S. Court of Appeals.[4] If the petition for review is not filed within this time period, the OSHRC may petition the Court for enforcement of the OSHRC's final order.[5]

Safety professionals should be aware that the filing of a petition for review does not constitute an automatic "stay" or stop the OSHRC's final order from pursuing the collection of fines and enforcing abatement requirements. The safety professional or legal counsel must file a motion for a "stay" with the OSHRC or the appropriate Court while this matter is on appeal. Absent a "stay," safety professionals should be aware that OSHA could cite the company or organization for a failure to abate or take other actions within the abatement period provided by OSHA in the citation while the Court review is pending.[6] If the company or organization ultimately prevailed in vacating the citation on appeal, the subsequent penalties would also be vacated.

The U.S. Court of Appeals is required to apply the "substantial evidence" test when reviewing the OSHRC's finding of facts to determine whether a violation should have been found.[7] This test is different than the "preponderance of the evidence" test utilized by the OSHRC for their review. Thus, if the Court finds that the OSHRC's finding of fact are supported by "substantial evidence," the Court will defer to the OSHRC decision. Additionally, safety professionals should be aware that the Court must find that the OSHRC abused its discretion before possessing jurisdiction to modify a penalty.[8] Additionally, safety professionals should be aware that the

---

[1] See *Brennan v. Gilles and Cottings, Inc.*, 504 F.2d 1255 (CA-4, 1974).

[2] See *McGowan v. Marshall*, 604 F.2d 885 (CA-5, 1979).

[3] 29 U.S.C. Section 660(a).

[4] Id.

[5] 29 U.S.C. Section 660(b).

[6] 29 U.S.C. Section 666(d).

[7] See *Brennan v. OSHRC*, 487 F.2d 438 (CA-8, 1973).

[8] See *Secretary of Labor v. OSHRC (Interstate Glass)*, 487 F. 2d 222 (8th Cir. 1975).

court will provide deference to OSHA's interpretations of their own rules so long as the interpretations are reasonable.[1]

Safety professionals and legal counsel should not take an appeal to the OSHRC and the federal courts lightly. Careful analysis of the citation as well as the costs, in terms of money, effort, and time, should be conducted prior to initiating this protracted appeals process as well as the probabilities of success. OSHA will be well prepared for any challenge; thus, safety professionals should be well prepared if this is the avenue of redress which is to be pursued.

---

[1] See *Secretary of Labor v. OSHA (CF&I Steel),* 111 S. Ct. 1171 (1991).

11 Operational Compliance

Say what you mean, mean what you say, but don't say it mean.

Anonymous

Achieving and maintaining compliance in a multi-facility operation or on the corporate level is a daunting task given the number of constantly changing variables, different job responsibilities and challenges, and the political climate of the company or organization. Corporate safety professionals who have achieved this level within their management hierarchy have achieved safety successes in the past and are now faced with the challenges of multiple locations, management of additional personnel, budget issues, staffing issues, and a myriad of other more complex and numerous issues that may have been encountered in the past. The management of the corporate level for safety and health is substantially different than at the plant level with the skills of leadership, delegation, and communication being essential components at this level.

Although the required duties and job responsibilities may vary from company to company, the skills and abilities required to perform at the multi-plant level or corporate level in order to appropriately plan, organize, direct, and control the multiple variables that are present within the safety function throughout the company requires careful development and management. If centralized management of the safety function is absent, each of the multiple entities will pursue individualized courses toward what they perceive are the company's safety objectives which can lead to exposures or gaps within the overall safety effort. Although some companies utilize a decentralized model for their management, the safety function is often most successful with a centralized model that provides more guidance and stability until such time as the safety professionals achieve a level of understanding and knowledge as well as a level of comfort with the direction and management of the plant-level safety programs. However, given the substantial level of turnover at the plant safety level due to promotion or departure, the centralized safety model offers a greater level of stability as the overall program develops and matures to a level at which compliance can be assured on a daily basis.

This being said, the corporate safety professional must be the leader and champion for the overall safety function as well as individual plant safety or facility programs and functions. The corporate safety professional sets the tone for the overall safety effort, protects his or her personnel within the corporate battles, manages and guides the overall safety efforts, as well as fights for the resources needed to effectively and efficiently manage the safety function at all levels. The corporate safety professional should be a person of action as well as a strategist utilizing the available elements to maximize the successes of the overall safety functions. And failures come with the territory. Corporate safety professionals should be prepared to "take the hit" for their personnel and program failures. Above all, the corporate safety professional must be trusted by his or her personnel as well as all levels of the management team and employees. The safety of each employee and the reputation of the company depend on the leadership and skill level of the corporate safety professional.

Corporate safety management from a distance requires management skills and abilities that can be different than those utilized successfully at the plant level. The corporate-level safety professional is not "in the trenches" every day and thus must acquire the pertinent level of knowledge through his or her personnel in the field. The information and knowledge transferred from the plant level to the corporate level must be truthful and accurate at all times. Depending on the pressures being applied at the plant level, the information flow and the accuracy of the information flow can be jeopardized. Trust is essential, and verification of the information is imperative in order to appropriately adjust the safety function and ensure effective and efficient management of the corporate-wide safety function.

Corporate safety professionals are normally "action" people. To achieve the corporate level, the corporate safety professional must have achieved a certain level of success at the plant or lower levels and possess the skills and abilities to achieve the corporate level. The skill set that served the safety professional at plant level and propelled the safety professional to the corporate level may not always be the same. Successful plant-level safety professionals are out of their offices and are primarily involved in "doing" activities such as inspections and training. The level of communications at the plant level is basically contained to employees and plant management with infrequent communications to the corporate level. Conversely, the corporate safety professional designs the programs rather than actually performs the activity, and the levels of communication are multiplied, ranging from communications with employees to the board of directors. The corporate safety professional should attempt to retain the skills that propelled him or her to the corporate level but add to these skills and level of knowledge which is required at the corporate level.

In the compliance area with the changing variables in the field operations, the frequency of the information as well as the type and verification of information is an essential component. Many corporate safety professionals require multiple levels of identification and verification as to the effectiveness of every element of each compliance program on a short-term and long-term basis. Given the fact that most companies or organizations possess a substantially large number of compliance programs operating simultaneously on a daily basis, any deficiency not identified can lead to substantial deficiencies within the overall program. To ensure compliance at all times, corporate safety professionals often install multiple layers of verification, including but not limited to daily inspections, weekly inspections, quarterly reviews, and outside audits, among other methods. Many safety professionals design their safety programs far above the minimum OSHA compliance levels in order to achieve the expected corporate-wide safety results.

Verification and accountability are essential for the successful management of the corporate safety function. Corporate safety professionals should install information systems that are timely and accurate as well as verifiable in order to effectively manage the multitude of variables as well as programs. The corporate safety professional's credibility and performance are based on the accuracy of the information provided from the plant or operations level. Additionally, it is important that the corporate safety professional be acquiring the "right" information from the plant or operational level and that no information is withheld or delayed. For example, if injury rates are part of the company's goals, corporate safety professionals may require a daily e-mail or report identifying each injury that occurred each day with an explanation of the injury and corrective action taken to prevent re-occurrence.

Providing the plant or operation level with fair, accurate, and attainable goals or objectives; providing the resources to achieve the goal or objectives; and holding the plant or operational entity accountable, with consequences, is an essential part of any corporate safety program. Although many studies have found that positive motivation provides greater results, it is essential within the safety function to provide the potential of negative reinforcement if the expected results are not achieved. A combination of positive and negative reinforcement has often achieved the expected results within the safety area.

In establishing corporate-wide safety goals and objectives, there are at least two primary schools of thought regarding the establishment of goals and objectives which should be considered prior to the establishment of safety goals. Conceptually, safety professionals should recognize whether a zero-tolerance goal and a progressive achievement goal would work best within their organizational structure and company philosophy. Basically, there are positive and negative aspects for each of these safety goal concepts which need to be explored and understood completely prior to implementation. Under a zero-tolerance safety goal or objective, no work-related injury or illness is acceptable. Conceptually, this is the goal or objective safety professionals are striving to achieve for their organization. However, if the company or organization is starting from a position at which a substantially large number of injuries or illnesses occur annually in the workplace, a zero-tolerance goal may be unachievable for the management team and workforce, and once the first injury occurs, the management team and workforce may lose focus on the goal that they perceive as unachievable. Conversely, a progress goal where a potentially achievable reduction in work-related injuries and illnesses is provided, such as a 25 percent reduction from last year's injuries and illnesses, the management team and workforce may be more motivated toward achievement of the goal. However, the safety professional and company or organization are accepting that they will incur work-related injuries and illnesses while striving toward the reduction goal. There is no one right way or wrong way to establish safety goals and objectives for a company or organization. A safety goal structure that may work for one company may not work for another company.

However, there are some common elements in any successful safety goal or objective program which are common to all. First, the use of positive and negative reinforcement is essential. On the positive side, there are a large number of safety incentive programs on the market ranging from T-shirts to automobiles depending on your budget. Positive reinforcement has been shown to produce better results than negative reinforcement; however, it is essential that the negative reinforcement not be removed from the overall safety goal program. The negative reinforcement is necessary for the percentage of the workforce who choose not to participate and is essential in the overall achievement and maintenance of your compliance effort. Second, a safety incentive program correlating with your safety goals and objectives is not a substitution for a strong foundational safety compliance program. If the foundation is not in place prior to initiating a safety incentive program, any progress toward the achievement of the safety goals or objectives will be short lived. Third, the overall company safety goals or objectives should be appropriately divided in order to provide departmental goals for which a unit or department can be held accountable. For example, if the company goal is 100 lost-time days, each of the four departments should be provided a departmental goal of 25 lost-time days. And each department or unit must be provided periodic feedback as to the achievement of its individual goal and be held accountable for the achievement of its department or unit goal.

Fourth, the selection of the appropriate category or categories from which to base your safety goals is essential. As pressure is being applied throughout the management hierarchy, smart managers will be looking for an edge or method through which to achieve this goal or objective. This is especially true if there is a monetary incentive or detriment tied to the achievement of the safety goal or objective. Safety professionals should carefully evaluate the categories selected for their safety goals and identify the loopholes and methods through which data on which the safety goals are established can be circumvented. For example, if the safety goal is established on the achievement of a requisite number of OSHA-recordable injuries and illnesses, the safety professional should recognize that there is the potential of inaccurate determinations given the specific and narrow definition of "recordable" injuries and illnesses identified in the OSHA record keeping standard. The information and data collected as part of the overall safety goals or objective program must be accurate at all times, and interpretations or other gray areas should be removed from the safety goals or objective system. Lastly, and most importantly, the management team must "buy in" and support the safety goals or objectives. With management support from the highest levels and the appropriate emphasis placed on the achievement of the safety goals or objectives, the company or organization is on its way to achieving the identified goals.

At the corporate or multi-plant level, assessments and constant evaluation at all levels of the compliance effort are essential in order to be able to identify the strengths and weaknesses in the overall program. Many companies and organizations require daily, weekly, monthly, quarterly, and annual examinations of various elements of their compliance efforts as well as an overall assessment of the functions of the entire program. These examinations can include daily safety inspections, weekly facility inspections, quarterly program audits, and annual program assessments that provide snapshot examinations of a particular program, such as daily personal protective equipment (PPE) inspections, through complete program audit of all programs throughout the company or organization. When developing such assessments, it is important to capture the exact data the instrument is intended to gather and ensure the accuracy of the data as the results are documented. Additionally, the data and information acquired through these assessments should be rated, if applicable, in some manner in order that the safety professional can provide a grade, percentage, or score that is easily understandable to the management team and workforce. As an example, presented below is a sample safety audit assessment instrument for three of the many possible categories that can be encompassed within a safety audit. In order to acquire a percentage, every element of every compliance program or other safety effort is provided a numerical top score. When conducting the safety audit assessment, the evaluation provides a numerical score up to the maximum allowable per category. When completed, the number of points acquired is divided by the number of points possible to acquire the percentage or score for the safety audit assessment. There are a number of different methods that can be utilized; however, the major point of emphasis is to provide a method that is easily understandable to your management team and workforce.

EXAMPLE

Safety Audit Assessment

Quarterly Report for _____ Quarter of _____ Year _____

Facility Name: _____

Total Points Available: _____   Audit Performed by: _____

Total Points Scored: _____   Signature: _____

Percentage Score _____   Date: _____

(total points scored divided by total points available)

Management Safety Responsibilities	Answer		Total Points	Score
1. Are the safety responsibilities of each management team member in writing?	Yes	No		10
2. Are the safety responsibilities explained completely to each team member?	Yes	No		10
3. Does each team member receive a copy of his or her safety responsibilities?	Yes	No		5
4. Has each team member been provided the opportunity to discuss his or her safety responsibilities and add input into the methods of performing these responsible acts?	Yes	No		10
Section total				**35**

Safety Goals	Answer		Total Points	Score
1. Has each member of the management team been able to provide input into the development of the operations safety goals?	Yes	No		5
2. Has each member of the management team been able to provide input into their department's goals?	Yes	No		10
3. Are goals developed in more than one safety area?	Yes	No		10
4. Are the goals reasonable and attainable?	Yes	No		10
5. Is there follow-up with feedback on a regular basis?	Yes	No		15
6. Is there a method for tracking the department's progress toward its goal?	Yes	No		15
7. Is the entire program audited on a regular basis?	Yes	No		10

	Answer		Total Points	Score
8. Does your management team fully understand the purpose of the Safety Goals Program?	Yes	No		10
9. Does your management team understand the OSHA recordable rate, loss time rate, and days lost rate (per 200,000 man-hours)?	Yes	No		10
10. Does your management team fully understand the provisions and requirements when the safety goals are not achieved on a monthly basis?	Yes	No		10
11. Is your management team provided with daily/ weekly feedback regarding the attainment of their safety goals?	Yes	No		10
Section total				**115**

Accident Investigations	Answer		Total Points	Score
1. Is your medical staff thoroughly trained in the completion of the Accident Investigation Report?	Yes	No		5
2. Are all supervisory personnel thoroughly trained in the completion of the Accident Investigation Report?	Yes	No		10
3. Are all management team members completing the Accident Investigation Report accurately?	Yes	No		5
4. Are the Accident Investigation Reports accurate, complete, and readable?	Yes	No		10
5. Are the Accident Investigation Reports being monitored for timeliness and quality?	Yes	No		10
6. Are management team members receiving feedback on the quality of the Accident Investigation Report?	Yes	No		10
7. Are management team members receiving feedback on safety recommendations identified on the Accident Investigation Report?	Yes	No		10
8. Is your Accident Investigation Report system computerized?	Yes	No		15
9. Is there follow-up on any items identified on the Accident Investigation Report to insure correction on the deficiency before there is a reoccurrence?	Yes	No		15
10. Are Accident Investigation Reports being discussed in staff meetings, line meetings, or safety committee meetings?	Yes	No		10
Section total				**100**

The data collected within the safety function should be as accurate as possible taking into account the differences among plant operations, state laws, and other factors. Safety professionals are encouraged to verify all data, especially in the area related to the OSHA record-keeping standard,[1] to ensure accuracy. Additionally, comparisons against other systems within the company or organization collecting parallel information, such as workers' compensation claims, can identify if inaccuracies potentially could be present within the data collection system and interpretation of the data for OSHA record-keeping purposes. For example, if the workers' compensation data reflect 100 claims filed for the operation and the OSHA 300 log data reflect 10 OSHA recordable injuries, there may be an inaccuracy within the plant- or operational-level system. Appropriate investigation and evaluation may be warranted.

Planning, directing, organizing, controlling, and managing a corporate-wide or multi-plant safety program often requires extraordinary effort during the development and implementation stages. Management commitment from all levels is an absolute necessity, and delegation of responsibilities, with a level of trust that the work will be completed on time, is essential. Safety professionals must keep in mind that "Rome wasn't built in a day," and there will be many trials and tribulations while striving to build a program. Remember, the safety efforts you are providing today in building a quality safety and health program can safeguard many employees for many years, permitting them to return home at the end of the day to their families and friends.

---

[1] 29 CFR 1904.

12 Multi-Employer Worksite Issues

I love my enemies for two reasons: They inspire me to recognize my weakness. They also inspire me to perfect my imperfect nature.

Sri Chinmoy

One area that often causes confusion and consternation for safety professionals is the OSHA Multi-Employer Citation Policy. Many companies or organizations utilize a variety of vendors and subcontractors to perform work for the entity ranging from custodial services to on-site construction. Additionally, on larger projects like a building construction, the company or organization may employ a general contractor to oversee and manage the substantial number of subcontractors on the worksite. Given this structure, it was often difficult for an OSHA inspector in the past to determine who was responsible for the violation and who should be issued the citation to acquire abatement of the violation.

To this end, OSHA adopted the Multi-Employer Citation Policy to assist OSHA compliance officers in determining and establishing who the responsible company or entity was for the violation. However, in establishing this policy, a company or organization can be cited for the violations of the vendor or subcontractors working on site or under the control of the company or organization despite presumed legal transference of liability for safety on the worksite via a written contract or other procedures. Safety professionals with vendors or contractors on site should become well acquainted with this important policy and adjust their activities to avoid or minimize exposure to the potential safety violations of these contractors.

Safety professionals should be aware that responsibility under the Multi-Employer policy depends solely on the safety professional, as an agent of the company, or other company representatives in the employer's role, and not on the job title. If the safety professional is found to meet the criteria of "creating, controlling, exposing, or correcting" on behalf of the vendor or contractor, then the safety professional has obligated his or her company for the alleged violation under OSHA's regulations.[1]

This policy, set forth in OSHA's Field Inspection Manual,[2] instructs the OSHA compliance officers to conduct a two-step analysis as follows:

Step 1 — Does the employer have responsibilities regarding compliance with OSHA standards?

- Decided by determining if the employer fits one or more of four of the following categories of employer: Exposing, Creating, Correcting, Controlling.

---

[1] Standard Interpretation of Russell B. Swanson, Director, Directorate of Construction, Dec. 13, 2001.
[2] CPL 2.103, Chapter III, Paragraph 6.c (Sept. 26, 1994).

- If the employer does not fit any of these categories, then it was not responsible for compliance with the standards.
- If the employer does fit one of these categories, then go to Step 2.

Step 2 — Did the employer take sufficient steps to meet the requirements of the standard?

- Extent of steps the employer was required to take depends on whether it was Exposing, Creating, Correcting, or Controlling employer under Step 1.
- If the employer took the steps required for its category, then it met its obligations and no citation is to be issued.
- If the employer failed to tale the steps required for its category, then it may be cited for a violation of the standard.[1]

Given this "test" utilized by OSHA as well as the OSHRC decisions identified below, many safety professionals have determined that there are two basic methods through which potential liability for violations by OSHA can be avoided. The first method is simply a complete "hands-off" by the safety professional regarding the compliance efforts of the subcontractor or vendor. This method avoids the "exposing, creating, correcting" and especially the "controlling" components of the policy. Conversely, other safety professionals simply assume the liability and exercise complete control over the safety for subcontractors and vendors working on site.

The primary reason for these two very distinct and opposing approaches is due to two companion cases decided by the OSHRC in 1976. In *Grossman Steel & Aluminum Corp.*,[2] the OSHRC found liability for a general contractor based upon the general contractor's supervisory authority and control of the workplace. In the companion case of *Anning-Johnson Company*,[3] the OSHRC found both the subcontractor and general contractor responsible for the violation involving the lack of a perimeter guard. In this case, the OSHRC found the general contractor as "controlling," and the subcontractor "knew or should have known of the hazard which his employees were exposed."[4]

The *Anning-Johnson and Grossman Steel* companion decisions, when analyzed, represent the position of the OSHRC, and thus OSHA, with regard to general contractors and subcontractors on multi-employer worksites. The *Anning-Johnson/Grossman Steel* analysis has been endorsed by five circuit courts of appeal,[5] and four distinct lines of cases have evolved from these decisions — namely, (1) the affirmative defense that the hazard was not created by the employer, (2) the employer did not know of the hazard despite due diligence, (3) the hazard was not controlled by the employer, and (4) the employer possessed control over the hazard. Given this

---

[1] Id.
[2] 4 OSH Cases 1185 (1976).
[3] 4 OSH Cases 1193 (1976).
[4] Id.
[5] See *Marshall v. Knutsun Const. Co.*, 566 F.2d 596 (8th Cir. 1977); *Beatty Equip. Leasing, Inc. v. Sec. of Labor*, 577 F.2d 534 (9th Cir. 1978); *Brennan v. OSHRC (Underhill Constr. Corp.)*, 513 F.2d 1032 (2d Cir. 1975); *Central of Ga. RR Co. v. OSHRC*, 576 F.2d 620 (5th Cir. 1978); *N.E. Tel. & Telegraph Co. v. Sec. of Labor*, 589 F.2d (1st. Cir. 1978).

analysis, OSHA adopted the Multi-Employer Worksite policy apportioning liability for the violation between the general contactor and the subcontractors (see Appendix J). In essence, OSHA will hold each employer on a multi-employer worksite primarily responsible for the safety of his or her own employees and the contractor whose employees are exposed to the violation are generally held responsible, even if another entity or contractor is contractually responsible for safeguarding against the violation or providing safety for the worksite. As noted above, OSHA has created the two-prong test for OSHA compliance officers to ascertain responsibility and thus liability under these circumstances.

Utilizing the OSHA two-prong test, if the employer did not create the hazard that is the basis of the violation and did not control the hazard in any way, potential liability for the violation can be avoided either by proving that the employer took the necessary steps to protect the employees or that the contractor or employer lacked the knowledge to recognize the hazard.[1] Given the above, safety professionals are left with the difficult decision as to either assume control for the safety of the contractor's employees working on site and thus assume the liability for any violations, or, alternatively, take a completely hands-off approach whereby the safety professional would possess no control over safety at the worksite and shift the liability for a violation to the contractor. It is important that safety professionals recognize the potential liabilities involved in these approaches and ensure that any contractor working on site meets all of the requirements and standards set for contractility as well as possesses all required insurances. Additionally, as part of the contractor's agreement, safety professionals should ensure that the contractor possesses a functioning safety and health program and possesses the ability to achieve and maintain compliance at the worksite at all times.

As noted in the article titled "Liability on Multi-Employer Worksites,"[2] OSHA's multi-employer worksite policy is applicable not only to construction but also to general industry. Additionally, the OSHA multi-employer policy has and is undergoing substantial challenges as can be seen in the *IBP*[3] and *Anthony Crane*[4] decisions. This article clearly points out the statutory basis, or lack thereof, which is the basis of liability under OSHA's policy, the "controlling" analysis which remains the basis for such liability, and the guidance provided in reducing the potential for liability by the employer through a "hands-off" approach to contractor safety.

Safety professionals, especially those working in the construction industry, should be aware of the recent decision in *Secretary of Labor v. Summitt Contractors, Inc.*,[5] wherein the OSHRC vacated an alleged scaffolding violation that was cited utilizing the OSHA Multi-Employer worksite policy. At issue was the conflict between Section 1910.12(a), wherein the section refers to the exposure of and duty owed to the employer's employees ("his employees") and the OSHA multi-employer worksite policy. This case also provided the history of the changes to the

[1] *OSHRC Field Manual*, supra N. 2

[2] Yohay, S.C., and Sapper, A.G., "Liability on Multi-Employer Worksites," *Occupational Hazards*, October, 1998, p. 28.

[3] *IBP, Inc. v. Herman*, 144 F.3d 861 (D.C. Cir. 1998).

[4] *Anthony Crane Rental, Inc. v. Reich*, 70 F.3d. 1298 (D.C. cir. 1995).

[5] OSHRC Docket No. 03-1633. Full text in Appendix E of this text.

multi-employer worksite policy. It should be noted that this case is currently on appeal by the Secretary of Labor.

In conclusion, safety professionals should be aware that their actions, although often well intentioned, can create liability for their companies or organizations under the OSHA Multi-Employer worksite policy or rule. Safety professionals should review this issue with their management team and decide which approach works best for their company. However, it is important that safety professionals select one approach or the other and "don't change horses mid-stream." With the OSHA Multi-Employer worksite policy as well as the identified cases, control of safety at the worksite is the key evaluator.

13 Workers' Compensation and Safety

It's great to be great, but it's greater to be human.

Will Rogers

There have been many books written about workers' compensation. There is only one reason why workers' compensation is included in this safety text — management does not "get it." Workers' compensation is a distinctively different program than safety and health. The management of safety and health is a proactive program predominately focused on prevention activities. Workers' compensation is a reactive program predominantly focused on providing benefits to those employees who have sustained a work-related injury or illness. Workers' compensation and safety and health are oil and water. However, often the management of companies and organizations simply do not see the difference and saddle the safety professional with both programs due to their inherent interaction at the point of the accident.

Although there are similarities between the management of a workers' compensation program and the management of a safety program, there are also many significant differences. Individual state workers' compensation programs are generally a no-fault, reactive mechanism through which to compensate employees with monetary benefits for injuries and illnesses that result after an accident has occurred. Conversely, safety and health programs are designed to be proactive programs with the emphasis on preventing employees from being injured in the first place. Safety professionals who are required to wear the dual hats of safety and health as well as workers' compensation are often placed in a precarious position with employees and must be able delineate which hat they are wearing at any given time. Safety professionals with dual responsibilities in these important areas must also be able to effectively manage the individuals, the situation, and the potential liabilities surrounding and involving the safety program as well as workers' compensation program.

The rising cost of workers' compensation for most employers has resulted in a significantly increased focus by management in this area. Employers, always cognizant of the bottom line, have found that their workers' compensation costs have significantly risen due to many factors including, but not limited to, increased injuries and illnesses, increased medical and rehabilitation costs, increased time loss and benefits, and other factors. With this increased focus, safety and health professionals are often thrust into the administrative world of workers' compensation with little or no training or education regarding the rules, regulations, and requirements. In the safety and health arena, many of the potential liabilities encountered in the area of workers' compensation are a direct result of acts of omission rather than commission. Safety and health professionals should understand the basic structure and mechanics of the workers' compensation system and the specific rules, regulations, and requirements under their individual state system.

Virtually every state's workers' compensation systems are fundamentally a no-fault mechanism through which employees who incur work-related injuries and illnesses are compensated with monetary and medical benefits. Either party's potential negligence is not an issue as long as this is the employer/employee relationship. In essence, workers' compensation is a compromise in that injured employees are guaranteed a percentage of their wages (generally two-thirds) and full payment for their medical costs when injured on the job. Companies, on the other hand, are guaranteed a reduced monetary cost for these injuries or illnesses and are provided a protection from additional or future legal action by the employee for the injury. In general, below is a synopsis of the usual perimeters of most states' workers' compensation programs. The typical workers' compensation system possesses the following features:

1. Every state in the United States has a workers' compensation system. There may be variations in the amounts of benefits, the rules, the administration, and so forth, from state to state. In most states, workers' compensation is the exclusive remedy for on-the-job injuries and illnesses.
2. Coverage for workers' compensation is limited to *employees* who are injured *on the job*. The specific locations of the work premises and what is considered "on the job" may vary from state to state.
3. Negligence or fault by either party is largely inconsequential. No matter whether the employer is at fault or the employee is negligent, the injured employee generally receives workers' compensation coverage for any injury or illness incurred on the job.
4. Workers' compensation coverage is automatic — employees are not required to sign up for workers' compensation coverage. By law, employers are required to obtain and carry workers' compensation insurance or be self-insured.
5. Employee injuries or illnesses that "arise out of and/or in the course of employment" are considered compensable. These definition phrases have expanded this beyond the four corners of the workplace to include work-related injuries and illnesses incurred on the highways, at various in- and out-of-town locations, and other such remote locales. These two concepts — "arising out of" the employment and "in the course of" the employment — are the basic burdens of proof for the injured employee. Most states require both. The safety and health professional is strongly advised to review the case law in his or her state to see the expansive scope of these two phrases. The injury or illness must "arise out of" the employment (i.e., there must be a causal connection between the work and the injury or illness) or must happen "in the course of" the employment (which relates to the time, place, and circumstances of the accident in relation to the employment). The key issue is a work connection between the employment and the injury or illness.
6. Most workers' compensation systems include wage-loss benefits (sometimes known as time-loss benefits) that are usually one-half to three-quarters of

the employee's average weekly wage. These benefits are normally tax free and are commonly called temporary total disability (TTD) benefits.

7. Most workers' compensation systems require payment of all medical expenses, including such expenses as hospital expenses, rehabilitation expenses, and prosthesis expenses.

8. In situations where an employee is killed, workers' compensation benefits for burial expenses and future wage-loss benefits are usually paid to the dependents.

9. When an employee incurs an injury or illness that is considered permanent in nature, most workers' compensation systems provide a dollar value for the percentage of loss to the injured employee. This is normally known as permanent partial disability (PPD) or permanent total disability (PTD).

10. In accepting workers' compensation benefits, the injured employee is normally required to waive any common law action to sue the employer for damages from the injury or illness.

11. If the employee is injured by a third party, the employer usually is required to provide workers' compensation coverage but can be reimbursed for these costs from any settlement that the injured employee receives through legal action or other methods.

12. Administration of the workers' compensation system in each state is normally assigned to a commission or board. The commission or board generally oversees an administrative agency located within state government which manages the workers' compensation program within the state.

13. The Workers' Compensation Act in each state is a statutory enactment that can be amended by the state legislatures. Budgetary requirements are normally authorized and approved by the legislatures in each state.

14. The workers' compensation commission or board in each state normally develops administrative rules and regulations (that is, rules of procedure, evidence, etc.) for the administration of workers' compensation claims in the state.

15. In most states, employers with one or more employees are normally required to possess workers' compensation coverage. Employers are generally allowed several avenues through which to acquire this coverage. Employers can select to acquire workers' compensation coverage from private insurance companies, from state-funded insurance programs, or become self-insured (that is, after posting bond, the employer pays all costs directly from its coffers).

16. Most state workers' compensation provides a relatively long statute of limitations. For *injury* claims, most states grant between 1 and 10 years in which to file the claim for benefits. For work related *illnesses*, the statute of limitations may be as high as 20 to 30 years from the time the employee first noticed the illness or the illness was diagnosed. An employee who incurred a work-related injury or illness is normally not required to be employed with the employer when the claim for benefits is filed.

17. Workers' compensation benefits are generally separate from the employment status of the injured employee. Injured employees may continue to

        maintain workers' compensation benefits even if the employment relationship is terminated, the employee is laid off, or other significant changes are made in the employment status.

18. Most state workers' compensation systems possess some type of administrative hearing procedures. Most workers' compensation acts have designed a system of administrative judges (normally known as administrative law judges [ALJ]) to hear any disputes involving workers' compensation issues. Appeals from the decision of the administrative law judges are normally to the workers' compensation commission or board. Some states permit appeals to the state court system after all administrative appeals have been exhausted.[1]

Safety professionals should be very aware that the workers' compensation system in every state is administrative in nature. In virtually every state, there is a substantial amount of required paperwork that must be completed in order for benefits to be paid in a timely manner. In most states, specific forms have been developed, and the forms are usually available on the individual state's Web site or through the state's workers' compensation office.

In general, the initial and often the most important form to initiate workers' compensation coverage in most states is the first report of injury or illness form. This form may be called a "First Report" form, an application for adjustment of claim, or may possess some other name or acronym like the SF-1 or Form 100. This form, often divided into three parts in order that information can be provided by the employer, employee, and attending physician, is often the catalyst that starts the workers' compensation system reaction. If this form is absent or misplaced, there is no reaction in the system, and no benefits are provided to the injured employee. Safety professionals can also utilize this claim form for comparison purposes with the OSHA 300 log to ensure accuracy.

In most states, information regarding the rules, regulations, and forms can be acquired directly from the state workers' compensation commission or workers' compensation board. Other sources for this information include insurance carriers, self-insured administrators, or state-fund administrators.

Safety professionals should be aware that workers' compensation claims can stretch for several years. Conversely, safety professionals are usually most familiar with the OSHA record keeping system where every year injuries and illnesses are totaled on the OSHA Form 300 log and a new year begins. With workers' compensation, once an employee sustains a work-related injury or illness, the employer is responsible for the management and costs until such time as the injury or illness reaches maximum medical recovery or the time limitations are exhausted. When an injury reaches maximum medical recovery, the employer may be responsible for payment of PPD or PTD benefits prior to closure of the claim. Additionally, in some states, the medical benefits can remain open indefinitely and cannot be settled or closed with the claim. Safety professionals should be aware that workers' compensation claims for a work-related injury or illness may remain open for several years (in

[1] Schneid, T.D., *Modern Safety and Resource Control Management*, John Wiley & Sons, New York, 2000.

comparison to OSHA's requirements) and thus require continued management and administration for the duration of the claim process in order not to lose a claim in the process.

Safety professionals should be aware that the rules regarding the management of workers' compensation can vary greatly. Some states allow the company to take the deposition of the employee claiming benefits, while others strictly prohibit depositions. Some states have a schedule of benefits and have permanent disability awards based strictly on a percentage of disability from that schedule. Other states require that a medical provider designate a percentage of functional impairment due to the injury or illness in accordance with the American Medical Association (AMA) Guidelines. Some states include other factors such as the employee's age, education, and work history in this formula.

Safety professionals should realize that most workers' compensation systems are no-fault systems that generally require the employer or the employer's insurance administrator to pay all required expenses no matter whether the employer or employee was at fault, whether the accident was the result of employee negligence or neglect, or whether the injury or illness was the fault of another employee. Most workers' compensation systems are designed to be liberally construed in favor of the employee.

Safety professionals who instinctively utilize a proactive method of identifying the underlying causes of accidents and immediately correcting the deficiency may find that management of the workers' compensation function can often be very time consuming, frustrating, and show little progress. In situations of questionable claims, safety professionals should be aware that in many states, the employee has the right to initiate a workers' compensation claim without even informing the company or safety professional. Additionally, this administrative procedure can be foreign to many safety professionals as well as stressful and frustrating to those accustomed to a more direct management style. Above all, the safety professional must realize that he or she must follow the prescribed rules, regulations, and procedures set forth under each state's workers' compensation system and any deviation thereof or failure to comply can place the company, the insurance carrier or administrator, and the safety and loss prevention professional at risk for potential liability.

Safety professionals should be aware that when an injured employee is represented by legal counsel, the direct lines of communication are severed between the safety professional and the employee and all communications must be through legal counsel. Additionally, a major component in the management of a workers' compensation program is the communications with the medical professionals who are treating the injured or ill employee. Safety professionals should be aware that this can be an area of potential miscommunication and conflict. The safety professional and the medical professional objectives are normally the same, namely to get the injured employee healed, but the methodology through which this objective is met can be a potential minefield. Safety professionals should make every effort to insure open and clear lines of communication with the medical community to avoid any such conflicts. The loss of trust between the safety professional and the medical community can ultimately lead to additional costs and substantial headaches.

Safety professionals responsible for the management of workers' compensation within the organization will find that an effective management system can control and minimize the costs related to this required administrative system while also maximizing the benefits to the injured or ill employee. Although the workers' compensation system is basically reactive in nature, safety professionals should develop a proactive management system through which to effectively manage the workers' compensation claims once incurred within the organization. Below is a basic guideline to implement an effective workers' compensation management system:

1. Become completely familiar with the rules, regulations, and procedures of the workers' compensation system in your state. A mechanism should be initiated to keep the professional updated with all changes, modifications, or deletions within the workers' compensation law or regulations. A copy of these laws and rules can normally be acquired from your state's workers' compensation agency at no cost. Additionally, the state bar association, universities, and law schools in many states have published texts and other publications to assist in the interpretation of the laws and rules.

2. A management system should be designed around the basic management principles of planning, organizing, directing, and controlling. Given the fact that most state workers' compensation programs are administrative in nature, appropriate *planning* can include, but is not limited to, such activities as acquiring the appropriate forms, developing status tracking mechanisms, establishing communication lines with the local medical community, and informing employees of their rights and responsibilities under the workers' compensation act. Organizing an effective workers' compensation system can include, but is not limited to, selection and training of personnel who will be responsible for completing the appropriate forms, coordination with insurance or self-insured administrators, acquisition of appropriate rehabilitation and evaluation services, and development of medical response mechanisms. The directing phase can include, but is not limited to, implementation of tracking mechanisms, on-site visitation by medical and legal communities, development of work-hardening programs, and installation of return-to-work programs. Controlling can include such activities as the establishment of an audit mechanism to evaluate case status and progress of the program, use of injured worker home visitation, acquisition of outside investigation services, among other activities.

3. Compliance with the workers' compensation rules and regulations must be of the highest priority at all times. Appropriate training and education of individuals working within the workers' compensation management system should be mandatory, and appropriate supervision should be provided at all times.

4. When an employee incurs a work-related injury or illness, appropriate medical treatment should be quickly provided. In some states, the employee possesses the first choice of a physician; in other states the employer has this choice. The injured or ill employee should be provided the best possible care in the appropriate medical specialty or medical facility as soon as feasible. Improper care in the beginning can lead to a longer healing period and additional costs.

5. Employers often fool themselves by thinking that if employees are not told their rights under the state workers' compensation laws that there is less chance that an employee will file a claim. This is a falsehood. In most states, employees possess easy access to information regarding their rights under workers' compensation through the state workers' compensation agency, through their labor organization, or even through television commercials. A proactive approach that has proven to be successful is for the safety and loss prevention professional or other representative of the employer to explain to the employee his or her rights and responsibilities under the workers' compensation laws of the state as soon as feasible following the injury. This method alleviates much of the doubt in the minds of the injured employee, begins or continues the bonds of trust, eliminates the need for outside parties being involved, and tends to improve the healing process.

6. The safety professional should maintain an open line of communication with the injured employee and attending physician. The open line of communications with the injured employee should be of a caring and informative nature and should never be used for coercion or harassment purposes. The open line of communications with the attending physician can provide vital information regarding the status of the injured employee and any assistance the employer can provide to expedite the healing process.

7. Timely and accurate documentation of the injury or illness and appropriate filing of the forms to insure payment of benefits is essential. Failure to provide the benefits in a timely manner as required under the state workers' compensation laws can lead the injured employee to seek outside legal assistance and cause a disruption in the healing process.

8. Appropriate, timely, and accurate information should be provided to the insurance carrier, organization team members, and others to insure that the internal organization is fully knowledgeable regarding the claim. There is nothing worse than an injured employee receiving a notice of termination from personnel while lying in the hospital because personnel was not informed of the work-related injury and counted the employee absent from work.

9. As soon as medically feasible, the attending physician, the insurance administrator, the injured employee, and the safety professional can discuss a return to light or restricted work. A prudent safety professional may wish to use photographs or videotape of the particular restricted duty job, written job descriptions, and other techniques in order to insure complete understanding of all parties of the restricted job duties and requirements. Once the injured employee is returned to restricted duty, the safety professional should insure that the employee performs only the duties agreed upon and within the medical limitations proscribed by the attending physician. An effective return-to-work program can be one of the most effective tools in minimizing the largest cost factor with most injuries or illnesses — namely, time-loss benefits.

10. In coordination with the injured employee and attending physician, a rehabilitation program or work hardening program can be used to assist the injured employee to return to active work as soon as medically feasible.

Rehabilitation or work hardening programs can be used in conjunction with a return-to-work program.

11. Where applicable, appropriate investigative methods and services can be used to gather the necessary evidence to address fraudulent claims, deny non-work-related claims, or address malingering or other situations.

12. A prudent safety professional should audit and evaluate the effectiveness of the workers' compensation management program on a periodic basis to insure effectiveness. All injured or ill employees should be appropriately accounted for, the status of each meticulously monitored, and cost factors continuously evaluated. Appropriate adjustments should be made to correct all deficiencies and to insure continuous improvement in the workers' compensation management system.[1]

For many safety professionals wearing the dual hats of safety and workers' compensation, the issue of trust lies at the heart of the management of these functions. Safety professionals must acquire the trust of their employees while serving as their liaison to management concerning all things safety related. Employees must trust the safety professional to "do the right thing" with regard to safety. Conversely, the management of the workers' compensation function can often place the safety professional at odds with individual employees over benefits, schedules, and related issues involved in the management of their claim. Additionally, the proactive management of safety is often at odds with the reactive and elongated nature of managing workers' compensation. Safety professionals with these commingled responsibilities should bifurcate these responsibilities as best as possible and remember the goals and objectives of each to be able to properly balance their daily activities. Although many companies and organizations believe safety and workers' compensation are the same, these are two distinctly different functions and should be separated to achieve maximum efficiency and effectiveness.

---

[1] Schumann, M.S., and Schneid, T.D., *Legal Liability: A Guide for Safety and Loss Prevention Professionals,* Aspen Publishers, Gaithersburg, MD, 1997.

Form 101
Revised 6/05

KENTUCKY
OFFICE OF WORKERS' CLAIMS
Application for Resolution of Injury Claim

Claim No. _____

....
Plaintiff   vs.

....
Social Security Number

....
Birth Date

....
Street Address

....
City/State/Zip Code

....
County

....
Phone Number

Filed:

....
Defendant/Employer

....
Street Address

....
City/State/Zip Code

....
Insurance Carrier

....
Street Address

....
City/State/Zip Code

....
Other Defendant

....
Street Address

....
City/State/Zip Code

Reason for Joinder:

....

....
Other Defendant

....
Street Address

....
City/State/Zip Code

Reason for Joinder:

....

....

I. Nature of Injury

1. Plaintiff states that on the _____ day of _____ 20 ___, he/she was injured within the scope and course of employment with defendant employer at _____

_____
(City/County/State)

2. Describe how the injury occurred: _____

_____

_____

3. Body part injured: _____

_____

4. State the date and means by which the plaintiff gave notice of injury to the employer: _____

_____

_____

5. Describe medical treatment, if any:_____

_____

_____

6. Name and address of physician whose report is attached: _____

_____

II. Personal Data

7. Name and address of last school attended: _____

_____

8. Highest grade completed in school: _____

9. GED awarded: _____ yes   _____ no

10. Professional or vocational degrees, certificates, or licenses: _____

11. Dependents:

Name	Date of Birth	Social Security Number	Relationship

12. Have you previously filed for or received workers' compensation benefits? ___ yes ___ no

If yes, give Office of Workers' Claims file number(s), dates, and nature of injury or disease and any award of benefits received: _____

_____

III. Employment Data

13. Is plaintiff currently working? _____ yes _____ no

14. Type of work performed at date of injury: _____

_____

15. Describe the physical requirements of job performed at date of injury: _____

_____

_____

_____

16. Weekly wage at date of injury: _____. Attach copy of any proof of wages, such as paycheck stub, W-2, and so forth.

17. Weekly wage currently earned: _____. Attach copy of any proof of current wages.

18. Name and address of current employer and description of job currently being performed:

_____

_____

19. Are you alleging a violation of a safety rule or regulation pursuant to KRS 342.165?
_____ yes _____ no

Notice: Any person who knowingly and with intent to defraud any insurance company or other person files a statement or claim containing any materially false information or conceals, for the purpose of misleading, information concerning any fact material thereto commits a fraudulent insurance act, which is a crime.

Plaintiff herein being duly sworn, states that the statements in this application and in Forms 104, 105, and 106 are true. This the _____ day of _____ 20___.

_____

Plaintiff's Signature

Subscribed and sworn to before me this _____ day of _____ 20___.

_____

Notary Public

My commission expires: _____ County: _____

Prepared and submitted by:

_____

Signature/Representative for Plaintiff

_____

Title

_____

Street Address

_____

City/State/Zip

_____

Telephone Number

INSTRUCTIONS FOR
COMPLETION OF FORMS 101, 102, AND 103

Form 101 — Application for Resolution of Injury Claim

1. All sections of this form must be completed, and must be accompanied by the following:
   a. Form 104 (Plaintiff's Employment History)
   b. Form 105 (Plaintiff's Chronological Medical History)
   c. Form 106 (Medical Waiver and Consent)
   d. Medical report describing and supporting the injury which is the basis of the claim.
   e. Proof of Wages, including W-2s, paycheck stubs, etc.
2. All information must be typewritten.

3. File the original of this form and sufficient copies for all named defendants with the Office of Workers' Claims, Prevention Park, 657 Chamberlin Avenue, Frankfort, Kentucky, 40601.
4. If you have no telephone number, please list a number at which you may be contacted.
5. If you have questions, call 1-800-554-8601.

Form 102 — Application for Resolution of Occupational Disease Claim, and Form 103 — Application for Resolution of Hearing Loss Claim

1. All sections of this form must be completed, and must be accompanied by the following:
   a. Form 104 (Plaintiff's Employment History)
   b. Form 105 (Plaintiff's Chronological Medical History)
   c. Form 106 (Medical Waiver and Consent)
   d. Medical report supporting the occupational disease
   e. Proof of Wages, including W-2s, paycheck stubs, etc.
   f. Social Security earnings record release form
2. This form may be filed in combination with an Application for Resolution of Injury Claim (Form 101) if both benefits are sought. Information provided should be current through the date application is signed by plaintiff.
3. All information must be typewritten.
4. File the original of this form and sufficient copies for all named defendants with the **Office of Workers' Claims**, Prevention Park, 657 Chamberlin Avenue, Frankfort, Kentucky, 40601.
5. If you have questions, call 1-800-554-8601.

Note: Special attention should be given to stating the correct name and address of the employer and insurance carrier. Otherwise, claim processing may be delayed.

14 Other Litigation Issues

Things don't turn up in this world until somebody turns them up.

James A Garfield

PRIVATE LITIGATION UNDER THE OSH ACT

Although there is no common law basis for actions under the OSH Act, OSHA regulations are used in many tort actions, such as negligence and product liability suits, as evidence of the standard of care and conduct to which the party must comply. Additionally, documents generated in the course of business that are required under the OSH Act are usually discoverable under the Freedom of Information Act (FOIA) and can be used as evidence of a deviation from the required standard of care.

According to Section 653(b)(4) of the OSH Act:

> Nothing in this Act shall be construed to supersede or in any manner affect any workmen's compensation law or to enlarge or diminish or affect in any other manner the common law or statutory rights, duties, or liabilities of employers and employees under any law with respect to injuries, diseases, or death of employees arising out of, or in the course of, employment.[1]

This language prevents injured employees or families of employees killed in work-related accidents from directly using the OSH Act or OSHA standards as an independent basis for a cause of action (i.e., wrongful death actions).[2] However, many federal and state courts have found that Section 653(b)(4) does not bar application of the OSH Act or OSHA standards in workers' compensation litigation or application of the doctrine of negligence or negligence per se to an OSHA violation.[3] See, for example, Cases 1567 (1st Cir. 1985) ("Our review of the legislative history of OSHA suggests that it is highly unlikely that Congress considered the interaction of OSHA regulations with other common law and statutory schemes other than workers' compensation. The provision is satisfactorily explained as intended to protect worker's compensation acts from competition by a new private right of action and to keep OSHA regulations from having any affect on the operation of the worker's compensation scheme itself."); *Frohlick Crane Serv. v. OSHR*, Cases 521 F.2d 628 (10th Cir. 1975); *Dixon v. International Harvester Co.*, 754 F.2d 573 (5th Cir. 1985); *Radon v. Automatic Fasteners*, 672 F.2d 1231 (5th Cir. 1982); and *Melerine v. Avondale Shipyards*, 659 F.2d 706, 10 O.S.H. Cases 1075 (5th Cir. 1981). These decisions distinguish between use of an OSHA standard as the basis for a standard of care in

---

[1] 29 U.S.C. §653 (b) (4).

[2] *Byrd v. Fieldcrest Mills*, 496 F.2d 1323, 1 O.S.H. Cases 1743 (4th Cir. 1974).

[3] *Pratico v. Portland Terminal Co.*, 783 F.2d 255, 12 O.S.H.

a state or federal common law action and the OSH Act or OSHA standards creating a separate and independent cause of action.

NEGLIGENCE ACTIONS

OSHA standards are most widely used in negligence actions. The plaintiff in a negligence action must prove the four elements: duty, breach of duty, causation, and damages. *Black's Law Dictionary* defines negligence "per se" as

> conduct, whether of action or omission, that may without any argument or proof as to the particular surrounding circumstances, either because it is in violation of a statute or valid municipal ordinance, or because it is so palpably opposed to the dictates of common prudence that it can be said without hesitation or doubt that no careful person would have been guilty of it.[1]

In simpler terms, if a plaintiff can show that an OSHA standard applied to the circumstances and the employer violated the OSHA standard, the court can eliminate the plaintiff's burden of proving the negligence elements of duty and breach through a finding of negligence per se.

The majority of courts have found that relevant OSHA standards and regulations are admissible as evidence of the standard of care,[2] and thus, violation of OSHA standards can be used as evidence of an employer's negligence or negligence per se. It should be noted, however, that courts have prohibited use of OSHA standards and regulations, and evidence of their violation, if the proposed purpose of the OSHA standards use conflicts with the purposes of the OSH Act,[3] unfairly prejudices a party,[4] or is meant to enlarge a civil cause of action.[5] The Fifth Circuit, reflecting the general application, approves the admissibility of OSHA standards as evidence of negligence but permits the court to accept or reject the evidence as it sees fit.[6]

In using OSHA standards to prove negligence per se, human resource (HR) professionals should be aware that numerous courts have recognized the OSHA standards as the reasonable standard of conduct in the workplace. With this recognition, a violation by the employer would constitute negligence per se to the employee.[7] A few other courts have held, however, that violations of OSHA standards can never constitute negligence per se because of Section 653(b)(4) of the Act.[8]

---

[1] *Black's Law Dictionary*, 5th ed., West Publishing Co., New York, 1983.

[2] Id., See also, *Teal v. E.I. Dupont de Nemours & Co.*, 728 F.2d 799, 11 O.S.H. Cases 1857 (6th Cir. 1984); *Johnson v. Niagara Machine & Works*, 666 F.2d 1223 (8th cir. 1981); *Knight v. Burns, Kirkley & Williams Construction Co.*, 331 So.2d 651, 4 O.S.H. Cases 1271 (Ala. 1976).

[3] *Cochran v. Intern. Harvester Co.*, 408 F. Supp. 598, 4 O.S.H. Cases 1385 (W.D. Ky. 1975)(OSHA standards not applicable where plaintiff worker was independent contractor); *Trowell v. Brunswick Pulp & Paper Co.*, 522 F. Supp. 782, 10 O.S.H. Cases 1028 (D.S.C. 1981) (motion in *Limine* prevented use of OSHA regulations as evidence).

[4] *Melerine v. Avondale Shipyards*, supra at n. 237.

[5] *Supra* at *n*. 240.

[6] *Spankle v. Bower Ammonia & Chem. Co.*, 824 F.2d 409, 13 O.S.H. Cases 1382 (5th Cir. 1987). (Trial Judge did not error in prohibiting OSHA regulations to be admitted which he thought were unfairly prejudicial under Fed. R. Evid. 403.)

[7] *Supra* at *n*. 237.

[8] 741 P.2d at 232.

In *Walton v. Potlatch Corp.*[1] the court set forth four criteria to determine whether OSHA standards and regulations could be used to establish negligence per se:

1. The statute or regulation must clearly define the required standard of conduct.
2. The standard or regulation must have been intended to prevent the type of harm the defendant's act or omission caused.
3. The plaintiff must be a member of the class of persons the statute or regulation was designed to protect.
4. The violation must have been the proximate cause of the injury.[2]

If the court provides an instruction on negligence per se rather than an instruction on simple negligence, the effect is that the jury cannot consider the reasonableness of the employer's conduct. In essence, the court has already established a violation that constituted unreasonable conduct on the part of the employer and that the conduct was prohibited or required under a specific OSHA standard. Thus, as a matter of law, the jury will not be permitted to address the reasonableness of the employer's actions.

OSHA STANDARDS AS DEFENSE

Under appropriate circumstances, HR professionals may be able to use OSHA standards and regulations as a defense. Simple compliance with required OSHA standards is not in itself a defense, and the use of OSHA standards as a defense has received mixed treatment by the courts. However, at least one court has held that violation of a state OSHA plan by an employee could be considered in determining the employee's comparative negligence in a liability case.[3] Use of OSHA standards and regulations to demonstrate an appropriate standard of care in third-party product liability actions, workers' compensation litigation, and other actions may be permitted and should be explored by personnel managers in appropriate circumstances.

The use of OSHA citations and penalties in tort actions has also received mixed treatment by the courts. In *Industrial Tile v. Stewart*,[4] the Alabama Supreme Court stated:

> We hold that it was not error to admit the regulation if the regulations are admissible as going to show a standard of care, then it seems only reasonable that the evidence of violation of the standards would also be admissible as evidence that the defendant failed to meet the standards that it should have followed. Clearly, the fact that Industrial Tile had been cited by OSHA for violating the standards, and the fact that Industrial Tile paid the fine, are relevant to the conduct of whether it violated the standards of care applicable to its conduct. It was evidenced from a number of witnesses that the crane violated the 10-foot standards. It seems to us that evidence that Industrial Tile

---

[1] *Wendland v. Ridgefield Construction Service*, 184 Conn. 173, 439 A.2d 954 (1981); *Hebel v. Conrail*, 273 N.E.2d 652 (Ind. 1985); *Cowan v. Laughridge Construction Co.*, 57 N.C. App. 321, 291 S.E.2d 287 (1982).

[2] *Walton v. Potlatch Corp.*, 781 P.2d 229, 14 O.S.H. Cases 1189 (Idaho 1989).

[3] *Zalut v. Andersen & Ass.*, 463 N.W.2d 236 (Mich. Ct. App., 1990).

[4] 388 So. 2d 171 (Ala. 1980).

paid the fine without objection was properly admitted into evidence as a declaration against interest.[1]

Other courts have found that OSHA citations and fines are inadmissible under the hearsay rule of the Federal Rules of Evidence.[2] However, this can be normally overcome easily by offering a certified copy of the citations and penalties to the court, under the investigatory report exception to the Federal Rules of Evidence.[3]

Investigation records and other documents gathered in the course of an OSHA inspection are normally available under the FOIA. As noted previously, if particular citations are deemed inadmissible, a certified copy of the citations and penalties is normally considered admissible under Section 803(8)(c) of the Federal Rules of Evidence and 28 U.S.C. Section 1733 governing admissibility of certified copies of government records. Although the issue of whether OSHA citations and penalties are admissible is determined by the court under the Rules of Evidence, personnel managers should be prepared for all documents collected or produced during an OSHA inspection or investigation to be presented to the court. Given the nature of these government documents and the methods of presenting OSHA documents under the Federal Rules of Evidence, it is highly likely in any type of related litigation that the other party will obtain the documents from OSHA and that they will be submitted for use at trial. Other information and documents, such as photographs, recordings, and samples, may also be admissible under the same theory. HR professionals should maintain as much control as possible over information gathered during an investigation or inspection (that is, trade secrets and speculation by management team members, and so forth) and be prepared for the information to become public through the FOIA and used by opponents in litigation or elsewhere.

In addition to direct litigation with OSHA and negligence actions, OSHA standards used as evidence of the standard of care and citations used to show a breach of the duty of care have also been used in product liability cases,[4] construction site injury actions against general contractors,[5] and toxic tort actions.[6] Other actions

---

[1] Id. at *Lowe v. General Motors*, 624 F.2d 1373 (5th Cir. 1980) (applied to National Traffic & Motor Vehicle Safety Act standards).

[2] Id.

[3] Fed. R. Evid., 28 U.S.C.A. §803.

[4] *Spangler v. Kranco*, 481 F.2d 373 (4th Cir. 1973); *Bunn v. Caterpillar Tractor Co.*, 415 F. Supp. 286 (W.D. Pa. 1976); *Scott v. Dreis & Krump Mfg. Co.*, 26 Ill. App. 3d 971, 326 N.E.2d 74 (1975); *Bell v. Buddies Super-Market*, 516 S.W.2d 447 (Tex. Civ. App. 1974); *Brogley v. Chambersburg Engineering Co.*, 452 A.2d 743 (Pa. Super. Ct. 1982). (Note that OSHA standards are usually used as evidence of acceptable standards of machine design, industrial standard of care, or of reasonable conduct by employer or industry.)

[5] "The general contractor normally has responsibility to assure that the other contractors fulfill their obligations with respect to employee safety which affects the entire site. The general contractor is well situated to obtain abatement of hazards, either through its own resources or through its supervisory role with respect to other contractors. It is therefore reasonable to expect the general contractor to assure compliance with the standards in-so-far as all employees on the site are affected. Thus, we will hold the general contractor responsible for violations it could reasonably have been expected to prevent or abate by reason of its broad supervisory capacity." *Secretary v. Grossman Steel & Aluminum Corp.*, 4 O.S.H. Cases 1185 (1976).

6 *See*, for example, *Hebel v. Conrail*, 475 N.E.2d 652 (Ind. 1985); *Sprankle v. Bower Ammonia & Chemical Co.*, 824 F.2d 409, 13 O.S.H. Cases 1382 (5th Cir. 1987). (Note that toxic tort cases can utilize various theories ranging from failure to warn under a strict liability or negligence theory to wanton misconduct.)

where OSHA standards and citations have been found admissible include[1] Federal Tort Claim Act actions,[2] against OSHA in the area of inspections, and actions under the Federal Employers' Liability Act.

MEDIA RELATIONS

Often simultaneously with a governmental agency inspection (especially in circumstances involving a work-related fatality, multiple injury situation, environmental release, or other catastrophic event) will be the arrival of the media. In this age of instant news via such services as CNN, it can be expected that a media-worthy "story" will attract the roving media to the location of the incident. HR professionals must be aware of this situation and prepare beforehand to properly address the media in order to protect their company from potential efficacy losses.

When developing your plan to address the media, prepare for the worst and hope for the best. With the onset of "gorilla journalism," you must be prepared for the media to place your company in the worst possible light, so you must prepare your plan to counteract, where possible, to minimize the possible damage to your company's image and good name. Remember, in most situations, your company has just incurred some traumatic situation whereby the stress level is extremely high and numerous abnormal activities, including governmental inspections, are taking place. Preparation beforehand minimizes the chances of a major blunder.

In many organizations, a written media plan addressing the various possible scenarios is prepared. These plans can include the following:

- Individual responsible for the information flow to the media and overall management of the media
- Centralized location for the media offering appropriate background
- Identified spokesperson for the company with appropriate dress, manner, and so forth
- Approval procedure from management and legal for all news releases
- Security and control of the media on site (i.e., keep them in one location)
- Availability of positive file footage

Management of the media during these extremely stressful situations is vital. For example, your company has incurred an accident in which there is a fatality. Do you want the fatally injured employee's family finding out that mom or dad is dead on the evening news? And additionally, the image sent via a 30-second sound bite can have a direct impact on your company's image, the sales of your products, your ability to attract employees, and a number of other factors. The person who is acting as your spokesperson "is" your company to the viewing public. *It is vital that you properly manage the image of your company during these intense time periods.*

---

[1] 20 U.S.C. §2671 *et seq.* See also *Blessing.*

[2] See, for example, *Blessing v. U.S.*, 447 F. Supp. 1160 (E.D. Pa. 1978). (Allegations of negligent OSHA inspections states a viable Federal Tort Claim Act claim under Pennsylvania law.); *Mandel v. U.S.*, 793 F.2d 964 (8th Cir. 1986).

ALTERNATIVE STRATEGIES

With most governmental agencies, there are alternatives or non-routine methods through which to address specific needs or unique situations. However, these unique alternatives are seldom advertised and usually must be requested by your company. Several alternatives to consider in the area of safety and health include the following:

State Education and Training Programs — In many states, specific education and training programs are provided, usually free of charge to employers, to assist the company in achieving and maintaining compliance with the requirements. These programs can include seminars, on-site consultation, on-site training, and on-site testing.

Variances — A variance is a request to be exempt from the particular standard or regulation. Under the OSH Act, there are four basic types of variances: permanent variances, temporary variances, experimental variances, and National Defense variances. The most often utilized are the permanent or temporary variances that apply to specific equipment, worksites, or situations and are not blanket exemptions from the regulations. Experimental variances are available when participating in a project approved by OSHA or the National Institute for Occupational Safety and Health (NIOSH). National Defense variances are seldom utilized today.

Interim Orders — OSHA regulations provide for interim orders that grant temporary relief from inspections or citations pending the outcome of a formal hearing regarding a variance application. Interim orders are normally included with variance petitions.

SUMMARY

Compliance with the numerous laws, standards, and requirements mandated by various governmental agencies has become an important part of most safety professionals' job functions. Safety professionals should be aware of the numerous regulatory requirements from other federal agencies, including, but not limited to, the Environmental Protection Agency (EPA), Food and Drug Administration (FDA), U.S. Department of Agriculture (USDA), Equal Employment Opportunity Commission (EEOC), as well as various state and local agencies. Each of these agencies usually possesses written rules and administrative procedures that are required to be followed. Given the potential detrimental effects to your company or organization, safety professionals must become proficient in the regulations applicable to their organization and acquire appropriate assistance. Safety professionals usually possess direct or indirect responsibility for this important area where achieving and maintaining compliance can be an enormous task. Proper management of this important function by the safety professional will provide a multitude of benefits for the employees, management team, and company. Failure to appropriately manage this important function is a recipe for disaster.

15 Other Laws Affecting the Safety Function

Nobody has a more sacred obligation to obey the law than those who make the law.

Jean Anouilh

As safety professionals are well aware, the safety function does not work in a vacuum. Although the OSH Act provides the primary framework of laws that govern the safety function, a myriad of other laws impact the safety function on a daily basis. Safety professionals should acquire a working knowledge of these laws and regulations in order to be able to recognize the interaction as well as the impact of these laws on safety-related activities, policies, and procedures. Below are some of the other laws and regulations that can impact the safety function.

AMERICANS WITH DISABILITIES ACT

The Americans with Disabilities Act of 1990 (known as the ADA) has opened a huge new area of regulatory compliance which will directly or indirectly affect most safety programs. In a nutshell, the ADA prohibits discrimination against qualified individuals with physical or mental disabilities in all employment settings. Given the breadth and scope of the ADA, this law has impacted most companies and organizations in a multitude of areas ranging from hiring to retirement.

Structurally, the ADA is divided into five titles, and all titles possess the potential of substantially impacting the safety function. Title I contains the employment provisions that protect all individuals with disabilities who are in the United States, regardless of national origin and immigration status. Title II prohibits discriminating against qualified individuals with disabilities or excluding qualified individuals with disabilities from the services, programs, or activities provided by public entities. Title II includes the transportation provisions. Title III addresses public accommodations and a broad expanse ranging from goods to facilities. Title III additionally covers transportation offered by private entities. Title IV addresses telecommunications. Title IV requires that telephone companies provide telecommunication relay services and television provide public service announcements produced or funded with federal money that include closed caption. Title V includes the miscellaneous provisions. This title noted that the ADA does not limit or invalidate other federal and state laws providing equal or greater protection for the rights of individuals with disabilities and addresses related insurance, alternate dispute, and congressional coverage issues.

Title I prohibits covered employers from discriminating against a "qualified individual with a disability" with regard to job applications, hiring, advancement, discharge, compensation, training, and other terms, conditions, and privileges of employment.

Section 101 (8) defines a "qualified individual with a disability" as any person

> who, with or without reasonable accommodation, can perform the essential functions of the employment position that such individual holds or desires ... consideration shall be given to the employer's judgment as to what functions of a job are essential, and if an employer has prepared a written description before advertising or interviewing applicants for the job, this description shall be considered evidence of the essential function of the job.[1]

The Equal Employment Opportunity Commission (EEOC) provides additional clarification as to this definition in stating "an individual with a disability who satisfies the requisite skill, experience and educational requirements of the employment position such individual holds or desires, and who, with or without reasonable accommodation, can perform the essential functions of such position."[2]

Although there is not a specific list of disabilities in the ADA, an individual has a disability if he or she possesses

1. a physical or mental impairment that substantially limits one or more of the major life activities of such individual;
2. a record of such an impairment;
3. is regarded as having such impairment.[3]

Safety professionals should be aware that for an individual to be considered "disabled" under the ADA, the physical or mental impairment must limit one or more "major life activities."[4] A "qualified individual with a disability" under the ADA is any individual who can perform the essential or vital functions of a particular job with or without the employer accommodating the particular disability.

Of particular concern to safety professionals is the treatment of the disabled individual, who, as a matter of fact or due to prejudice, is believed to be a direct threat to the safety and health of others in the workplace. To address this issue, the ADA provides that any individual who poses a *direct threat* to the health and safety of others that cannot be eliminated by reasonable accommodation may be disqualified from the particular job.[5] The term "direct threat" is defined by the EEOC as "a significant risk of substantial harm to the health and safety of the individual or others that cannot be eliminated by reasonable accommodation."[6] The determining factors safety professionals should consider in making this determination include the duration of the risk, the nature and severity of the potential harm, and the likelihood the

---

[1] ADA Section 101 (8).
[2] EEOC Interpretive Rules, 56 Fed. Reg. 35 (July 26, 1991).
[3] Subtitle A, Section 3(2).
[4] 28 CFR Section 41.31.
[5] ADA Section 103.
[6] EEOC Interpretive Rules, supra n. 2.

potential harm will occur.[1] Additional guidance in this area can be acquired in the EEOC interpretive rules found on the EEOC Web site (www.eeoc.gov).

Safety professionals should also note that Title I addresses the issue of reasonable accommodation. Section 101 (9) defines a "reasonable accommodation" as

a. "making existing facilities used by employees readily accessible to and usable by the qualified individual with a disability" and includes:

b. job restriction, part-time or modified work schedules, reassignment to a vacant position, acquisition or modification of equipment or devices, appropriate adjustments or modification of examinations, training materials, or policies, the provisions of qualified readers or interpreters and other similar accommodations for … the QID.[2]

The EEOC further defines "reasonable accommodation" as

1. Any modification or adjustment to a job application process that enables a qualified individual with a disability to be considered for the position such qualified individual with a disability desires, and which will not impose an undue hardship on the … business; or

2. Any modification or adjustment to the work environment, or to the manner or circumstances which the position held or desired is customarily performed, that enables the qualified individual with a disability to perform the essential functions of that position and which will not impose an undue hardship on the … business; or

3. Any modification or adjustment that enables the qualified individual with a disability to enjoy the same benefits and privileges of employment that other employees enjoy and does not impose an undue hardship on the … business.[3]

The above is just a brief overview of Title I of the ADA which can impact the safety professional. Title I also addressed controlled substance testing, medical examinations, as well as other areas impacting the safety function.

Title II of the ADA addresses public entities. This title covers the provision of services, programs, activities, transportation, and employment by public entities. A public entity under Title II includes the following:

- A state or local government.
- Any department, agency, special-purpose district, or other instrumentality of a state or local government.
- The National Railroad Passenger Corporation (Amtrak) and any commuter authority as this term is defined in section 103(8) of the Rail Passenger Service Act.[4]

---

[1] Id.

[2] ADA Section 101(9).

[3] EEOC Interpretive rules, supra n. 2.

[4] ADA Section 201.

Title III is focused on the accommodations in public facilities and makes it unlawful for public accommodations not to remove architectural and communication barriers from existing facilities and transportation barriers from vehicles "where such removal is readily achievable."[1]

Of importance to most safety professionals, Title III also requires "auxiliary aids and services" be provided for the qualified individual with a disability, including, but not limited to, interpreters, readers, amplifiers, and other devices (not limited or specified under the ADA), to provide the qualified individual with a disability with an equal opportunity for employment, promotion, and so forth.[2] This section also addresses modification of existing facilities to provide access to the qualified individual with a disability and requires all new facilities to be readily accessible and usable by the qualified individual with a disability.

Of less importance to safety professionals, **Title IV** requires all telephone companies to provide "telecommunications relay service" to aid the hearing and speech impaired qualified individual with a disability.

Title V assures that the ADA does not limit or invalidate other federal or state laws that provide equal or greater protection for the rights of individuals with disabilities. Title V also addresses specific exclusions from protection under the ADA.

Safety professionals should be aware that individuals' extended protection under the ADA includes all individuals associated with or having a relationship to the qualified individual with a disability. This inclusion is unlimited in nature, including family members, individuals living together, and an unspecified number of others.[3] The ADA extends coverage to all "individuals," and thus the protection is provided to all individuals, legal or illegal, documented or undocumented, living within the boundaries of the United States regardless of their status.

Although the ADA is far more expansive than can be covered in this text, safety professionals should become familiar with the applicable aspect of the ADA as it relates to the safety function and ensure that programs, policies, and decisions do not discriminate against protected individuals. Additional information regarding the ADA and its requirements can be found on the Equal Employment Opportunity Commission's Web site (www.eeoc.gov).

FAMILY AND MEDICAL LEAVE ACT

Another law impacting the safety function is the Family and Medical Leave Act (or FMLA).[4] Although this is primarily a human resource function, the FMLA impacts the safety function in the areas of safety personnel and staffing. The key provisions of this law for safety professionals are Title I and Title II. Under Title I, an eligible employee is defined as being an employee who has been employed at least 12 months with the company or organization and who has provided at least 1250 hours of service during the 12 months. Safety professionals should be aware that companies or

---

[1] ADA Section 302.
[2] ADA Section 3(1).
[3] ADA Sections 102(B)(4) and 302(b)(1)(e).
[4] 29 U.S.C. Sections 2601–2654.

organizations with fewer than 50 employees within a 75-mile radius of the worksite are excluded from coverage.

Title I also includes the broadly construed definitions of the terms "parent," "son or daughter" (includes biological, adopted, or foster child or legal ward), but does not define "spouse." "Serious health condition" is defined as "an illness, injury, impairment, or physical or mental condition" involving either inpatient care or continuing treatment by a health-care provider.

Eligible employees are entitled to 12 unpaid workweeks of leave during any 12-month period for three basic reasons: (1) the birth or placement for adoption or foster care of a child (within 12 months of the birth or placement), (2) serious health condition of a spouse, child, or parent, or (3) the employee's own serious health condition.

In general, the company or organization may require that the eligible employee provide certification of a serious health condition of him- or herself or a family member. Certification includes the date of the serious health condition, the duration of the condition, appropriate facts regarding the condition, and a statement that the eligible employee needs to care for the spouse, child, parent, or self. For cases of intermittent leave, the dates of the leave should also be noted. The company or organization may require a second opinion be obtained by the employee at his or her own expense. The second opinion may not be provided by the health-care provider employed by the company or organization. In the event of a conflict in opinions, a third and final opinion can be required at the expense of the company or organization.

The employee who completes a period of leave is to be returned either to the same position or to a position equivalent in pay, benefits, and other terms and conditions of employment. This leave cannot result in the loss of any previously accrued seniority or employment benefits, but neither of these benefits is required to accrue during the leave. Health benefits are to continue throughout the leave. The company or organization may be required to pay the health coverage premiums but may recover these premiums if the employee fails to return to the job following the leave. As with most laws, the company or organization is prohibited from discriminating against the employee for use of the Family and Medical Leave Act.

Safety professionals can find additional information regarding the FMLA on the U.S. Department of Labor Web site (www.dol.gov). Safety professionals should be aware that legislation is currently pending to provide protection to the families of military personnel and other changes to the FMLA.

FAIR LABOR STANDARDS ACT

Safety professionals should be aware that the Fair Labor Standards Act (FLSA) addresses the issues of minimum wage and overtime pay as well as record keeping and child labor standards. The FLSA applies to most companies and organizations that have employees who are engaged in interstate commerce, producing goods for interstate commerce, or handling, selling, or working on goods or materials that have been moved in or produced for interstate commerce and have an annual dollar volume of business of over $500,000.

Some employees are exempt from the Act's overtime pay provisions or both the minimum wage and overtime pay provisions under specific exemptions provided in the law. Because these exemptions are generally narrowly defined, safety professionals should carefully check the exact terms and conditions at www.dol.gov or by contacting local offices of the U.S. Department of Labor, Wage and Hour Division.

Section 13(a)(l) of the FLSA exempts executive, administrative, professional, and outside sales employees from the minimum wage and overtime requirements of the FLSA, provided they meet certain tests regarding job duties and responsibilities and are compensated "on a salary basis" at not less than stated amounts. Subject to certain exceptions set forth in the regulations, in order to be considered "salaried," employees must receive their full salary for any workweek in which they perform any work without regard to the number of days or hours worked. This rule applies to each exemption that has a salary requirement. (Outside sales employees and certain licensed or certified doctors, lawyers, and teachers have no salary requirement. For certain computer-related occupations under the professional exemption, they need not be paid a salary if they are paid on an hourly basis at a rate not less than $27.63 per hour.)

The FLSA requires covered employers to pay employees the identified minimum wage. Of importance to safety professionals, the FLSA does not limit the number of hours in a day or week, and the employee may be required to work a week (except for workers at least 16 years of age). The FLSA does, however, require that the employee be paid one and one-half times his or her regular rate of pay for all hours worked in excess of 40 hours in a workweek. Companies or organizations are required to maintain records on wages, hours, and other related items established under the Department of Labor's Wage and Hour regulations.

The FLSA provides protection for employees against discrimination for filing a complaint or participating in legal proceedings under the Act. The FLSA also prohibits shipment of goods in interstate commerce which were produced in violation of the minimum wage, overtime, child labor, or special minimum provisions. Enforcement of the FLSA is provided by the Wage and Hour Division investigators from the Department of Labor housed at various locations throughout the United States. When investigators find a violation, they usually recommend changes to the employment practices to bring the organization into compliance and request the back pay for the employees; however, a variety of remedies are available. Willful violations may be referred to the Department of Justice for criminal prosecution and the violator may be fined up to $10,000. A second conviction can result in imprisonment and willful and repeat violations and result in civil penalties up to $1000 per violation. The Secretary of Labor may also bring a suit for back pay and an equal amount of liquidated damages and also obtain injunctive relief. Employees are also provided a cause of action for back pay and liquidated damages, including attorney fees and costs, if the Department of Labor does not bring suit.

As with OSHA, safety professionals should be aware that if an FLSA violation with monetary penalty is assessed, the company or organization must file an exception to this determination within 15 days of receipt. The exception is usually referred to an Administrative Law Judge (ALJ) for a hearing and determination as to the appropriateness of the penalty. If an exception is not filed in a timely manner, the penalty becomes final. Additional information regarding the FLSA can be found on the U.S. Department of Labor Web site (www.dol.gov).

TITLE VII OF THE CIVIL RIGHTS ACT

Safety professionals should be aware that Title VII is the seventh title in the broad civil rights legislation. Title VII of the Civil Rights Act of 1964, as amended by the Civil Rights Act of 1991 and other intermediate acts (hereinafter referred for the purposes of this text as "Title VII"), is the primary law that prohibits discrimination based upon race, sex, color, creed, religion, and national origin in the workplace. Title VII is applicable to all public and private-sector employers as well as labor unions, employment agencies, and joint labor–management organizations. Virtually all companies and organizations are covered under Title VII.

A company or organization is subject to Title VII if it has fifteen employees on the payroll on each working day of at least 20 weeks in the current or preceding calendar year. Safety professionals should be aware that employment agencies and labor organizations also are covered by Title VII.

Safety professionals should be aware that individuals who believe that they have been discriminated against can file a Title VII charge with the EEOC. A charge must be filed within 180 days after the alleged unlawful employment practice occurred. Before processing a charge, EEOC must notify the employer of the allegations and give state and local fair employment agencies at least 60 days to resolve it. EEOC investigates the charge after the state or local agency has acted or 60 days have elapsed. During investigations, EEOC can require employers to provide relevant records or evidence.

If the investigation reveals no evidence of discrimination, EEOC issues a letter of determination stating that there is no reasonable cause to believe that discrimination occurred. If EEOC finds support for the discrimination charge, it must try to resolve the dispute informally through conciliation proceedings. If this effort fails, the commission can file suit against the employer in a U.S. district court.

The act allows employees and job applicants who have Title VII complaints to bring suit directly in a U.S. district court after their charges have been on file with EEOC for 180 days and the commission has not dismissed them. Complainants also retain their private right to sue for 90 days after receiving a no cause determination by EEOC.

Safety professionals should be aware that the Civil Rights Act of 1991 contains a number of provisions that broaden or clarify existing civil rights laws. In brief, the Act covers the following:

- Prohibits all racial discrimination in the making and enforcement of contracts.
- Provides for damages and jury trials in cases of intentional employment discrimination.
- Shifts the burden of proof to employers in cases of disparate impact.
- Clarifies that any discrimination practice is unlawful, even if other lawful factors motivated the action.
- Prohibits certain consent order challenges.
- Extends discrimination protection to U.S. citizens working abroad for American companies.
- Prohibits discrimination in employment testing.

- Requires the EEOC to establish a Technical Assistance Training Institute and conduct educational and outreach activities.
- Expands the right to challenge discriminatory seniority systems.
- Provides for the same interest against the federal government for delay in payments of awards that is provided in cases involving private employers.
- Increases the time for filing a Title VII suit against the federal government from 30 days to 90 days after receipt of a notice of right to sue.
- Authorizes the awarding of expert witness fees.
- Establishes a "Glass Ceiling Commission" to study and make recommendations on eliminating barriers to the advancement of women and minorities.
- Extends coverage of the major civil rights laws to Senate employees and presidential appointees.
- Extends coverage of Title VII to employees of the House of Representatives.

The 1991 Act amended Title VII by including the following:

- Requiring employers to demonstrate that a challenged practice is job related and consistent with business necessity.
- Extending employees the right to challenge seniority systems when adopted, when they are actually harmed by the application of a system.
- Prohibiting consent decree challenges by certain individuals after court approval.
- Clarifying that any discriminatory practice is unlawful, even if lawful factors also motivated the action.
- Including expert witness fees in recoverable attorneys' fees.
- Extending protection from discrimination to U.S. citizens employed abroad by American companies.
- Prohibiting the adjusting of test scores, use of different cut-off scores, or altering of the results of employment-related tests on the basis of race, color, religion, sex, or national origin.

Safety professionals should also be aware that the 1991 Act clarified that the 1866 law prohibited not only racial bias that occurs in the hiring process and in promotions, but also on-the-job harassment, and other post-hiring conduct.

The 1991 act also expanded the categories of intentional bias victims who can obtain compensatory and punitive damages. Previously, only victims of intentional racial and ethnic bias could receive compensatory and punitive damages. The 1991 act extends such coverage to victims of intentional sex, religious, and disability discrimination under Title VII, ADA, and the Rehabilitation Act.

In intentional discrimination cases, compensatory damages can be recovered from private employers, state and local governments, or the federal government. However, punitive damages can be sought only from private employers, and the complainant must demonstrate that the employer engaged in the discriminatory practice "with malice or with reckless indifference."

Further, most compensatory and all punitive damages are capped according to the size of an employer's workforce in each of 20 or more weeks in the current or preceding year.

Damages cannot exceed the following:

- $50,000 for employers with more than 14 and fewer than 101 employees.
- $100,000 for employers with more than 100 and fewer than 201 employees.
- $200,000 for employers with more than 200 and fewer than 501 employees.
- $300,000 for employers with more than 500 employees.

Back pay, interest on back pay, or other relief authorized under 42 U.S.C. Section 2000e-5(g) is not included in compensatory damages, and caps do not apply to past pecuniary losses such as medical bills. An individual seeking compensatory or punitive damages can demand a jury trial, but courts are prohibited from informing the jury about cap limitations. No compensatory or punitive damages can be awarded in reasonable accommodation cases where the employer demonstrates that it has made a good faith effort to reasonably accommodate the person with the disability.

Safety professionals should be aware that the 1991 Act amended ADA by extending protection from disability discrimination to U.S. citizens employed by American companies.

The 1991 Act also deleted a provision in ADA that allowed the running of the limitations period for filing suit to be suspended up to one year while EEOC attempted to resolve the charge. A new provision was added that requires EEOC to notify complainants when a charge is dismissed or otherwise terminated. Suit then must be filed within 90 days of receipt of that notice.

Safety professionals should be well versed in the requirements of Title VII and ensure that all programs, policies, and decisions are free from discrimination. Additional information regarding Title VII can be found on the EEOC Web site (www.eeoc.gov).

AGE DISCRIMINATION IN EMPLOYMENT ACT

With the aging population in the United States and individual workforces aging, safety professionals should be cognizant of the protections afforded current employees and applicants under the Age Discrimination in Employment Act (ADEA). This legislation, when combined with the protections afforded under related laws such as the ADA, can pose substantial challenges to safety professionals in avoiding potential areas of possible discrimination within the safety function as well as in the hiring and training processes.

The ADEA of 1967[1] was enacted by Congress for the purpose of promoting employment of individuals over the age of 40 and prohibiting arbitrary age discrimination by employers. Safety professionals should recognize that ADEA protects individuals who are between the ages of 40 and 70 from being discriminated against or discharged because of their age.[2] The ADEA provides coverage to

---

[1] 29 U.S.C.A. Section 621, et. seq.
[2] Id.

virtually all private-sector employers, employment agencies, and labor organizations.[1] An employer is defined under the ADEA as "one engaged in an industry affecting commerce and having 20 or more employees for each working day in each of the twenty or more calendar weeks during the current or preceding calendar year." Safety professionals should be aware that the ADEA prohibits companies and organizations from discriminating against employees, members, referrals, applicants, or other covered individuals as a means of retaliation because the employee has made a charge of discrimination, has testified, has assisted, or has participated in any manner in any investigation, proceeding, or litigation under the ADEA.[2] Additionally, companies and organizations are prohibited from printing or publishing or causing to be printed or published any notice or advertisement related to employment which indicates a preference, limitation, specification, or discrimination based upon age.[3] Safety professionals should be aware that the following activities and actions are prohibited by the ADEA:

1. Failing or refusing to hire or to discharge any individual or otherwise discriminate against any individual with respect to his or her compensations, terms, conditions, or privileges of employment because of such an individual's age.
2. Limiting, segregating, or classifying his or her employees in any way that would deprive or tend to deprive any individual of employment opportunities or otherwise adversely affect the status of the employee because of age.
3. Reducing the wage rate of any other employee or employees in order to comply with the ADEA.[4]

Safety professionals should be aware that the statute of limitations for an ADEA claim are relatively confusing and guidance may be needed from legal counsel. An individual who has been discriminated against can file an EEOC charge. The individual must file within 180 days from the alleged unlawful acts, unless the misconduct occurred in a state where the age anti-discrimination group or agency permits a 300-day statute of limitations.[5] The ADEA charge must be filed with the state agency or the EEOC in writing. An investigative procedure and conciliation would mirror that as set forth under Title VII of the Civil Rights Act. Safety professionals should also be cognizant that a person, as well as the EEOC, may seek to bring a civil action against the company or organization in any court of competent jurisdiction if the following conditions are met prior to the filing of the action:

1. The employee has waited 60 days after filing the charge with the EEOC or state agency before initiating his or her private lawsuit.
2. The Age Discrimination Charge is filed within the specified statute of limitations.

---

[1] Id.
[2] 29 USC Section 623(d).
[3] 29 USC Section 623(e).
[4] 29 USC Section 623(a).
[5] 29 U.S.C.A. § 626(d)(2).

3. The lawsuit by the employee is filed within the 2- or 3-year statute of limitations.

The ADEA provides that an individual is entitled to a trial by jury for most remedies except for injunctive relief.[1] Safety professionals should realize that, like with most legal actions, a wide variety of possible relief is permitted under an ADEA action, including permanent injunctions, reinstatement or promotion, judgments compelling employment, monetary damages, and, in willful situations, liquidated damages.[2] Unlike OSHA and related actions, safety professionals need to make their management team aware that attorney's fees can be awarded to the prevailing party.

Safety professionals should be aware that there are a variety of different defenses available depending on the circumstances. These defenses include providing deference based on reasonable factors other than age; the bona fide occupational qualification (BFOQ) defense; a bona fide seniority system; the employer's good faith reliance on a written administrative order; and procedural or technical deficiencies in the ADEA claim itself.

EMPLOYMENT AND LABOR LAWS

Safety professionals should be aware that there are a myriad of labor- and employment-related laws, both federal and state, that may directly or indirectly affect their company or organization. At the federal level and in most states, laws have been adopted in the labor and employment area addressing specific employees working within their jurisdiction. Safety professionals should take careful note of the laws that are applicable to their programs as well as to their company on the federal, state, and even local levels to insure compliance.

As a general rule, safety professionals should realize that most employees in the American workforce are considered "at will"; thus, the employee can be terminated at will for good cause, bad cause, or no cause at all. The general exceptions to this rule are when an employee is governed by a collective bargaining agreement (union contract) or the individual possesses a personal contract for employment. The courts have generally found that these contractual relationships remove the particular party from the at-will doctrine. In addition to the above exceptions, many courts in the United States have developed additional exceptions from the at-will doctrine including, but not limited to, the following:

1. Implied Contract Exception
   - Safety professionals should note that many courts have recognized a cause of action where the employer, or an agent of the employer, has promised through a personnel handbook or other document that the employee would not be discharged until given a fair chance to perform.

---

[1] 29 U.S.C.A. Section 217.
[2] U.S.C.A. § 216(b) and (c).

The court recognizes this exception through a handbook as constituting an implied in fact employment contract.[1]

2. Covenant of Good Faith and Fair Dealing

- The courts have recognized that the employer may have an implied obligation to deal fairly with an employee and not act arbitrarily in a discharge situation. Safety professionals should note that most of these cases involve employees who have spent an extended period of time with the employer and then are terminated under the at-will doctrine.[2]

3. Independent Consideration Exception

- Although not recognized in many states, safety professionals should be aware of the independent consideration exception to the at-will doctrine. In this exception, the employer's oral promise not to terminate the employee except for good cause and when services are performed in addition to regular required employment services is utilized after the employee is terminated.[3]

4. Public Policy Exceptions

- Utilized most often in the safety area, safety professionals should be aware that most states recognize that an employer may not terminate an employee for pursuing a workers' compensation claim, complaint to OSHA, EEOC claim, or other public policy rights.[4]

5. Presumption of Term Exception

- In many states an oral contract for a yearly salary raises the inference of a year-to-year oral employment contract.[5]

6. Promissory Estoppel Exception

- Some courts have recognized a promissory estoppel exception to the at-will doctrine holding that an employee may recover moving expenses where he or she detrimentally relied upon a prospective employer's promise of a job.[6]

7. Intentional Infliction of Emotional Distress Exception

- Safety professionals should be aware that some courts have found that where an employer's conduct is so outrageous as to go beyond all possible bounds of decency resulting in severe emotional distress to an employee, an action will be permitted to address this wrongdoing.[7]

8. Refusal to Commit Unlawful Acts

- Safety professionals should be aware that most states recognize that where an employee refuses to commit an unlawful act, and the employer

---

[1] See, for example, *Walker v. Northern San Diego County Hospital District*, 135 Cal. App. 3d. 896, 185 Cal. Rep. 617 (1982).

[2] See, for example, *Pugh v. See's Candies, Inc.*, 116 Cal. App. 3d. 311, 171 Cal. Rep. 917 (1981); *Cancellier v. Federated Department Stores*, 672 F. 2d. 1312 (Ca-9, 1982).

[3] See, for example, *Alvarez v. Dart Industries, Inc.*, 55 Cal. App. 3d. 91, 127 Cal. Rep. 222 (1976).

[4] See, for example, *Firestone Textile Company Divisions v. Meadows*, 666 S.W. 2d 730 (Ky. 1983); *Tammany v. Atlantic Richfield Co.*, 27 Cal. 3d 167, 610 P. 2d. 1330, 164 Cal. Rep. 839 (1980).

[5] See, for example, *Moore v. YWCA*, Case No. 84-CA-1508-MR (Ky. Court of Appeals, 1985).

[6] See, for example, *Lorson v. Falcon Coach, Inc.*, 522 P. 2d. 449 (Kansas 1974).

[7] See, for example, *Harris v. First Federal Savings & Loan Assoc. of Chicago*, 473 N.E. 2d. 457 (Ill. Court of Appeals, 1984).

discharges the employee, that this may constitute an exception to the at-will doctrine.[1]

Safety professionals should recognize that, in addition to the at-will doctrine, there are numerous labor- and employment-related laws that have been enacted on the federal and state levels which directly impact the safety function. The Welsh-Healey Act, 41 U.S.C. § 35, enacted in 1936, addresses the situation of government contracts in excess of $10,000. Employees working for contractors governed by this Act must pay the minimum wage and overtime benefits as specified by law. There are numerous exceptions to the Welsh-Healey Act.[2] The Department of Labor is the agency that enforces the Welsh-Healey Act, and failure to comply may permit the government to blacklist the contractor from further contracts with the U.S. government in the event of non-compliance.[3]

The David-Bacon Act, enacted in 1931, requires the payment of minimum wages and fringe benefits to laborers on federal public works contracts in excess of $2000.[4] Safety professionals should be aware that the provisions of the David-Bacon Act normally apply to construction activities as distinguished from manufacturing activities and require the employer to comply with wage and hour provisions. The Department of Labor is responsible for the administration and enforcement of the David-Bacon Act.

Safety professionals should be aware that there are a large number of labor-related laws that may come into play depending on the circumstances. The Copeland Act, sometimes known as the "anti-kickback act," prohibits government contractors from compelling or inducing the employee to give back part or all of his or her compensation. The Miller Act provides that government contractors working on public works contracts must furnish a bond to protect the payment of wages to all employees.[5] The Contract Work Hours and Safety Standards Act provides that employers pay time and one-half for hours worked in excess of 8 hours in any one day or 40 hours in any one week and additionally requires that the employer provide a safe and healthy work environment for employees.

Of particular interest for safety professionals with workers' compensation and related responsibilities, the Consumer Credit Protection Act (Title III) provides restrictions set on the amount of an individual's earnings which may be deducted in any one week as a result of a garnishment proceeding.[6] The Consumer Credit Protection Act establishes a maximum part of the aggregate disposable income of any individual which may be subject to garnishment to be no more than 25 percent of the employee's disposable earnings in any one week, or the amount by which the disposable earnings for that week exceed 30 times the federal minimum hourly wage as set forth under the Fair Labor Standards Act.[7] Safety professionals should be aware that exceptions can be made from a court order on any bankruptcy as set up under

---

[1] See, for example, *Peterman v. International Brotherhood of Teamsters, Local 396*, 174 Cal. App. 2d 184, 344 P. 2d 25 (1959).

[2] CFR Sections 50-201.603.

[3] 41 U.S.C.A. Section 37.

[4] 40 U.S.C. Section 270(a); 32 C.F.R. Section 10.101-110.

[5] 40 U.S.C.A. Section 329.

[6] 15 U.S.C.A. Section 1671, *et. seq.*

[7] Id.

the Bankruptcy Act or any debt due for the payment of state or federal taxes. The Consumer Credit Protection Act further provides that no employer may discharge any employee by reason of the fact that his or her earnings are subject to garnishment for any one indebtedness.[1]

Safety professionals need to be aware that the above are just a few of the myriad of labor-related laws that are "on the books." There are numerous other labor and employment acts on the federal and state levels which address specific industries and specific situations. In the area of employment practices, many states have adopted anti-discrimination laws; protections for political freedom, jury duty, sterilization, investigating consumer reports, arrest record, access to personnel files, confidentiality of medical information, employment under false pretense, polygraph restrictions; protection of military personnel; protection from blacklisting and fingerprinting; whistle blowing statutes; and alcohol rehabilitation, plant closure, equal pay, and anti-reprisal statutes. Other rights that may be available under state labor laws include collective bargaining rights of public employees, higher education employees, non-right-to-work policies, yellow duck contract laws, strike replacement laws, and anti-injunction statutes. Safety professionals are urged to become familiar with the particular laws of their jurisdiction as well as the federal laws that apply to their specific industry as well as company or organization.

PRIVACY LAWS

Safety professionals may be faced with privacy issues ranging from locker rooms to medical records. Privacy issues are often commingled with other laws, such as ADEA, Title VII, and state laws prohibiting discrimination based upon race, sex, color, religion, age, national origin, disability, and related protected areas. The protections afforded on the federal level tend to be narrow in scope and address specific issues of privacy. Additionally, individual states have developed privacy laws that address privacy issues encountered in their jurisdiction.

On the federal level, there are eight primary privacy laws that include the following:

1. Freedom of Information Act (FOIA)
   - For safety professionals, the FOIA can be a conduit through which information about your programs and company can be disseminated to the public. The FOIA requires federal agencies to make its rules, opinions, orders, records, hearings, and proceedings available to the general public and, in litigation or other activities, information provided to a governmental agency may be subject to an FOIA disclosure.
   - The federal agency is required to make available for public inspection and copying information such as opinions and orders made from cases heard by the agency as well as policies, interpretations, manuals, and related information. Safety professionals should know that the FOIA does not apply to certain documents such as individual employee's personnel or medical records, and other related documents or information

[1] 15 U.S.C.A. § 1674.

exempted under the Sunshine Act [5 U.S.C.A. Section 552 (b)(1) through (9)]. If a federal agency fails to comply with the FOIA, the requesting party may bring an action in the appropriate U.S. District Court.

2. Privacy Act of 1974
   - The Privacy Act was designed to protect individual rights of privacy from invasion from actions by federal governmental agencies. This law requires federal agencies to maintain records of individuals in an accurate and relevant manner to permit individual citizens to gain access to records and information about themselves and to refuse access about individuals to other persons or agencies except in very specific circumstances.
   - The Privacy Act requires every federal agency to develop and maintain a system of records to provide the individual requesting the review of his or her personal records in a timely manner. Additionally, if the federal agency refuses to correct a record after review by the individual, the federal agency must provide a final determination within 30 days after the request is made by the individual. Under the exceptions in this law, a federal agency may provide records of an individual, without his or her consent or request, to other agencies, like the Census Bureau, or to others under the FOIA.

3. Federal Sunshine Act
   - The Federal Sunshine Act is a companion law to the Privacy Act of 1974. This statute provides that meetings held by federal agencies that have two or more participants appointed by the president be open to the public. The primary purpose of this act is to provide the public access to the decision-making process of the federal government.
   - Certain federal agency meetings are exempt from this law — namely, meetings involving law enforcement, military meetings involving classified materials, and meetings involving other confidential or classified materials. All federal agencies are required to promulgate regulations for implementing this law, and if the agency should refuse to comply, the individual adversely affected may bring an action in the appropriate U.S. District Court.

4. Fair Credit Reporting Act
   - The Fair Credit Reporting Act provides protection for applicants and employees from the company or organization acquiring consumer reports unless the employer provides a written notice to the applicant or employee. This written notice to an applicant or employee must disclose the nature and scope of the investigation which can include information as to the individual's character, reputation, characteristics, and lifestyle.
   - The Fair Credit Reporting Act also requires the company or organization, if employment is denied based upon the investigative report, to inform the applicant as to the information that was the basis of the decision and provide the name and address of the reporting agency. An individual harmed by the willful or negligent failure to comply with the

Fair Credit Reporting Act is provided a cause of action in the appropriate U.S. District Court.

5. Immigration and Nationality Act of 1972
   - As most safety professionals are aware, this area is under intense scrutiny at this time. The Immigration and Nationality Act addresses the admission of immigrant aliens and temporary foreign workers for employment in the United States and only permits a limited number of immigrants when there is a showing that American workers are not available for the work in question and the wages and working conditions will not adversely affect similar employed U.S. workers.

6. Equal Access to Justice Act
   - The Equal Access to Justice Act provides for the award of attorney fees, expert witness fees, and other costs against the United States for individuals, corporations, partnerships, labor, or other parties in defending against governmental actions. The act provides that a governmental agency conducts an adversary proceeding to award reasonable fees for the attorneys, witnesses, and other reasonable costs to the prevailing party. The prevailing party must submit an application for fees within 30 days from the final disposition, and certain parties, such as corporations with a net worth over $5 million, are ineligible for this award.

7. Omnibus Crime Control and Safe Streets Act
   - This law prohibits *deliberate* interception of wire and oral communications between two individuals. This law prohibits employers, as well as other parties, from deliberately intercepting confidential information without the consent of the individuals. A private cause of action is provided to the violated parties.

8. Civil Service Reform Act of 1978
   - Of importance to safety professionals, the Civil Service Reform Act of 1978 (also known as the federal "whistle-blowers" law) prohibits federal agencies and officials from retaliating against an applicant or employee for disclosing information believed to be in violation of any law, abuse of authority, waste of funds, mismanagement, or danger to public health or safety. This law empowers the Office of Special Counsel to investigate claims of retaliation against federal employees or applicants.

Safety professionals should be aware that the list above does not include all of the privacy-related laws that could impact their company or organization. As seen throughout this text, may other laws possess prohibitions against retaliation as well as other protections against violation of individual privacy. Review by local counsel for all situations involving charges or retaliation or invasion of privacy is essential. Most labor laws can be found on the U.S. Department of Labor Web site (www.dol.gov).

16 Ethics in the Safety Profession

People think of the Golden Rule as something mild and innocuous, like a baby lamb. But when they suffer an infringement of it, they think they've been mauled by a panther.

Francis Wren

What is the Code of Professional Conduct for safety professionals? Is it a voluntary code or a mandatory code? How is it enforced? Who is the governing group who enforces the code? What qualifications must a safety professional possess? What educational level is required to be a safety professional? What certifications are required? Is ethical behavior required in the safety profession? These are just some of the questions that surround the safety profession and can often lead to issues and conflicts within and outside of the safety professional's organization.

Let us examine how an individual enters the safety profession. In general, and there are exceptions, the modern safety profession started with the enactment of the Occupational Safety and Health Act of 1970. American companies saw a need for individuals with specialized expertise to manage the compliance area to avoid the penalties that could be provided by OSHA. At this time, there were few colleges and universities providing educational programs in this unique area, and thus individuals were either transferred within the organization to manage this area, were hired from the outside, or the safety responsibilities were added to the employee's existing job responsibilities. The individuals tasked with safety responsibilities primarily utilized the OSHA standards as the "safety bible" and learned safety "on the fly."

As the profession expanded, colleges and universities began developing programs to train safety professionals, and numerous public and private entities began to offer specialized training to assist the safety professional. Professional-safety-related organizations began to emerge to provide the safety professional with a shared experience opportunity and certifications and often provided a voluntary Code of Professional Conduct for their members. In addition, OSHA expanded opportunities for continuing education for safety professionals through their outreach efforts and regional training centers.

With this emergence of the safety profession, the duties and responsibilities of the average safety professional also expanded. Depending on the organization, the safety professional may be tasked with additional responsibilities in security, workers' compensation, training, wellness, human resources, insurance, environmental concerns, and a myriad of other job responsibility combinations. Correlating with these added responsibilities, the safety professional may now be privy to files that include confidential information, such as medical files, personnel files, and insurance information. The safety professional may be responsible for disciplining employees, denying workers' compensation claims, and other related functions that can run adverse to the safety function.

When a safety professional acts or performs in an unethical or unprofessional manner, what happens, and what are the negative connotations for the safety professional, the company, and the profession? In many circumstances when unethical or unprofessional behavior is identified, the individual company provides any disciplinary action, up to and including discharge, to the safety professional. At this point in time, there are literally no repercussions from the profession itself.

Correlating with professional conduct is the issue of education. What proscribed level of education is required to be a safety professional? Is there any proficiency testing to become a safety professional? At this point in time, there are no educational requirements to enter the safety profession. The individual may be a supervisor on Friday and become a safety manager on Monday with little or no training or education. The newly crowned safety professional often must sink or swim by his or her own wits. There is no minimum required level of education at this point in time, and we have professionals with educational levels ranging from high school to doctoral levels.

In this chapter, we will explore the pivotal question as to how we ensure the ethical and professional conduct of the individuals serving within the safety profession. Additionally, we will explore common ethical dilemmas confronted by safety professionals and identify several possible options through which professional and ethical behavior among our ranks can be created and maintained.

First, what constitutes ethical behavior in business today? In recent years, we have experienced the ethical failures of several major businesses, including such companies as Enron and Adelphia, which have led to indictments and convictions for several top executives and failure of these companies costing shareholders and employees billions of dollars. Conversely, primarily as a result of the Sarbanes-Oxley Act in 2002 which addressed oversight for publicly traded companies, many companies have instituted policies and procedures to attempt to address unethical behavior and practices within their organizations.

However, given the broad spectrum of potential unethical behaviors or practices, what really constitutes unethical behavior? Is it the golden rule to not "lie, cheat or steal" from or for the employer, or is it broader in scope? Can personal behaviors be unethical but within the scope of the company's policies? Is there a line that personal behaviors cannot cross? Who sets the standard as to whether a behavior is ethical or unethical?

Second, can ethical behavior be taught and learned within the business environment? As stated, "ethical beliefs are shaped by their personal experiences, peer pressure, family and cultural and religious standards."[1] Is it possible to take a wide variety of diverse individuals with varying backgrounds, religious beliefs, family beliefs, and a myriad of different experiences and provide a level of training and learning to achieve a minimal level of acceptable ethical behavior? Of professional conduct? If so, what type of training and educational experience should be provided to these individuals to achieve this minimum level of acceptable behavior?

[1] Nichols, N., Nichols Jr., G.V., and Nichols, P.A., "Professional Ethics, The Importance of Teaching Ethics to Future Professionals," *Professional Safety*, July 2007, p. 37.

Third, can ethical behavior, or the lack thereof, actually be policed within the business environment and within the safety profession? Utilizing the medical and legal professions as models, these professions possess educational requirements, examination and testing requirements, continuing education requirements, as well as the ability to sanction members and remove them from the profession. Conversely, at this point in time, the safety profession possesses no educational requirements, no examination requirements, no licensing requirements, no continuing education requirements, and no ability to sanction professionals. The only guidance and education provided as of today to safety professionals in the area of ethics and professional conduct are the codes of professional conduct provided by the professional organizations, of which safety professionals are not required in any way to become members of or participate in any manner.

Let us start our search with the most basic of questions: What is ethical behavior? Ethical behavior, not unlike the U.S. Supreme Court's definition of pornography, is that you know it when you see it. Conversely, you also can identify unethical behavior when you see it. But where can we draw the line when it comes to ethical or unethical behavior when it involves the safety function? One often utilized test designed for the business community which has application in the safety profession is as follows:

1. *Is It Legal?* Would the activity violate any federal, state, or local law or law of any kind or company policy?
2. *Is It Balanced?* Would the parties involved believe the action taken was fair and served all involved in a positive manner now and in the future?
3. *Is It Right?* Is this a decision in which all parties can be proud?[1]

Although this test can provide guidance, safety professionals often face tough issues in the performance of their jobs which often fall within gray areas with regard to legality, ethics, and morality. The circumstances and individuals involved in the situation place the safety professional in difficult positions of making a decision where there is not a solid or acceptable alternative pathway. Safety professionals must often follow their own moral compass in making these decisions. Is fairness as well as impartiality, candor, and fidelity to trust, dignity, compassion, and courage also to be evaluated in this assessment by the safety professional? Does the safety professional owe the duty of ethical behavior to the profession? The company? The company shareholders or owners? His or her superior? Or the employees?

In theory, all organizations want their employees, especially their safety professionals and members of the management team, to possess the highest moral and ethical standards and to perform in an ethical manner as they represent the company. Some companies, as exemplified by Texas Instruments, Inc., provide guidance to their employees to assist them in making ethical decisions on the job. Texas Instruments, Inc., provided their employees with a business card that contains the following thought-provoking questions:

---

[1] Blanchard, K., and Peale, N.V., *The Power of Ethical Management*, William Morrow, New York, 1988.

1. Does it comply with our values?
2. If you do it, will you feel bad?
3. How would it look in the newspaper?
4. If you know it's wrong, don't do it!
5. If you're not sure, ask.
6. Keep asking until you get an answer.[1]

Although there is much written about business ethics, the safety profession is substantially different and provided unique ethical issues that can be perplexing for professionals. For example, when is a lie not a lie, and what constitutes a lie? This should be a simple answer. The same answer your mother told you for years when you stretched the truth. But in a plant environment, the safety professional, as a member of the management team, may not be able to simply tell the truth in all circumstances. For example, the safety professional, as a member of the management team, is discussing a layoff that involves a large number of employees. If an employee, let us say a friend of the safety professional, asks the safety professional if the rumors of a layoff are true because he wants to buy a new bass boat, can or should the safety professional tell the truth? If the safety professional tells his friend the truth, the safety professional will be violating the confidence of the management team (and will probably lose his job). If the safety professional does not tell his friend/employee the truth, the friend will make a major financial mistake and the friendship is in jeopardy. Should the safety professional just "play dumb" and answer "I don't know," which is, in essence, a lie? Whichever direction the safety professional decides, this is a "no win" situation. Can simple silence constitute an ethical violation?

Although there is no simple solution, safety professionals can often avoid ethical dilemmas by avoiding "off the cuff" or "shooting from the hip" decision making. Safety professionals should evaluate their individual situation as well as their personal position in advance in order to determine where personal perimeters are located and determine what "lines" the safety professional absolutely would not cross under any circumstances. This pre-determination can permit the safety professional to establish his or her individual code of conduct that can be relied upon when confronted with difficult decisions.

Additionally, several professional organizations, such as the American Society of Safety Engineers, National Safety Management Society, Institute of Professional Safety and Health, and others provide voluntary Codes of Professional Conduct for their members. These codes of conduct are often broad in scope in order to address many of the concerns of their members; however, they can serve as a framework for the development of a safety professional's personal code of conduct.

Below is a sample code of conduct that can assist safety professionals in analyzing and determining their personal code of conduct or ethical rules. This sample code is provided only for guidance. Safety professionals are encouraged to add other

---

[1] Boone, L.E., and Kurtz, D.L., *Contemporary Business*, 11th ed., Thompson, South-Western, Mason, Ohio, 2005. Also see *Professional Ethics*, FN 1.

issues or areas that may be pertinent to their individual situation and analyze each of these situations to determine their individual perimeters.

SAMPLE

Individual Code of Conduct

1. I will assist in maintaining the integrity and competence of the safety profession.
2. I will maintain my individual competency in the safety profession.
3. I will maintain my own integrity and the integrity of the safety profession.
4. I will always tell the truth.
5. I will never provide materially false statements.
6. I will never deliberately fail to disclose material facts or documents.
7. I will never falsify any documents or statements.
8. I will not engage in illegal conduct.
9. I will not engage in acts involving moral turpitude.
10. I will not engage in conduct involving dishonesty, fraud, deceit, or misrepresentation.
11. I will not engage in any conduct that adversely affects or reflects on my fitness as a safety professional.
12. I will fulfill my duties and responsibilities to my company, organization, management team, and employees to the best of my abilities.
13. I will work with law enforcement and other authorities within the boundaries of the law.
14. I will ensure the proper handling of all monies.
15. I will ensure the confidentiality of all appropriate records.
16. I will promote the safety, health, and welfare of my employees.
17. I will preserve all information provided to me in confidence.
18. I will exercise independent judgment within the bounds of my job function.
19. I will inform my employer of all potential violations of the law and encourage him or her to comply with the law. If my employer chooses to violate the law, I will take appropriate actions.
20. I will not abuse alcohol, controlled substances, prescription medication, or other mind-altering substances.
21. I will not discriminate against any party.
22. I will not conceal or knowingly fail to disclose information.
23. I will not lie or provide perjured testimony or false evidence.
24. I will avoid the appearance of impropriety.
25. I will not misappropriate funds of an individual or my company.
26. I will maintain complete documents and records.
27. I will address situations in a timely and appropriate manner.
28. I will not attempt to improperly influence the judgment of employees, management, or others.
29. I will always strive to take the "high road" in every circumstance.
30. I will always do the best I can possibly do in every situation.[1]

As discussed later in this chapter, there is no mandatory code of conduct required for all safety professionals. Thus, the guidance provided to safety professionals at this point in time is simply voluntary in nature, and there is no enforcement, outside

---

[1] Schneid, T.D., *Modern Safety and Resource Control Management*, John Wiley & Sons, New York, 2000.

of company-related policies, for any conduct- or ethics-related violation by a safety professional. Safety professionals who are presented with a potential conflict or ethical issue should establish and utilize their personal code of conduct to identify and establish personal parameters. However, a systematic method through which to analyze the potential ethical issue and manage the specific problem is also necessary. The following guideline is provided to assist safety professionals in this analysis:

- STOP! Slow the situation down and ensure that you have all of the facts.
- DO NOT make snap judgments or decisions without all of the information and evidence.
- Review all of the information and evidence in detail.
- Identify possible options.
- Identify possible ramifications of each option.
- Identify short-term impacts. Identify long-term impacts.
- Are there any options that can resolve the situation?
- Are there legal considerations?
- Is there someone you trust that you can talk to regarding the conflict?
- Who is being injured through the conflict?
- Do you possess a duty to respond in any way?
- Identify your options.
- Select the appropriate option.
- Can you live with your decision?
- Is your decision aligned with your personal code of conduct?
- Is your decision within the bounds of your company's code of conduct?
- Is your decision within the bounds of the law?
- Is the decision correct given the circumstances?
- Plan your actions. How are you going to address the situation?
- Initiate your actions.
- Re-evaluate as necessary.[1]

One common area of potential conflict for safety professionals is the area of sensitive information. As noted above, safety professionals are often privy to internal company information as well as individual employee information, such as medical records, which is not common knowledge and is often personal in nature. In most circumstances, the information is in regard to the specific company or organization or individual issues. Safety and resource control professionals should exercise extreme caution with regard to the confidentiality of this information and how, if at all, this information is utilized within the job function. For example, the safety professional is told by a supervisor that he or she has acquired a sexually transmitted disease from someone who is not his or her spouse during a recent business trip. Does the safety professional owe a duty to the supervisor who came to him or her for assistance? Does the safety professional owe a duty to the company and potentially to the public if the disease can contaminate the product? The implications in this type of situation can be catastrophic to many parties, including the company (workers' compensation

---

[1] Id.

claim, governmental regulations regarding contamination of the product, etc.); the spouse (confirmation of infidelity, etc.); the general public (contaminated products, etc.); as well as the individual. This type of situation places the safety professional in the very delicate position of being required to act upon the circumstance in an expedited manner because of the nature of the situation, but the implications of the safety professional's action can be devastating to many involved parties.

Company-established safety goals can often be an area ripe with ethical issues for safety professionals. Given the monetary or other rewards provided by the company, safety professionals may be asked to neglect to file a claim form, conveniently lose the form, or hold the claim for a specified period of time in order for the management team or employees to receive their safety reward. As an example, say the company's safety goals are established on the number of workers' compensation claims. On December 29 at 6:00 p.m., an employee incurs a laceration that may or may not require sutures. If the employee is sent for sutures, this will constitute a claim, and the plant will not receive their safety "bonus" constituting several thousand dollars for each team member. The plant manager, who gets the biggest bonus and is your boss, asks you to not send the employee to the hospital and to try to take care of the laceration at the plant. Or, in the alternative, the plant manager asks you to not send the employee to the hospital until after midnight when the claim is recorded in the New Year (and the safety bonus time period has elapsed). Do you see any ethical issues with these scenarios?

Safety professionals can be confronted with situations where the requested action can be illegal as well as unethical. Safety professionals can be placed in positions where the ultimate decision that belongs to his or her boss may result in a violation of the law as well as an ethical violation such as falsifying OSHA records. These types of situations place the safety professional in a very difficult position. If the safety professional does not adhere to the direction of his or her boss, there may be employment implications — namely, termination. Conversely, if the safety professional performs the requested illegal act, he or she could be in violation of the law. Many companies have attempted to assist not only safety professionals but all employees with a protected avenue through which to communicate with corporate or an outside entity without fear of repercussions. Additionally, under several laws and regulatory schemes, employees, including safety professionals, can be protected for reporting illegal activities to the governing agency. However, safety professionals should exercise extreme caution in situations involving whistle blowing due to the immediate employment implications as well as potential civil and criminal actions that may result.

A more common dilemma for safety professionals is the conflict between creating a safety program that is beneficial to the company as well as all employees and the conflict with specific policies, programs, or procedures. This can result in a detrimental situation for aspects of the safety program or individual employee. Safety professionals often work in a multi-level world where they must interact with the executive management group, the middle management or supervisors, and the employees on a daily basis. Additionally, safety professionals may be acquiring direction from the corporate level, the plant level, and many different departments. This diversity within the job function can create ethical conflicts for the safety professional.

Ethical conflicts and violations of the voluntary codes of conduct are inevitable in the safety profession. Individuals in the safety profession must take a leadership role and develop and adhere to a personal code of conduct given the lack of guidance from the profession. Absent any type of mandatory guidance and enforcement, safety professionals are often on an island and are left to their own devices in determining what is ethical and what is not on a daily basis.

The second issue concerns whether ethics can be taught and learned in a business environment.

If we assume that ethics and professional conduct can be taught and learned, which I believe it can in most circumstances, then what is the best way through which to transfer this knowledge to the safety professional? Should the methods and curricula for an entry-level safety professional be different than those for a seasoned safety professional? Taking our guidance in this area from the legal profession, there appears to be a minimum of three levels of education to consider: initial education for new or beginning safety professionals, continuing education for all safety professionals, and case studies or identification of situations in which other safety professionals have encountered ethical conflicts.

For the initial educational component, we should assume that every participating safety professional bring with him or her a history of personal experiences, ethical beliefs, cultural beliefs, religious beliefs, and additional experiences and beliefs that serve as the framework from which we begin their education. Within the time period provided, the educational experience can include, but is not limited to, identification and discussion of the core values, review of current organizational codes of conduct, common expectations of professional conduct, company policies regarding ethical and professional conduct, common ethical dilemmas, and case studies of ethical situations previously encountered by safety professionals. Although substantially more can be added to this initial educational experience, the primary goal of this initial experience is to ensure that the new or entry-level safety professional is aware of the ethical and professional conduct expectations of his or her employer and the profession as well as alert them to common conflicts and the methods through which to address the various conflict issues.

The second prong of the educational experience is the continuous education component. Within this component, entry-level as well as experienced safety professionals would receive traditional classroom training, written materials, and other communications regarding current trends, recent ethical or professional conflict situations, new laws or legislation addressing ethics or professional conduct, legal actions involving ethical or professional conduct issues, as well as case studies involving ethics or professional conduct issues. Again, utilizing the legal system's model, safety professionals would be required to acquire a proscribed number of classroom hours in this subject matter and would be provided written materials for future reference.

The third prong of the educational experience would include periodic reinforcement directly to the safety professional in the form of case studies, updates, e-mail messages, and other forms of communications in order to maintain ethics and professional conduct in the forefront of his or her mind. Although this sounds simplistic, in the hectic day of a safety professional, a simple reminder can spur thought prior to performing any action that can be construed as unethical or unprofessional.

Education, in my opinion, is the key to ensuring ethical and professional conduct within the safety professional ranks. Errors within the ethical or professional conduct area result from lack of knowledge, lack of awareness, work pressures and stress, peer or superior pressure, controlled substances (including alcohol), or outside pressures or stress. Given this assumption, I believe the vast majority of the safety professionals wish to act in an ethical or professional manner, but this is not to say there is not a maverick or two hidden within the ranks. Most safety professionals aspire to work at the highest level and standard of professional conduct, but even the best safety professional can err depending on the circumstances. Through education, the safety professional can be made aware of the pitfalls and learn from the errors of others to continuously improve and aspire to greater heights.

The third question is can ethical behavior, or the lack thereof, actually be policed within the business environment and within the safety profession? Although this may be a bit controversial, the safety profession, in my opinion, is not a profession as it is currently operating. The safety profession does have several very good professional organizations; however, membership in these organizations is voluntary. At this point in time, there are no educational requirements to enter the safety profession, and the current spectrum ranges from high school to doctoral level. An individual may be a supervisor on Friday and, with a change of the color of a hard hat, instantly becomes a safety manager on Monday. There is no mandatory testing for competency within the safety profession, although there are several voluntary competency programs available through the voluntary professional organizations. And lastly and most importantly, the safety profession possesses no licensing or enforcement within the profession. When a safety professional errs within the job function, the only enforcement is through internal disciplinary action, or lack thereof, within the company in which the safety professional is employed. Given the above, is the safety profession really a profession within every sense of the word? Do people in general see the safety professional on the same level as a physician? A lawyer? Do employees view the safety professional as a glorified supervisor or as a professional in the area of safety?

Again utilizing the existing professionals of physician and lawyer as a template, the safety profession should consider instituting at least minimum requirements in the following areas:

Educational Level — Should there be a minimum level of education required to become a safety professional? Should a safety professional possess at least a bachelor's degree in a related field to be considered for entry into the profession?

Testing — Should an individual who wished to enter the safety profession be required to successfully pass an examination that tests the minimum level of competencies in the related subject matter?

Licensing — Should an individual who wishes to enter the safety profession be required to apply for and acquire a license issued by a governing agency or body? Should a background check be required? Is acquisition of the license a prerequisite to employment in the field? Should there be reciprocity between states? International?

Continuing Education — Should safety professionals be required to continue their education and growth in the profession throughout their career? Should safety professionals be required to take a requisite number of educational hours per year, including required hours of ethics and professional conduct?

Mandatory Enforcement — If and when a safety professional goes outside of the bounds of acceptable professional conduct, should there be a method of sanctioning the safety professional? Should the same agency or board providing licensure be tasked with enforcement?

Although the above is interesting in theory, this type of licensure and education endeavor for the safety profession will take new legislation on the federal or state levels in order to become reality. It is my hope that this type of legislation can be a proactive measure to improve the quality of the safety profession and not reactive legislation after a major disaster involving the safety profession. The safety profession has grown and improved substantially since its inception and is performing more than adequately in today's American workplaces. However, given the various responsibilities of today's safety professional, the new technologies, and the myriad of laws and regulations, it is important that the safety profession look to the future and ensure that quality, competency, and professional conduct are maintained at the highest levels.

17 Emerging Issues in Safety

News is a rough draft of history.

Philip L. Graham

Safety professionals should always have one eye on the horizon to identify new and emerging trends in the profession, new OSHA standards, and new risks that have not been identified previously. In the safety profession, new "things" are always happening in a wide variety of areas which require careful assessment and analysis in order to properly prepare and protect our company or organization. To this end, below are listed several issues that may have applicability to your company or at least should be "on your radar screen."

OSHA has not had a raise in the monetary penalties for more than 18 years. The maximum monetary penalty under the OSH Act in 1970 was $10,000. With the Omnibus Budget Bill in 1989, the maximum penalty was increased to a maximum of $70,000. Given the current economy and time span since the last increase, prudent safety professionals may be on the lookout for an increase in the maximum penalties in budget, OSHA, or related legislation.

From a horizontal standard point of view, the most expansive potential standard currently sitting on the shelf is the previously proposed ergonomics standard. Although there is currently an OSHA guideline on this subject, and OSHA has cited companies and organizations under the general duty clause for ergonomically related hazards, there is a substantial likelihood that we will see an ergonomics standard proposed by OSHA in the near future. It is anticipated that this standard will involve the known risk factors of force, posture, and repetition as well as the assessment methods identified in the OSHA guideline.

Given the recent risks regarding Severe Acute Respiratory Syndrome (SARS), avian influenza, and other related airborne pathogens and influenzas as well as the information flow regarding the potential of pandemic effects, it is not out of the realm of possibility that OSHA will expand their current efforts in this area to include a standard requiring companies and organizations to prepare for the impact of such diseases. OSHA currently has a publication titled, "Pandemic Influenza Preparedness and Response Guidance for Healthcare Workers and Healthcare Employers"[1] that can serve as the foundation for a standard. This extensive publication covers a number of areas that are health-care related and encompasses areas such as disaster planning that have implications for private-sector companies and organizations.

Another area of increased interest that is ripe for a standard is the area of workplace violence. In 2005, there were 564 homicides in the workplace out of a total of 5702 fatal work injuries and illnesses.[2] Workplace violence is the second leading

---

[1] OSHA Publication 3328-05 (2007).
[2] OSHA Web site — workplace violence.

cause of death on the job (only behind motor vehicle accidents) and the leading cause of workplace deaths for females.[1] As can be seen from the information in Appendix H, OSHA as well as the National Institute for Occupational Safety and Health (NIOSH) and several of the state plans have published guidance in this emerging area. Given the sheer number of deaths involved on a national basis, the probability of a standard emerging in this area is likely in the near future.

With the American workforce aging, safety for older workers may emerge as a new area of concern for safety professionals. Although older workers may be more experienced on the job, considerations and accommodations may be necessary to ensure their safety, including, but not limited to, the areas of lifting, hearing, seeing, walking, and related functions. Considerations may also be necessary in the training function as well as equipment operations, vehicle operations, and related areas.

Correlating to the aging of our workforce, areas such as on-site childcare and eldercare are fast emerging due to the current structure of the American family. Safety professionals, if they are not already, may be responsible for the safety at day-care centers and after-school centers that are located at or near the worksite. Additionally, given the fact that today's worker belongs to the "sandwich generation" (for example, workers "sandwiched" between young children and elderly parents), the need for on-site eldercare is fast emerging. These changes in the workplace can create new and novel safety concerns for the safety professional.

With the expansion of computers and their uses at and outside of the workplace, issues involving cybercrime are fast emerging in the American workplace. With sensitive business data electronically pulsating through telephone lines and in the air, new viruses, bugs, worms, and spyware are emerging which can detrimentally affect the flow of vital information. Safety professionals need to be cognizant of this potential risk and ensure appropriate precautions are taken to minimize or eliminate this risk in their workplaces. Although it is not perceived that this will be an OSHA-related risk, the impact on the company or organization requires vigilance.

Correlating with computers, the issue of Internet use and e-mail has fast emerged as a risk factor for consideration in the workplace. Many companies have initiated policies and procedures to prohibit the use of the Internet to acquire pornography and related information, and e-mail policies have been initiated to prevent against discrimination and workplace harassment. However, safety professionals should be aware of the security issues involving potential theft of intellectual property, internal theft, and related issues that can impact a company or organization.

One area often overlooked is the risk of loss involving executives and other key officials within your company or organization. What would happen if you lost the top leadership tier of your company's management in an accident? Does your company possess a succession plan? To protect executives in their travels throughout the world, safety professionals may consider providing protections such as providing a 5-minute airpack in case the executive is in a hotel fire, requiring executives to stay below the seventh floor when residing in a hotel, instituting policies where top executives do not fly together on the same aircraft, and related protections. Additionally,

---

[1] NIOSH Fact Sheet, June 1997.

given the increase in executive kidnappings in certain parts of the world, executive training and special protections may be considered.

Last is the issue of an increased risk of domestic as well as foreign terrorism. We are living in a new era after 9/11 and the Oklahoma City bombing. Safety professionals are encouraged to assess their company's risk of such terrorism and take the appropriate precautions to reduce these risks through deterrents or other means.

The above are but a few of the emerging risks for safety professionals today. Safety professionals are encouraged to assess all of the potential risks as well as the changes in the safety arena and take the appropriate steps to safeguard your entire workforce no matter where they are located on this planet. Our world is constantly changing, and safety professionals need to assess and adapt in order to be prepared and to prepare their workforce for these changes.

Appendix A

Rights and Responsibilities under the OSH Act

Men are entitled to equal rights — but to equal rights to unequal things.

Charles James Fox

Safety professionals are well aware that their companies or organizations are responsible for creating and maintaining a safe and healthful work environment for their employees. Additionally, as described in detail throughout this text, safety professionals have numerous responsibilities, as the agent of their company or organization, before, during, and after an OSHA inspection. Additionally, safety professionals should be aware that the OSH Act provides additional rights, as described below, which safety professionals should be aware of and ensure that these rights are not violated.

Safety professionals should be aware that every employee possesses the right to contact OSHA, usually with a complaint, regarding workplace safety. The OSH Act provides employees working for companies or organizations the basic rights to file complaints with OSHA, to be protected against discrimination for reporting violations of the OSH Act, and even the right to refuse unsafe work without retaliation. Safety professionals should be aware that in addition to the protections provided under the OSH Act, many states as well as other federal laws also provide this protection against retaliation and discrimination based on filing a complaint.

Specifically, the OSH Act states that "no person shall discharge or in any manner discriminate against any employee," so that the discrimination prohibitions "are not limited to actions taken by employers against their employees."[1] Safety professionals should be aware that the term "employee" is liberally construed and extended to applicants for employment as well as traditional employees. Safety professionals should note that many public-sector employees of states or other political subdivisions are excluded from the discriminatory protection provided under the OSH Act.[2] Additionally, safety professionals should be aware that protection against discrimination can extend to labor organization representatives working on the company premises.[3]

Specifically, safety professionals should be aware that the OSH Act prohibits the company or organization from discharging or otherwise discriminating against an employee who has filed a complaint, instituted or testified in any proceeding, or

---

[1] 29 CFR Section 1977.4.
[2] 29 CFR Section 1977.5.
[3] See *Marshall v. Kennedy Tubular Products,* 5 OSH Cases 1467 (W.D. Pa. 1977).

otherwise exercised any right afforded by the OSH Act.[1] The OSH Act also specifically gives employees the right to contact OSHA and request an inspection without retaliation from the company or organization if the employee believes a violation of a health or safety standard threatens physical harm or creates an imminent danger.[2] Additionally, safety professionals should be aware that employees who exercise their right to contact OSHA with a complaint can also remain anonymous to the company but usually must provide their name and required information to OSHA to acquire the above-noted protections against discrimination.[3]

In addition to the complaint and inspection protections, employees are protected against discrimination under the OSH Act when testifying in proceedings under or related to the OSH Act. Generally, employees who testify in hearings regarding inspections, hearings involving employee-contested abatement dates, employee-initiated hearings for promulgating new standards, hearings involving employee applications for modifying or revoking variances, employee-based judicial challenges to OSHA standards, and employee appeals from decisions by the Occupational Safety and Health Review Commission (OSHRC) are covered under this protection.[4] Safety professionals should be aware that an employee "need not himself directly institute the proceedings" but may merely set "into motion activities of others which result in proceedings under or related to the Act" to acquire the protections.[5] Additionally, this protection to employees is extended to proceedings instituted or caused to be instituted by the employee, as well as "any statement given in the course of judicial, quasi-judicial, and administrative proceedings, including inspections, investigations, and administrative rule making or adjudicative functions."[6]

Safety professionals should become acquainted with the specific procedures OSHA must follow when addressing a charge of discrimination under the OSH Act in order to be able to properly defend these allegations. First, the charges must be brought by an employee of your company or organization. Second, any employee who believes he or she has been discriminated against may file a complaint with the Secretary within thirty days of the alleged violation which the Secretary will then investigate.[7] When an employee has filed a complaint, the Secretary must notify the employee as to whether an action will be filed on his or her behalf in federal court. Safety professionals should be aware that OSHA may bring discrimination action against corporate officers as individuals or against the corporation itself and the officers in their official capacities.[8]

---

[1] 29 CFR Section 1977.5.

[2] OSH Act Section 8(f)(1); 29 USC Section 657(f)(1).

[3] 29 CFR Part 1977.

[4] OSH Act Section 6(b) and (d), 10(c); 29 USC Sections 655(b–f).

[5] 29 CFR Section 1977.10(b).

[6] 29 CFR Section 1977.11.

[7] 29 CFR Section 1977.15. (Note that the purpose of the 30-day limitation is "to allow the Secretary to decline to entertain complaints that have become stale." This relatively short period can be tolled under special circumstances and has no effect on other causes of action.)

[8] See *Donovan v. RCR Communications*, 12 OSH Cases 1427 (M.D. Fla 1985) (as individuals); *Moore v. OSHRC*, 591 F.2d 991 (4th Cir. 1979) (officers in official capacities).

One common right provided under the OSH Act which employees commonly invoke is the right to review information. Under the OSH Act, employees have the following rights:

- Review copies of appropriate standards, rules, regulations, and requirements that the employer is required to have available at the workplace.
- Request information from the employer on safety and health hazards in the workplace, appropriate precautions to take, and procedures to follow if the employee is involved in an accident or is exposed to toxic substances.
- Gain access to relevant employee exposure and medical records.
- Request an OSHA inspection if they believe hazardous conditions or violations of standards exist in the workplace.
- Accompany an OSHA compliance officer during an inspection tour, or have an authorized employee representative do so.
- Respond to questions from an OSHA compliance officer.
- Observe any monitoring or measuring of hazardous materials and see the resulting records, as specified under the OSH Act and required by OSHA standards.
- Review or have an authorized representative review the employer's Log of Work-Related Occupational Injuries and Illnesses (OSHA 300) at a reasonable time and in a reasonable manner.
- Object to the timeframe set by OSHA for the employer to correct a violation by writing to the OSHA area director within 15 working days from the date the employer receives the citation.
- Submit a written request to the National Institute for Occupational Safety and Health (NIOSH) for information on whether a substance in the workplace has potentially toxic effects in the concentration being used, and, if requested, have their names withheld from the employer.
- Be notified if the employer applies for a variance from an OSHA standard, and have an opportunity to testify at a variance hearing and appeal the final decision.
- Be advised of OSHA actions regarding a complaint, and request an informal review of any decision not to inspect the site or issue a citation.
- File a complaint if punished or discriminated against for acting as a "whistleblower" under the OSH Act or 13 other federal statutes for which OSHA has jurisdiction, or for refusing to work when faced with imminent danger of death or serious injury and there is insufficient time for OSHA to inspect.[1]

To clarify the employee's right to refuse to perform unsafe or unhealthy work, safety professionals may look to the case law for guidance. The U.S. Supreme Court, in addressing this issue in *Whirlpool Corp. v. Marshall*,[2] stated: "circumstances may exist in which the employee justifiably believes that the express statutory arrangement

---

[1] *OSHA: Employee Workplace Rights*, OSHA publication 3021-08R (2003); Also see Appendix A of this text.
[2] 445 U.S. 1, 100 S. Ct. 883, 63 L.Ed.2d 154 (1980).

does not sufficiently protect him from death or serious injury." Given this right, safety professionals should be aware that this employee right may surface when "(1) the employee is ordered by the employer to work under conditions that the employee reasonably believes pose an imminent risk of death or serious bodily injury, and (2) the employee has reason to believe that there is not sufficient time or opportunity either to seek effective redress from the employer or to apprise OSHA of the danger."[1]

An additional right not specifically identified in the above-noted list of employee rights is the employee's right to acquire training from the company or organization. As identified on the OSHA Web site, employees have the following rights:

- Get training from your employer on chemicals you are exposed to during your work and information on how to protect yourself from harm. Employers must establish a comprehensive, written hazard communication program (Chemical Hazard Communications). Your employer must label chemical containers, make material safety data sheets with detailed hazard information available to employees, and train you about the health effects of the chemical you work with and what the employer is doing and what you can do to protect yourself from these hazards.
- The program must list the hazardous chemicals in each work area, how the employer will inform employees of the hazards of non-routine tasks (for example, the cleaning of reactor vessels), and hazards associated with chemicals in unlabeled pipes and how the employer will inform other employees at a multi-employer worksite of the hazards to which their employees may be exposed.
- Getting training from your employer on a variety of other health and safety hazards and standards that your employer must follow. These include lockout-tagout, bloodborne pathogens, confined spaces, construction hazards and a variety of other subjects.[2]

Safety professionals should be aware that the twenty-three state plan states and territories as well as the two public employee plans provide the same rights and protections, or even additional rights to employees, as provided under the OSH Act. Safety professionals should become familiar with the employee rights under their state plan, if applicable, and be able to fully and completely explain these rights to their management team.

As safety professionals are well aware, OSHA requires a posting of employee rights in every workplace. Additionally, in this age of instantaneous communications, employees are well aware of their rights in the workplace through television, the Internet, and traditional communication methods. Safety professionals should acquire a working knowledge of employee rights in the workplace and ensure these rights are respected and guarded throughout the safety programs and efforts.

---

[1] Id.
[2] OSHA Web site — *Worker Rights under the Occupational Safety and Health Act of 1970.*

Office of Investigative Assistance

THE WHISTLEBLOWER PROGRAM

The OSH Act is designed to regulate employment conditions relating to occupational safety and health and to achieve safer and more healthful workplaces throughout the nation. The Act provides for a wide range of substantive and procedural rights for employees and representatives of employees. The Act also recognizes that effective implementation and achievement of its goals depend in large measure upon the active and orderly participation of employees, individually and through their representatives, at every level of safety and health activity.

To help ensure that employees are, in fact, free to participate in safety and health activities, Section 11(c) of the Act prohibits any person from discharging or in any manner discriminating against any employee because the employee has exercised rights under the Act.

These rights include complaining to OSHA and seeking an OSHA inspection, participating in an OSHA inspection, and participating or testifying in any proceeding related to an OSHA inspection.

OSHA also administers the whistleblowing provisions of thirteen other statutes, protecting employees who report violations of various trucking, airline, nuclear power, pipeline, environmental, and securities laws.

A person filing a complaint of discrimination or retaliation will be required to show that he or she engaged in protected activity, the employer knew about that activity, the employer subjected him or her to an adverse employment action, and the protected activity contributed to the adverse action. Adverse employment action is generally defined as a material change in the terms or conditions of employment. Depending upon the circumstances of the case, "discrimination" can include:

- Firing or laying off
- Blacklisting
- Demoting
- Denying overtime or promotion
- Disciplining
- Denial of benefits
- Failure to hire or rehire
- Intimidation
- Reassignment affecting prospects for promotion
- Reducing pay or hours

The fourteen statutes enforced by OSHA and the regulations governing their administration are listed below:

- Section 11(c) of the Occupational Safety and Health Act of 1970 (OSHA)
- The Surface Transportation Assistance Act of 1982 (STAA)
- The Asbestos Hazard Emergency Response Act of 1986 (AHERA)
- The International Safe Container Act of 1977 (ISCA)
- The Safe Drinking Water Act of 1974 (SDWA)
- The Federal Water Pollution Control Act of 1972 (FWPCA)
- The Toxic Substances Control Act of 1976 (TSCA)
- The Solid Waste Disposal Act of 1976 (SWDA)
- The Clean Air Act of 1977 (CAA)
- The Comprehensive Environmental Response, Compensation and Liability Act of 1980 (CERCLA)
- The Energy Reorganization Act of 1974 (ERA)
- The Wendell H. Ford Aviation Investment and Reform Act for the 21st Century (AIR21)
- Section 806 of the Corporate and Criminal Fraud Accountability Act of 2002 (CCFA) (Sarbanes-Oxley Act)
- Section 6 of the Pipeline Safety Improvement Act of 2002 (PSIA)

REGULATIONS

- 29 CFR Part 1977 — Discrimination Against Employees Exercising Rights under the Williams-Steiger Occupational Safety and Health Act of 1970
- 29 CFR Part 1978 — Rules for Implementing Section 405 of the Surface Transportation Assistance Act of 1982
- 29 CFR Part 1979 — Procedures for the Handling of Discrimination Complaints under Section 519 of the Wendell H. Ford Aviation Investment and Reform Act for the 21st Century
- 29 CFR Part 1980 — Procedures for the Handling of Discrimination Complaints under Section 806 of the Corporate and Criminal Fraud Accountability Act of 2002
- 29 CFR Part 1981 — Procedures for the Handling of Discrimination Complaints under Section 6 of the Pipeline Safety Improvement Act of 2002
- 29 CFR Part 24 — Procedures for the Handling of Discrimination Complaints under Federal Employee Protection Statutes

FILING A COMPLAINT

If you believe your employer has discriminated against you because you exercised your safety and health rights, contact your local OSHA Office right away. Most discrimination complaints fall under the OSH Act, which gives you only thirty days to report discrimination. Some of the other laws have complaint-filing deadlines that differ from OSHA's, so be sure to check.

You can telephone, fax, or mail your OSHA complaint. OSHA conducts an in-depth interview with each complainant to determine the need for an investigation. If evidence supports the worker's claim of discrimination, OSHA will ask the employer to restore the worker's job, earnings, and benefits. If the employer objects, OSHA may take the employer to court to seek relief for the worker. The procedures for investigations of discrimination complaints are contained in the OSHA Whistle-blower Investigations Manual:

- Whistleblower Investigations Manual
- Revised Interim Guidelines for Handling Privacy Act Files and FOIA Requests
- Policy for Approval of Settlements with Future Employment Waivers

Appendix B

Guide to Review Commission Procedures: Occupational Safety and Health Review Commission (August 2005)

TABLE OF CONTENTS

SECTION 1 — INTRODUCTION

THE REVIEW COMMISSION

The Occupational Safety and Health Review Commission ("Review Commission") is an independent agency of the U.S. Government that was established by the Occupational Safety and Health Act of 1970 ("Act") to be like a court that resolves certain disputes under the Act. The Review Commission is composed of three members who are appointed by the President of the United States and confirmed by the Senate for six-year terms. It employs Administrative Law Judges to hear cases.

The Act was passed by Congress to "assure safe and healthful working conditions for working men and women." The Act also established another agency, the Occupational Safety and Health Administration ("OSHA"), which is part of the U.S. Department of Labor, to enforce the law. OSHA issues regulations setting occupational safety and health standards that an employer must follow. As part of its enforcement responsibilities, OSHA may also conduct an inspection of a workplace. If OSHA's inspectors find what they believe are unsafe or unhealthy conditions, they may issue a **citation** to an employer. A citation includes allegations of workplace safety or health violations, proposed penalties, and proposed dates by which an employer must correct the alleged hazardous conditions.

If the cited employer or any of its employees or an employee representative disagrees with the citation, they may then file a timely **notice of contest**. The Review Commission (which is **completely independent** of OSHA) then comes into the picture to resolve the dispute over the citation.

PURPOSE OF THIS GUIDE

This Guide is intended to inform employers, employees, and other interested persons about Review Commission proceedings. It provides an overview of the proceedings conducted before the Administrative Law Judges and the Commission Members and it is primarily intended to assist persons in defending their business or their employer's business after having contested an OSHA citation. It will also be useful to other persons who desire a general overview of the Review Commission and its procedures.

The Review Commission also publishes a Guide to Simplified Proceedings and an Employee Guide to Review Commission Procedures that may be obtained at the Review Commission Web site (www.oshrc.gov), or by writing or calling: Executive Secretary, U.S. Occupational Safety and Health Review Commission, 1120 20th Street, N.W., 9th Floor, Washington, DC 20036-3457.

RULES OF PROCEDURE

The Review Commission's Rules of Procedure are published in Part 2200 of Title 29, Code of Federal Regulations ("C.F.R."). These Rules may be available in a local library. They can also be obtained at the Review Commission Web site (www.oshrc. gov) or by contacting the Review Commission's Office of The Executive Secretary at the address or telephone number above. References to the Rules in this Guide state "See Rule" and the appropriate number. (For example, "See Rule 4" refers to 29 C.F.R. § 2200.4.)

This guide is intended to provide an overview of the Review Commission's procedures and it is not intended to be a substitute for the Rules of Procedure, which are followed in the Review Commission's proceedings in deciding cases. Parties to cases should review the Rules and follow them in proceedings before judges and the Commission members.

USING THIS GUIDE

This guide describes many of the documents and steps in proceedings before the Commission members and judges. Throughout this Guide, important terms are shown in **bold italics** and many are included in the Glossary.

PARTIES MAY REPRESENT THEMSELVES

The Review Commission's Rules do not require that a **party** — an employer, a union, or affected employee(s) — be represented by a lawyer. However, proceedings before the Review Commission are legal in nature. Certain legal formalities must be followed. OSHA will be represented by lawyers from the **Solicitor of Labor's** Office, the employer may be represented by a lawyer, and the decision in the case may have consequences beyond the amount of the penalty. For example, a decision may require corrective actions at a worksite. Parties to cases should consider carefully whether to hire a lawyer to represent them in their case.

TIME IS OF THE ESSENCE

Many of the documents parties are required to file, such as those needed to disagree with an OSHA citation or proposed penalty, must be filed within a specific time period. Failure to file documents as required could result in a citation becoming a final order without an opportunity to appeal. Therefore, parties to cases must respond promptly to communications received from the judge, the Commission, or any of the other parties to the dispute.

SAMPLE LEGAL DOCUMENTS

The Appendixes contain forms and sample correspondence that may be used or referred to in preparing a case. These are mentioned as appropriate throughout the Guide.

QUESTIONS REGARDING PROPER PROCEDURE

Parties to cases having questions regarding the Commission's procedures in cases pending before a judge should call the judge's office. At other stages of the proceedings, inquiries should be directed to the Executive Secretary's Office at (202) 606-5400. Commission employees cannot give legal advice or advise a party how to proceed. However, they can provide information about the Rules of Procedure and the Commission's methods of processing cases.

SECTION 2 — PRESERVING RIGHTS AND CHOOSING A PROCEEDING

OSHA Citation

Cases that come before a Review Commission judge arise from inspections conducted by OSHA, an agency of the United States Department of Labor. When OSHA finds what it believes to be a **violation** at a worksite, it will notify the employer in writing of the alleged violation and the period of time it thinks reasonable for correction by issuing a written **citation** to the employer.

The period of time stated in the citation for an employer to correct the alleged violation is the **abatement period**. OSHA likely will also propose that the employer pay a monetary penalty.

The Act requires that the employer **immediately post a copy of the citation** in a place where **affected employees** will see it, to have legal notice of it. An affected employee is an employee who has been exposed to or could be exposed to any hazard arising from the cited violations.

Employer's Notice of Contest

If an employer disagrees with any part of the OSHA citation — the alleged violation, the abatement period, or proposed penalty — **it must notify OSHA in writing of that disagreement within 15 working days** (Mondays through Fridays, excluding Federal holidays) of receiving the citation. This written notification is referred to as a **notice of contest,** and if it is filed late with OSHA, the employer is not usually entitled to have the dispute resolved by the Commission.

The notice of contest must be **delivered in writing to the Area Director of the OSHA office that mailed the citation**. The Area Director's name and address will be listed on the citation. **A notice of contest must not be sent to the Commission.**

Informal Conference with OSHA

If a citation is issued, an employer may schedule an informal conference or engage in settlement discussions with the OSHA Area Director, **but this does not delay the 15 working day deadline** for filing a notice of contest. Thus, if an informal conference is conducted that does not result in a written settlement agreement, if a notice of contest is not filed within the 15 working day deadline all citation items must be abated and all penalties must be paid.

Content and Effect of Notice of Contest

The notice of contest is a statement that an employer intends to contest (1) the alleged violations, (2) the specific abatement periods, and/or (3) the penalties proposed by OSHA. The notice should state in detail those matters being contested. (See Appendixes 1A, 1B.)

For example, if there are two citations and the employer wishes to contest only one of them, the citation being contested should be identified. If there are six different items alleged as violations in a single citation and the employer wishes to contest items 3, 4, and 6, those items should be specified.

If the employer wishes to contest the entire penalty, or only the amount for one citation or specific items of one citation, or only the abatement period for some or all of the violations alleged, this should also be specified.

For any item (violation) not contested, the abatement requirements must be fully satisfied and any related penalty must be paid to the Department of Labor. If the employer contests whether a violation occurred, the abatement period and the proposed penalty for that item is suspended until the Commission issues a final decision.

Notice of Docketing

The OSHA Area Director sends the notice of contest to the Commission. The Executive Secretary's Office then notifies the employer that the case has been received and assigns a docket number. This docket number must be printed on all documents sent to the Commission.

Employee Notification

At the time the employer receives the Notice of Docketing that the case has been filed and given a docket number, the Commission will furnish a copy of a notice to be used to inform affected employees of the case. A pre-printed postcard is sent to the employer with this notice; the employer returns the postcard to the Commission to inform it that affected employees have been notified.

Employees May Contest Abatement Period

Unions or **affected employees** wishing to participate in a dispute may file a notice of contest (see Appendix 1C) challenging the reasonableness of the period of time given to the employer for abating (correcting) an alleged violation.

Even if the employer does not contest the citation, unions or affected employees can object to the abatement period. **This must be done within 15 working days of the employer's receipt of the citation.** The notice of contest should state that the signer is an affected employee or a union that represents affected employees and that the signer wishes to contest the reasonableness of the abatement period.

The employee or the union must **mail the notice of contest to the Area Director of the OSHA office that mailed the citation, not the Commission**. The Area Director's name and address will be listed on the citation. (See Section 10 of the Act and Rules 20, 22 and 33.)

Employees May Elect Party Status

Employees may also elect party status to a case by filing a written notice of election at least ten days before the hearing. A notice of election filed less than ten days prior to the hearing is ineffective unless good cause is shown for not timely filing the notice. It must be served on all other parties in accordance with Rule 7. (See Rule 20.)

Party Requests for Simplified Proceedings

Cases heard by Administrative Law Judges may proceed in one of two ways: conventional proceedings or simplified proceedings. Each method is described in detail in Sections 3 and 4 of this Guide. The Chief Administrative Law Judge may designate a case for simplified proceedings soon after the notice of contest is received at the Review Commission. Parties may also request simplified proceedings within 20 days of the date on the notice of docketing. If a case is not designated for simplified proceedings, conventional proceedings are in effect.

Choosing Simplified Proceedings or Conventional Proceedings

Simplified proceedings are appropriate for cases that involve less complex issues and for which more formal procedures used in conventional proceedings are deemed unnecessary to assure the parties a fair and complete contest. Simplified proceedings are covered in Section 4 of this Guide and the Commission has developed a Guide to Simplified Proceedings that is published on the Commission's Web site (www.oshrc.gov) or may be obtained by writing or calling: Executive Secretary, U.S. Occupational Safety and Health Review Commission, 1120 20th Street, N.W., 9th Floor, Washington, DC 20036-3457, (202) 606-5400.

SECTION 3 — AN OVERVIEW OF HEARINGS CONDUCTED UNDER CONVENTIONAL PROCEEDINGS

This section describes the major features of the Commission's hearings conducted under the Conventional Proceedings method as opposed to hearings conducted under Simplified Proceedings. Simplified Proceedings are explained briefly in Section 4 and in a separate guide that should be consulted by those persons interested in that method of hearing cases.

The Complaint

Within 20 calendar days of receipt of the employer's notice of contest, the Secretary of Labor must file a written **complaint** with the Commission. A copy must be sent to the employer and any other parties. The complaint sets forth the alleged violation(s), the abatement period, and the amount of the proposed penalty. See Appendix 2A for an example of a complaint. (See Rule 34.)

The Answer

The employer must file a written **answer** to the complaint **with the Commission within 20 calendar days** after receiving the complaint from the Secretary of Labor. The answer must contain a short, plain statement denying allegations of the complaint that the employer wishes to contest. **Any allegation not denied by the employer is considered to be admitted.** In addition, if the employer has a specific defense it wishes to raise, such as (1) the violation was due to employee error or failure to follow instructions, or (2) compliance with a standard was infeasible, or (3) compliance with a standard posed an even greater hazard, the answer **must** describe that defense.

If the employer fails to file an answer to the Complaint on time, its Notice of Contest may be dismissed, and the Citation and Penalties may become final. The answer must be filed with the Commission by mailing it to: Executive Secretary, U.S. Occupational Safety and Health Review Commission, 1120 20th Street N.W., 9th Floor, Washington, DC 20036-3457 or to the judge, if the case has been assigned to one. A copy of the answer must also be sent to the Secretary of Labor. See Appendix 3. (See Rule 34.)

DISCOVERY

Discovery is a method used whereby one party obtains information from another party or person before a hearing. Discovery techniques in Commission cases include (1) written questions, called interrogatories; (2) oral statements taken under oath, which are depositions; (3) asking a party to admit the truth of certain facts, called requests for admissions; and (4) requests that another party produce certain documents or objects for inspection or copying. In conventional proceedings, any party can use these discovery techniques without the judge's permission, except for depositions, which require that that parties agree to take depositions or that the judge order the taking of depositions after a party files a motion requesting permission to do so. (See Rules 51–57.)

SCHEDULING ORDER OR CONFERENCE

In conventional cases, discovery takes place after the answer is filed and before the hearing date. After the answer to the complaint is filed, the judge will issue an order setting a schedule for the case and may also hold a conference with the parties to clarify the issues, consider settlement, or discuss other ways to expedite the hearing. (See Rule 51.)

WITHDRAWAL OF NOTICE OF CONTEST

A party wishing to withdraw its notice of contest to all or parts of a case may do so at any time. The **Notice of Withdrawal** must be served on all affected employees and all other parties. A copy must also be sent to the judge. See example at Appendix 8. The withdrawal terminates the proceedings before the Commission with respect to the citation or citation items covered by the notice of withdrawal. (See Rule 102.)

SETTLEMENT

The Commission encourages the Settlement of cases. Cases can be settled at any stage. The Secretary of Labor and the employer must agree to the settlement terms, and the affected employees or their union must be shown the settlement before it will be approved.

Any party can also request that a Settlement Judge be appointed to help facilitate a settlement. (See Rule 100.)

HEARINGS

Hearings are governed by Rules 60-74. The parties will be notified of the time and place of the hearing at least 30 days in advance. The employer must post the hearing notice if there are any employees who do not have a representative and served on all unions representing affected employees. The hearing is usually conducted as near the work place as possible.

At the hearing, a Commission Judge presides. The hearing enables the parties to present evidence on the issues raised in the complaint and answer. Each party to the proceedings may call witnesses, introduce documentary or physical evidence, and cross-examine opposing witnesses. In conventional proceedings, the Commission follows the Federal Rules of Evidence. Under these rules, evidence is only admitted into the record if it meets certain criteria that are designed to assure that the evidence is reliable and relevant.

HEARING TRANSCRIPTS

A transcript of the hearing will be made by a court reporter. A copy may be purchased from the reporter.

POST-HEARING BRIEFS

After the hearing is completed and before the judge reaches a decision, each party is given an opportunity to submit to the judge proposed findings of fact and conclusions of law with reasons why the judge should decide in its favor. Proposed findings of fact are what a party believes actually happened in the circumstances of a case based upon the evidence introduced at the hearing. Proposed conclusions of law are how a party believes the judge should apply the law to the facts of a case. The statement of reasons is known as a **brief**. (See Rule 74.)

JUDGE'S DECISION AND PETITION FOR DISCRETIONARY REVIEW

After hearing the evidence and considering all arguments, the judge will prepare a decision based upon all of the evidence placed in the hearing record and mail copies of that decision to all parties. The parties then can object to the judge's decision by filing a **Petition for Discretionary Review.** (See Appendix 6 for an example.) **Instructions for submitting such a petition will be stated in the judge's letter transmitting the decision and in a Notice of Docketing of Administrative Law Judge's Decision issued by the Executive Secretary's Office.** See Rule 91 for further information on filing **Petitions for Discretionary Review.**

DECISIONS FINAL IN 30 DAYS

If a Commissioner does not order review of a judge's decision, it becomes a final order of the Commission 30 days after the decision has been filed. If a Commissioner does direct review, it will ultimately issue its own written decision and that decision becomes the final order of the Commission.

Any party who is adversely affected by a final order of the Commission can appeal to a United States Court of Appeals. However, the courts usually will not hear appeals from parties that have not taken advantage of all possible appeal rights earlier in the case. **Thus, a party who failed to file a petition for review of the judge's decision with the Commission likely will not be able to later appeal that decision to a court of appeals.**

SECTION 4 — SIMPLIFIED PROCEEDINGS: AN OVERVIEW FOR EMPLOYERS AND EMPLOYEES

WHAT ARE SIMPLIFIED PROCEEDINGS?

Simplified Proceedings are designed to resolve small and relatively simple cases in a less formal, less costly, and less time-consuming manner. **The Commission's Chief Administrative Law Judge ("Chief Judge") or the judge assigned to your case notifies you that your case will be heard under Simplified Proceedings.** Even though the legal process is streamlined, **the proceedings are still a trial** before an Administrative Law Judge with sworn testimony and witness cross-examination.

MAJOR FEATURES OF SIMPLIFIED PROCEEDINGS

Under Simplified Proceedings:

1. Early discussions among the parties and the Administrative Law Judge are required to narrow and define the disputes between the parties.
2. Motions, which are requests asking the judge to order some act to be done, such as having a party produce a document, are discouraged unless the parties try first to resolve the matter among themselves.
3. Disclosure. The Secretary is required to provide the employer with inspection details early in the process. In some cases, the employer will also be required to provide certain documents, such as evidence of their safety program, to the Secretary.
4. Discovery, which is the written exchange of information, documents and questionnaires between the parties before a hearing, is discouraged and permitted only when ordered by the judge.
5. Appeals of actions taken by the judge before the trial and decision, such as asking the Commission to rule on the judge's refusal to allow the introduction of a piece of evidence, called interlocutory appeals, are not permitted.
6. Hearings are less formal. The Federal Rules of Evidence, which govern other trials, do not apply. Each party may present oral argument at the close of the hearing. Post-hearing briefs (written arguments explaining your position in the case), will not be allowed except by order of the judge. (See Rule 209(e).) In some instances, the judge will render his or her decision "from the bench," which means the judge will state at the end of the hearing whether the evidence and testimony proved the alleged violations and

will state the amount of the penalty the employer must pay, if a violation is found.

Cases Eligible for Simplified Proceedings

It is possible that not all relatively small cases eligible for Simplified Proceedings will be selected. (See Rules 202 and 203(a).) **The Chief Judge will assign cases for Simplified Proceedings or, if your case is not selected, you may request that it be chosen.** Cases appropriate for Simplified Proceedings are those with one or more of the following characteristics:

- Relatively simple issues of law or fact with relatively few citation items
- Total proposed penalty of not more than $30,000
- A hearing that is expected to take less than two days
- A small employer whether appearing with or without an attorney

Cases having willful or repeated violations or that involve a fatality are not deemed appropriate for Simplified Proceedings.

Employee or Union Participation

Affected employees or their unions who file a notice of contest may also request Simplified Proceedings. Unions or an affected employee (ones exposed to the alleged health or safety hazard) wishing to participate in a dispute may file a notice of contest (see Appendix 1C) challenging the reasonableness of the period of time given to the employer for abating (correcting) an alleged violation. Even if the employer does not contest the citation, unions or affected employees can object to the abatement period. **This must be done in writing within 15 working days of the employer's receipt of the citation.** You might consider Simplified Proceedings if you or your local union wish to avoid the time and expense of a full-blown hearing. You might also participate by electing party status after the employer files a notice of contest, but must do so promptly.

When affected employees or their unions contest the time allowed for abatement, and the employer does not contest the citation, the employer may in turn elect to participate. Once the abatement date has been contested, other employees or unions may likewise elect to participate.

An employee or a union must **mail a written notice of contest to the Area Director of the OSHA office that issued the citation, not the Commission**. First-class mail will be sufficient for this purpose. The Area Director's name and address will be listed on the citation. This process is governed by Section 10 of the Act and Commission Rules 20, 22 and 33.

Should You Ask for Simplified Proceedings?

If you are an employer, have received an OSHA citation, have filed a notice of contest, and the total proposed penalties in the citation are between $20,000 and $30,000, the Chief Judge may designate your case for Simplified Proceedings. If the penalties are $20,000 or less, you may file a request for Simplified Proceedings provided that there is no allegation of willfulness or a repeat violation, and the case does not involve a fatality.

You must file your request within 20 days of docketing of your case by the Executive Secretary's Office. The request must be in writing and it is sufficient if you state: "I request Simplified Proceedings." The Chief Judge or the assigned judge will then rule on your request.

Your case may be appropriate for Simplified Proceedings but that does not necessarily mean that your particular interests are best served by requesting Simplified Proceedings. In addition to considering time and expense, you should base your decision on the facts of your case, the nature of your objections to the citation, what you will try to show the judge at the hearing, the amount of paperwork involved if your case proceeds under conventional proceedings as compared to Simplified Proceedings, and whether you have legal representation.

You should also remember that, in most circumstances, your interests may be best served if you can reach a fair and equitable settlement of your case with OSHA before a hearing. Either way, Simplified Proceedings or conventional, the proceedings are legal and the Secretary of Labor will most likely be represented by an attorney. You have the right to represent yourself or to be represented by an attorney or by anyone of your choosing.

COMPLAINT AND ANSWER

Once your case is selected for Simplified Proceedings, the complaint and answer are not required. However, until an employer is notified that a case has been designated for Simplified Proceedings, conventional procedures should be followed and an answer must be filed. (See Rule 205(a).)

BEGINNING SIMPLIFIED PROCEEDINGS

You need not give any reasons for requesting Simplified Proceedings. A letter saying simply "I request Simplified Proceedings," and indicating the Docket Number assigned to your case, is sufficient. (See Appendix 4.) The letter must be sent to: Executive Secretary, U.S. Occupational Safety and Health Review Commission, 1120 20th Street, N.W., 9th Floor, Washington, DC 20036-3457.

NOTIFYING OTHER PARTIES

It is required that a copy of your request for Simplified Proceedings be sent to the Regional Solicitor of the Department of Labor office for your region. The address is on your Notice of Docketing. All employee representatives, including an employee union, that have elected party status must also be sent a copy of your request for Simplified Proceedings. **A brief statement indicating to whom, when, and how your request was served on the parties** in the case **must be received with the request for Simplified Proceedings**. An example of such a "Certificate of Service" follows. (See Rule 203(b).)

Example: I certify that on October 1, 2004, a copy of my request for Simplified Proceedings was sent by first class mail to Jane Doe, Office of the Solicitor, U.S. Department of Labor, 123 Street, City, State Zip Code and to John Doe, President, Local 111,
GHI International Union, 456 Street, City, State Zip Code. (See Appendix 2B.)

Objections to and Discontinuing Simplified Proceedings

Should you decide to object to the Chief Judge's assignment of your case to Simplified Proceedings or another party's request for Simplified Proceedings, all you need
to do is file a brief written statement with the judge assigned to your case or, if the
case has not been assigned to a judge, with the Chief Judge, explaining why your
case is inappropriate for Simplified Proceedings. The judge is required to rule on
a request for Simplified Proceedings within 15 days. Therefore, you must file your
objections as soon as possible.

If you disagree with another party's request to discontinue Simplified Proceedings and you want your case to continue under Simplified Proceedings rules, you
have seven days to file a letter explaining why you disagree. (See Rule 204(b).)

If it appears that a case is inappropriate for Simplified Proceedings, the use of
this method may be discontinued by the judge at his or her discretion. A party may
also request that the judge discontinue Simplified Proceedings. The request must
explain why the requesting party believes that the case is inappropriate for Simplified Proceedings. **If you agree with another party's request to discontinue Simplified Proceedings, you should submit a letter saying so.** When all parties agree
that a case is inappropriate for Simplified Proceedings, the judge is required to grant
the request. If the judge orders that a case be taken out of Simplified Proceedings,
the case will proceed under the Commission's conventional procedures.

Pre-Hearing Conference

Soon after the parties exchange the required information, the judge will hold a pre-
hearing conference to either reach settlement in the case or to find out which factual
and legal issues the parties agree on. This discussion may be conducted in person but
is usually conducted by a telephone conference call. The purpose of the pre-hearing
conference is to settle the case or, if settlement is not possible, to determine what
areas of dispute must be resolved at a hearing. Even if a settlement of the entire case
cannot be reached, the parties are required to attempt agreement on as many facts and
issues as possible. The discussion will include the following topics: (See Rule 207.)

1. **Narrowing of Issues.** The parties will be expected to discuss all areas in
   dispute and to resolve as many as possible. Where matters remain unresolved, the judge will list the issues to be resolved at the hearing.
2. **A Statement of Facts.** The parties are expected to agree on as many of the
   facts as possible. Examples of these facts may include the size and nature
   of the business, its safety history, details of the inspection, and the physical
   nature of the worksite.
3. **A Statement of Defenses.** You will be required to list any specific defenses
   you might have to the citation. The burden is on the Secretary to establish

that each violation occurred. However, you should be prepared to tell the judge all reasons why you believe that the Secretary's allegations are wrong.

You might also have what is called an "affirmative defense." An affirmative defense is a recognized set of circumstances in which an employer will be found not in violation even though the employer did not comply with the cited standard. For example, you may believe that the alleged violation was the result of an employee acting contrary to a work rule that has been effectively communicated and enforced. Or, you may think that compliance with the standard was impossible or infeasible, or would have resulted in a danger to employees that was greater than the danger that the standard was designed to prevent.

You should be aware that **the burden of proving an affirmative defense is on you, the employer**. Therefore, if you argue that the violation was the result of employee misconduct, at the hearing you will have to prove to the judge that you had an effectively communicated and enforced work rule. As will be discussed later, if you raise an affirmative defense, the judge may require you to provide the Secretary of Labor with certain documents before the hearing regarding the defense. For example, if you claim that an employee violated a written work rule, you will probably be required to provide the Secretary with a copy of your company's safety rules.

It is critical that you set forth your defenses at the pre-hearing conference. You may be prohibited from later asserting any defenses not raised at the pre-hearing conference. Remember, even if your defense does not excuse the violation, the judge may find it relevant in determining the penalty amount.

4. **Witnesses and Exhibits.** The parties are expected to list the witnesses they intend to call if there is a hearing, and to list any documents or physical evidence they intend to introduce to support their positions. For example, you should list any photographs that you believe show the existence of a safety device that the Secretary claims you failed to provide.

Review of the Judge's Decision

Any party dissatisfied with the judge's decision may petition the Commission for review of that decision.

No particular form is required for the petition (see Appendix 6). However, it should clearly explain why you believe that the judge's decision is in error on either the facts or the law or both. **Review of a judge's decision is at the discretion of the Commission. It is not a right.** (See Rules 91 and 210.)

Your petition should be filed no later than 20 days after issuance of the judge's written decision. Under the law, the Commission cannot grant any petition for review more than 30 days after the judge's decision is filed. Therefore, **your petition must be filed as soon as possible to obtain maximum consideration.**

The Commission will notify you whether your petition has been granted (see Appendix 7). If it is granted, your case will then proceed under the Commission's conventional rules.

SECTION 5 — OTHER IMPORTANT THINGS TO KNOW

APPEARANCES IN COMMISSION PROCEDURES

Any employer, employee, or union that initially files a notice of contest is automatically a party to the proceedings. Affected employees or their union may also choose to participate as a party where the employer has filed a notice of contest. Any party may appear in a Commission proceeding either personally, through an attorney, or through another representative. (See Rule 22.) Such a person need not be an attorney. However, all representatives of parties must either enter an appearance by signing the first document filed on behalf of the party or intervenor, or thereafter by filing an entry of appearance. (See Rule 23.)

Every party to the case must serve every other party or representative with copies of every document it files with the Commission or judge. Service is made by first-class mail, electronic transmission, or personal delivery. (See Rule 7(c).)

All notices the Commission sends to the parties will list the name and address of all parties or their representatives. (See Rule 22.) Parties must do the same.

PENALTIES

OSHA only **proposes** amounts which it believes are appropriate as penalties. These proposals automatically become penalties assessed against the cited employer when the enforcement action is **not** contested. Once a **citation or proposed penalty** is contested, the amount of the penalty for that citation, if any, will be decided by the Commission or a judge.

When a case goes to hearing before a Review Commission judge, the employer's evidence and argument on what penalty, if any, should be assessed, receives the same consideration as the evidence and argument of the Secretary of Labor.

The four factors that the law requires the Commission to consider in determining the appropriateness of civil penalties are

- the size of the business of the employer being charged;
- the gravity of the violation;
- the good faith of the employer;
- the employer's history of previous violations.

The amounts that may be assessed as civil penalties by the Commission under Section 17 of the Act are as follows:

- For a serious or non-serious violation: up to $7000
- For violations committed willfully or repeatedly: up to $70,000

- For failure to correct a violation within the period permitted: up to $7000 for each day it remains uncorrected.

Private (Ex Parte) Discussions

Parties to cases before the Commission may not communicate **ex parte** (without the knowledge or consent of the other parties) with respect to the merits of a case with the judge (except a Settlement Judge), a Commissioner, or any employee of the Commission. In other words, no participant, directly or indirectly, may discuss the case or make any argument about a matter in a case to any of these people unless done in the presence of the other participants who are given an equal opportunity to present their side, or unless it is done in writing and copies are sent to all other parties. Violation of this rule may result in dismissal of the offending party's case before the Commission. This prohibition does not, however, preclude asking questions about the scheduling of a hearing or other matters that deal only with procedures. (See Rule 105.)

Petition for Modification of Abatement

An employer who does not contest a **citation** is required to correct all of the violations within the **abatement period** specified in the **citation**. If the Commission upholds a contested citation, the employer must then correct the violation, with the **abatement period** starting on the date of the Commission's final order. If the employer has made a good faith effort to correct a violation within the **abatement period** but has not been able to do so because of reasons beyond his or her control, the employer may file a **Petition for Modification of Abatement (PMA)**. This petition is filed with the **OSHA** area director and should be filed no later than the end of the next working day following the day on which abatement was to have been completed. It must state why the abatement cannot be completed within the given time. The PMA must be posted in a conspicuous place where all affected employees can see it or near the location where the violation occurred. The PMA must remain posted for ten days. The Secretary of Labor may not approve a PMA until the expiration of 15 working days from its receipt.

At the end of the 15-day period, if the Secretary of Labor, affected employees, or their union object to the petition, the Secretary of Labor is required to forward the PMA to the Commission. After notice by the Commission to the employer and the objecting parties of its receipt of the PMA, each objecting party has 10 calendar days in which to file a response to the PMA setting out the reasons for opposing it. Proceedings before the Commission are conducted in the same way as notice-of-contest cases, except that they are expedited. The employer must establish that abatement cannot be completed for reasons beyond the employer's control, and has the burden of proving the petition should be granted. In cases of this kind, the employer is called the Petitioner, the Secretary of Labor is called the Respondent. (See Rules 37 and 103.)

EXPEDITED PROCEEDINGS

In certain situations, time periods allowed for certain procedures are shortened. The Commission's Rules of Procedure provide that an **Expedited Proceeding** may be ordered by the Commission. If an order is made to speed up proceedings, all parties in the case will be specifically notified. All **Petitions for Modification of Abatement** and all **employee contests** are automatically expedited. (See Rule 103.) Expedited proceedings are different from Simplified Proceedings. (See Rule 103.)

MAINTAINING COPIES OF PAPERS FILED WITH THE JUDGE

In order that Affected Employees may have the opportunity to be kept informed of the status of the case, the employer must keep available at some convenient place copies of all pleadings and other documents filed in the case so they can be read at reasonable times by Affected Employees.

SECTION 6 — DESCRIPTIVE TABLE OF CONVENTIONAL PROCEEDINGS FOR CONTESTING AN OSHA CITATION

EVENTS COMMON TO ALL PROCEEDINGS

- Employer files **notice of contest** with OSHA office that mailed citation — within 15 working days of receiving the citation.
- Employer receives notification (**notice of docketing**) from Commission of case, docket number and forms to notify employees.
- Employer **posts notification** to employees of case in progress.
- Union and/or affected employees may **contest reasonableness of abatement period**; notice of contest is sent to citing OSHA office within 15 working days of the employer's receipt of the citation.
- If the Chief Administrative Law Judge has not assigned the case for Simplified Proceedings, and if a party has not requested Simplified Proceedings within 20 days of the notice of docketing and the request is not granted, conventional proceedings will be used. (See Rule 203.)

EVENTS PERTAINING TO CONVENTIONAL PROCEEDINGS

The Employer:

- Receives a **complaint** from OSHA's attorneys.
- Files an **answer** to the complaint within 20 calendar days of receiving the complaint.
- Discusses **discovery** techniques with the judge when applicable.
- Participates in a **conference call** to discuss issues and a possible settlement.
- Engages in discovery; exchanges interrogatories and depositions.
- **Discusses settlement** in another conference call with OSHA and judge. If not settled, then

- prepares for and participates in the **hearing**;
- may purchase a copy of the hearing **transcript** and may choose to submit a **brief** to the judge.
- Judge issues a **decision**. (If dissatisfied, any party may ask for **Commission review** of the decision.)

SECTION 7 — DESCRIPTIVE TABLE OF CONTESTING AN OSHA CITATION AND CHOOSING SIMPLIFIED PROCEEDINGS

EVENTS COMMON TO ALL PROCEEDINGS

- Employer files **notice of contest** with OSHA office that mailed citation — within 15 working days of receiving the citation.
- Employer receives notification (**notice of docketing**) from Commission of case, docket number and forms to notify employees.
- Employer **posts notification** to employees of case in progress.
- Union and affected employees may **contest reasonableness of abatement period**; notice of contest is sent to citing OSHA office within 15 working days of citation's posting at workplace.
- If the Chief Administrative Law Judge has assigned the case for Simplified Proceedings, or if a party has requested Simplified Proceedings and the request is granted, Simplified Proceedings will be in effect. (See Rule 203.)

EVENTS PERTAINING TO SIMPLIFIED PROCEEDINGS

If all disputed issues not resolved at the pre-hearing conference, then parties

- list **witnesses** and **exhibits**;
- prepare for and participate in the **hearing**;
- present **oral arguments** at the close of the hearing;
- may purchase a copy of the hearing **transcript**;
- decide whether to request permission to file a **brief**.

JUDGE ISSUES DECISION

If dissatisfied, any party may ask for **Commission review** of the decision.

SECTION 8 — DESCRIPTIVE TABLE OF EVENTS PERTAINING TO REVIEW OF AN ADMINISTRATIVE LAW JUDGE'S DECISION

If an employer is dissatisfied with an administrative law judge's decision and wishes to seek review by the Commission members, the employer

- receives judge's decision; dissatisfied with the outcome;

- files petition for discretionary review of the judge's decision;
- receives notification from Commission that case is or is not directed for review.

If the case is not directed for review, the judge's decision is a final order of the Commission and the employer may file a petition for review in a Court of Appeals.

If the case is directed for review, all parties do as follows:

- Receive a request from Commission for briefs on review.
- File briefs on review before Commission.
- Receive Commission decision that may supercede the judge's decision and affirm, modify or reverse it. In some cases, the judge's decision may be remanded for further proceedings.
- Files petition for review in Court of Appeals if dissatisfied with Commission decision.

See also Rules 90–96

GLOSSARY

Abatement Period Period of time specified in citation for correcting alleged workplace safety or health violation.

Affected Employee An affected employee is one who has been exposed to or could be exposed to any hazard arising from the cited violations — that is, the circumstances, conditions, practices, or operations creating the hazard.

Answer Written document filed in response to a complaint, consisting of short plain statements denying the allegations in the complaint which the employer contests.

Authorized Employee Representative A labor organization, such as a union, that has a collective bargaining relationship with the employer and represents affected employees or may be an affected employee(s) in cases where unions do not represent the employees.

Brief A written document in which a party states what the party believes are the facts of the case and argues how the law should be applied.

Certificate of Service A document stating the date and manner in which the parties were served (given) a document. See Appendix 2B for sample certificate. (Also see definition of "service.")

Citation Written notification from OSHA of alleged workplace violation(s), proposed penalty(ies), and abatement period.

Complaint Written document filed by the Secretary of Labor detailing the alleged violations contained in a citation.

Conventional Proceedings Typical Review Commission proceedings, which are similar to, but less formal than, court proceedings.

Discovery The process by which one party obtains information from another party prior to a hearing.

Exculpatory Evidence Information that may clear one of a charge or of fault or of guilt; in the context of OSHRC cases, information that might help the employer's case.

Exhibit Something (*e.g.*, a document, video, etc.) that is formally introduced as evidence at a hearing.

File To send papers to the Commission Executive Secretary, or to the judge assigned to a case, and to give copies of those papers to the other parties in the case.

Interlocutory Appeal An appeal of a judge's ruling on a preliminary issue in a case that is made before the judge issues a final decision on the full case. These types of appeals are infrequently made and are infrequently allowed. One example of an issue often raised in an interlocutory appeal is whether certain material that a party wants kept confidential, such as an employer's trade secrets or employee medical records, should become part of the public record in a case.

Motion A request asking that the judge direct some act to be done in favor of the party making the request or motion.

Notice of Appearance A written letter informing the Review Commission of the name and address of the person or persons who will represent a party (that is, the employer or a union or OSHA) in a case.

Notice of Contest Written document disagreeing with any part of an OSHA citation.

Notice of Docketing Written document from the Review Commission's Executive Secretary telling an employer, the Secretary of Labor, and any other parties in a case that the case has been received by the Commission and given an OSHRC docket number.

Notice of Withdrawal A written document from a party withdrawing its notice of contest or the citation and thus terminating the proceedings before the Commission. (See Appendix 8.)

Party The Secretary of Labor, anyone who files a notice of contest, or a union or affected employee(s) that requests party status.

Petition for Discretionary Review A written request from a party in a case asking the Commission in Washington, DC, to review and change the judge's decision. The grounds on which a party may request discretionary review are (1) it believes the judge made findings of material facts which are not supported by the evidence; (2) it believes that the judge's decision is contrary to law; (3) it believes that a substantial question of law, policy, or abuse of discretion is involved; or (4) it believes that a prejudicial error was committed.

Pro Se Latin for "without an attorney."

Secretary of Labor The head of the U.S. Department of Labor. OSHA is part of that Department.

Service Sending by first-class mail or personal delivery a copy of documents filed in a case to all parties in the case. See Definition of "Certificate of Service." (See Rule 7.)

Settlement An agreement reached by the parties resolving the disputed issues in a case.

Simplified Proceedings Review Commission proceedings that are less formal than Conventional Proceedings and designed for smaller and relatively simple cases. A complaint and answer are not required and discovery occurs only if the judge permits it.

Solicitor of Labor The U.S. Department of Labor's chief lawyer who has offices throughout the country. Lawyers from these offices represent the Secretary of Labor and OSHA in Review Commission cases.

APPENDIXES/SAMPLE LEGAL DOCUMENTS

This section is not intended to be a manual of forms, and the sample legal documents here are limited in number. The sample legal documents are intended for illustration to familiarize the reader with the general nature of some of the documents received and issued. Many of the documents received by the Commission, such as those in Appendixes 2, 3, and 6 (Complaint, Answer, and Petition for Discretionary Review), vary significantly from case to case.

APPENDIX 1 — NOTICE OF CONTEST

APPENDIX 1A. NOTICE OF CONTEST TO CITATION AND PROPOSED PENALTIES

XYZ Corp.
123 Street
City, State Zip Code

February 26, 2004

ABC, Area Director
Occupational Safety and Health Administration
U.S. Department of Labor, Federal Building
City, State Zip Code

Dear Mr. ABC:

This is to notify you that XYZ Corp. intends to contest all of the items and penalties alleged in the Citation and Proposed Penalty, received February 20, 2004, and dated February 19, 2004 (a copy is attached).

Very truly yours,

XYZ, President

APPENDIX 1B. NOTICE OF CONTEST TO PROPOSED PENALTIES ONLY

XYZ Corp.
123 Street
City, State Zip Code

September 14, 2004

ABC, Area Director
Occupational Safety and Health Administration
U.S. Department of Labor, Federal Building
City, State Zip Code

Dear Mr. ABC:

I wish to contest the amount of the Proposed Penalties of $1200 issued September 9, 2004, based on the violations cited by you during your recent inspection.

Sincerely,

XYZ, President
General Manager

APPENDIX 1C. NOTICE OF CONTEST BY EMPLOYEE REPRESENTATIVE

GHI International Union
456 Street
City, State Zip Code

June 9, 2004

ABC, Area Director
Occupational Safety and Health Administration
U.S. Department of Labor, Federal Building
City, State Zip Code

Dear Mr. ABC:

We have been authorized by the employee representative, GHI International Union, to file this notice of contest to the OSHA citations issued on June 2, 2004, against the employer, XYZ Co. The abatement dates of June 27, 2004, for Items No. 1 and No. 3 of the non-serious citation, and January 5, 2005, for Item No. 1 of the serious citation, are unreasonable and will continue to expose workers to safety hazards.

Sincerely,

JKL, Director
Safety Department
GHI International Union

APPENDIX 2 — COMPLAINT AND CERTIFICATE OF SERVICE

APPENDIX 2A. COMPLAINT

U. S. Occupational Safety and Health Review Commission, Secretary of Labor, Complainant, v. OSHRC Docket No. 99-9999 XYZ Co., Respondent

COMPLAINT

This action is brought to affirm the Citations and Notifications of penalty issued under the Occupational Safety and Health Act of 1970, 29 U.S.C. § 651, et seq., hereinafter the Act, of violations of §5(a) of the Act and the Safety and Health Regulations promulgated thereunder.

I

Jurisdiction of this action is conferred upon the Commission by §10(a) of the Act.

II

Respondent, XYZ Co., is an employer engaged in a business affecting commerce within the meaning of §3(5) of the Act.

III

The principal place of business of respondent is at 123 Street, City, State, Zip Code, where it was engaged in retail sales as of the date of the alleged violations.

IV

The violations occurred on or about June 9, 2004, at 123 Street, City, State, Zip Code (hereinafter "workplace").

V

As a result of an inspection at said workplace by an authorized representative of the complainant, respondent was issued three Citations and Notifications of Penalty pursuant to §9(a) of the Act.

VI

The Citations and Notifications of Penalty, copies of which are attached hereto and made a part hereof as Exhibits "A," "B," and "C" (consisting of one page each) identify and describe the specific violations alleged, the corresponding abatement dates fixed, and the penalties proposed.

VII

On or about July 29, 2004, by a document dated July 26, 2004, the complainant received notification, pursuant to §10(a) of the Act, of respondent's intention to contest the aforesaid Citations and Notifications of Penalty.

VIII

The penalties proposed, as set forth in Exhibits "A," "B," and "C" are appropriate within the meaning of §17(j) of the Act. The abatement dates fixed were and are reasonable.

WHEREFORE, cause having been shown, complainant prays for an Order affirming the Citations and Notifications of Penalty, as aforesaid.

JKL, Attorney
Office of the Solicitor
U.S. Department of Labor, Federal Building
City, State Zip Code

APPENDIX 2B. CERTIFICATE OF SERVICE*

I certify that the foregoing Complaint was served this 19th day of August, 2004, by mailing true copies thereof, by first-class mail to:

XYZ
XYZ Corp.
123 Street
City, State Zip Code

PQR
Attorney

* A similar document must accompany all other documents requiring a certificate of service.

APPENDIX 3 — ANSWER

U. S. Occupational Safety and Health Review Commission, Secretary of Labor, Complainant, v. OSHRC Docket No. 99-9999 XYZ Corp., Respondent

ANSWER

I, II, III

Respondent admits Paragraphs I, II, and III.

IV

Respondent denies Paragraph IV.

V

Respondent neither admits nor denies the allegations at Paragraph V.

VI

Respondent denies Paragraph VI.

VII

Respondent neither admits nor denies the allegations at Paragraph VII.

VIII

Respondent denies the allegations at Paragraph VIII. The penalties are excessive under § 17(j) of the Act based upon the small size of the employer, which has only twelve employees, and the low gravity of the alleged violations.

IX

Respondent pleads the affirmative defense of "greater hazard." Abatement of the alleged violations will increase the safety risk to employees. Respondent also pleads the affirmative defense of "unpreventable employee misconduct." The alleged conditions were the result of unauthorized actions by certain employees which resulted in the conditions referred to in the alleged violations.

RESPONDENT

By _____

Attorney
XYZ Corp.
123 Street
City, State Zip Code

APPENDIX 4 — REQUEST FOR SIMPLIFIED PROCEEDINGS

XYZ Corp.
123 Street
City, State Zip Code

March 26, 2004

Executive Secretary
U.S. Occupational Safety and Health
Review Commission
1120 20th Street, N.W., 9th Floor
Washington, DC 20036-3457

Dear Executive Secretary;

I request Simplified Proceedings. The Review Commission Docket Number assigned to my case is 99-9999.

Very truly yours,

XYZ, President

APPENDIX 5 — NOTICE OF DECISION

NOTICE OF DECISION

In Reference to:

Secretary of Labor v. XYZ Corp.
OSHRC Docket No. 99-9999

1. Enclosed is a copy of my decision. It will be submitted to the Commission's Executive Secretary on January 3, 2004. The decision will become the final order of the Commission at the expiration of thirty (30) days from the date of docketing by the Executive Secretary, unless within that time a member of the Commission directs that it be reviewed. All parties will be notified by the Executive Secretary of the date of docketing.

2. Any party that is adversely affected or aggrieved by the decision may file a petition for discretionary review by the Review Commission. A petition may be filed with the Judge within ten (10) days from the date of this notice. Thereafter, any petition must be filed with the Review Commission's Executive Secretary within twenty (20) days from the date of the Executive Secretary's notice of docketing. See Paragraph No. 1. The Executive Secretary's address is as follows:

   Executive Secretary
   Occupational Safety and Health Review Commission
   1120 20th Street, N.W., 9th Floor
   Washington, DC 20036-3457

3. The full text of the rule governing the filing of a petition for discretionary review is 29 C.F.R. 2200.91. It is appended hereto for easy reference, as are related rules prescribing post-hearing procedure.

MNO
Administrative Law Judge
December 1, 2004

APPENDIX 6 — PETITION FOR DISCRETIONARY REVIEW

U.S. Occupational Safety and Health Review Commission, Secretary of Labor, Complainant, v. OSHRC Docket No. 99-9999 XYZ Corp., Respondent

Petition for Discretionary Review

Comes now Respondent, XYZ Corp., being aggrieved by the Decision and Order of the Administration Law Judge in the above-styled matter, and hereby submits its Petition for Discretionary Review pursuant to 29 CFR 2200.91-Rule 91, Rules of Procedure of the Occupational Safety and Health Review Commission.

Statement of Portions of the Decision and Order to which Exception Is Taken

1. XYZ Corp. takes exception to that portion of the Decision and Order wherein the Administrative Law Judge held XYZ Corp. in serious violation of the standard published at 29 CFR 1926.28(a) as alleged in Serious Citation 1, Item 1, in finding that XYZ's employee John Jones was exposed to the alleged violation. (Judge's Decision at pp. 8–12.)

2. XYZ Corp. takes exception to that portion of the Decision and Order pertaining to Serious Citation 1, Item 1, wherein the Administrative Law Judge held that action of employee John Jones was not unpreventable employee misconduct. (Judge's Decision at pp. 13–17.)

Statement of Reasons for which Exceptions Are Taken

1. In his Decision, the Administrative Law Judge failed to follow the test set forth for the Fifth Circuit's Decision in *Secretary of Labor v. RPQ Corp.* for determining the existence of employee exposure. The testimony at transcript pages 25 to 45 clearly shows that John Jones was not in the zone of danger because he was on a work break and outside of the definition of the zone.

2. The evidence of record supports XYZ's position that the actions taken by employee John Jones were unpreventable. The Commission has set forth the test for determining unpreventable employee misconduct at *Secretary of Labor v. ROM Corp.* The testimony of XYZ's employees at transcript pp. 46 to 59 met all of the requirements of ROM Corp. to prove John Jones's actions were unpreventable.

For the reasons stated herein, XYZ Corp. hereby submits that the Occupational Safety and Health Review Commission should direct review of the Decision and Order of the Administrative Law Judge.

Respectfully submitted,

By_____

Attorney for XYZ Corp.

123 Street
City, State Zip Code
Tel. No. (999) 999-9999

APPENDIX 7 — DIRECTION FOR REVIEW

U.S. Occupational Safety and Health Review Commission Secretary of Labor, Complainant, v. OSHRC Docket No. 99-9999 XYZ Corp., Respondent

DIRECTION FOR REVIEW

Pursuant to 29 U.S.C. § 66(j) and 29 C.F.R. § 2200.92(a), the report of the Administration Law Judge is directed for review. A briefing order will follow.

COMMISSIONER
Dated:

APPENDIX 8 — NOTICE OF WITHDRAWAL

U.S. Occupational Safety and Health Review Commission Secretary of Labor, Complainant, v. OSHRC Docket No. 99-9999 XYZ Corp., Respondent

RESPONDENT'S WITHDRAWAL OF NOTICE OF CONTEST

Respondent, XYZ Corp., by the undersigned representative, hereby withdraws its Notice of Contest in the case with the docket number above, pursuant to 29 CFR 2200.102 of the Rules of Procedure for the Commission.

XYZ
XYZ Corp.
123 Street
City, State Zip Code
March 30, 2004

Appendix C
SAMPLE Corporate Compliance Program Checklist

To assess whether your company is in need of a comprehensive compliance program in the area of safety and resource control and other areas of potential risk, the following list of questions is provided in order to assess your current position. **Every "no" answer should send a signal that a potential risk is at hand and a program is needed.**

1. Does the board of directors put a high priority on human resources, safety and health, environmental, resource control, discrimination areas, and other regulatory compliance requirements?
2. Has the company adopted policies with regard to compliance with governmental regulations and other laws having a direct bearing on the operations?
3. Has the company established and published a Code of Conduct and distributed copies to employees?
4. Has your company employed an individual who will be directly responsible for human resources, safety and health, and governmental compliance? Are these individuals properly educated and prepared to manage these functions?
5. Has the company formally developed programs to involve employees such as safety committees?
6. Does your company possess all of the necessary resources to effectively develop and maintain an effective compliance program?
7. Are your required personnel, safety and health, environmental, and other compliance programs in writing?
8. Are corporate officers and managers and supervisors sensitive to the importance of governmental compliance?
9. Are your employees involved in your safety and resource control and related efforts?
10. Are your corporate officers and managers committed to your safety and resource control compliance efforts? Do they provide the necessary resources, staffing, and so forth to effectively perform the function successfully?
11. Does the company conduct periodic compliance and other legal audits to detect compliance failures? Are deficiencies or failures corrected in a timely manner?
12. Has the company established a "hotline" or other mechanisms through which to facilitate reporting of compliance failures such as can occur with human resource, safety and resource control, environmental, and so forth?

13. Are all employees properly trained in the required aspects of your compliance programs?
14. Is your training properly documented (i.e., will your documentation prove beyond a shadow of a doubt that a particular employee was trained in a particular regulatory requirement)?
15. Do you possess a new employee orientation program?
16. Does the orientation program for new employees include review of personnel policies, safety and resource control policies, codes of conduct, and other policies and procedures?
17. Are employees provided additional training? Other on-the-job training?
18. Does your company conduct compliance training sessions to sensitive managers and rank-and-file employees to their legal responsibilities?
19. Does your company provide information and assistance to employees regarding their rights and responsibilities under the individual state's workers' compensation laws?
20. Does your company communicate human resource, safety and resource control compliance issues to employees with posting, newsletters, brochures, and manuals?
21. Does your company go beyond the "bare bones" compliance requirements to create an appropriate work environment?
22. Does your company "keep up" with the new and revised laws, standards, and other regulatory compliance requirements?
23. Does your company appropriately discipline employees for failure to follow rules, policies, and regulations? Is this discipline fair and consistent?
24. Is creating a good working environment a high priority for your officers and directors?
25. Is your company proactive in the areas of human resources, safety and resource control, environmental, and so forth?

Issues for consideration in the development of a Corporate Code of Conduct:

1. Content of a Corporate Code of Conduct. The most common laws covered in codes of conduct include labor law, antitrust law, business ethics, conflicts of interest, corporate political activity, environmental law, safety and health laws, employee relations law, securities laws, and so forth. Special care should be provided in the area of individual and personal rights.
2. Distribution to officers, directors, and employees. Most companies distribute their codes of conduct to directors, officers, and employees when they join the company and on a yearly basis. The company may want to document that the director, officer, and employee have read, understood, and will adhere to the code.
3. Provide training and education to promote compliance. In-house seminars can be targeted to small groups of key managers to sensitize them to their legal responsibilities. Some companies use guidebooks as memos to communicate the importance of compliance to their employees. Newsletters and memos can be distributed to managers on a periodic basis to remind

them about their legal responsibilities, advise them of the developments of the law, and give them preventive law tips.

4. Enforcing the Code of Conduct. Compliance programs should include disciplinary procedures to punish violations of Code of Conduct. Compliance programs are not effective if they are not enforced. Sanctions for violation can include verbal warnings, written warnings, suspension, demotion, discharge, and referral to law enforcement agencies. In addition to sanctions, the disciplinary procedures should include provisions for protecting whistle-blowers and for investigation allegations of illegal conduct.

5. Monitor compliance through legal audits. Legal audits can include, but are not limited to, the following:
   a. Assemble legal audit team
   b. Educate and train legal audit team
   c. Assign legal team responsibilities
   d. Develop a site-specific audit instrument
   e. Conduct site visitation and facility inspection
   f. Conduct employee interviews
   g. Conduct records search
   h. Develop and use employee questionnaires
   i. Review record retention procedures and policies
   j. Develop an audit report
   k. Present audit report to board of directors and officers
   l. The legal audit mechanism, like the internal safety and health audit discussed above, produces documentation that may be the subject of discovery requests in civil or criminal litigation. This type of evaluation may also produce sensitive data that the company seeks to keep confidential. Extreme caution should be exercised to preserve privilege and confidentiality.

Appendix D
The U.S. Equal Employment Opportunity Commission

DISCRIMINATORY PRACTICES

Under Title VII of the Civil Rights Act of 1964, the Americans with Disabilities Act (ADA), and the Age Discrimination in Employment Act (ADEA), it is illegal to discriminate in any aspect of employment, including

- hiring and firing;
- compensation, assignment, or classification of employees;
- transfer, promotion, layoff, or recall;
- job advertisements;
- recruitment;
- testing;
- use of company facilities;
- training and apprenticeship programs;
- fringe benefits;
- pay, retirement plans, and disability leave; and
- other terms and conditions of employment.

Discriminatory practices under these laws also include

- harassment on the basis of race, color, religion, sex, national origin, disability, or age;
- retaliation against an individual for filing a charge of discrimination, participating in an investigation, or opposing discriminatory practices;
- employment decisions based on stereotypes or assumptions about the abilities, traits, or performance of individuals of a certain sex, race, age, religion, or ethnic group, or individuals with disabilities; and
- denying employment opportunities to a person because of marriage to, or association with, an individual of a particular race, religion, national origin, or an individual with a disability. Title VII also prohibits discrimination because of participation in schools or places of worship associated with a particular racial, ethnic, or religious group.

Employers are required to post notices to all employees advising them of their rights under the laws the Equal Employment Opportunity Commission (EEOC) enforces and their right to be free from retaliation. Such notices must be accessible, as needed, to persons with visual or other disabilities that affect reading.

Note: Many states and municipalities also have enacted protections against discrimination and harassment based on sexual orientation, status as a parent, marital status, and political affiliation. For information, please *contact the EEOC District Office nearest you.*

OTHER DISCRIMINATORY PRACTICES UNDER FEDERAL EQUAL EMPLOYMENT OPPORTUNITY (EEO) LAWS

Title VII

Title VII prohibits not only intentional discrimination, but also practices that have the effect of discriminating against individuals because of their race, color, national origin, religion, or sex.

National Origin Discrimination

- It is illegal to discriminate against an individual because of birthplace, ancestry, culture, or linguistic characteristics common to a specific ethnic group.
- A rule requiring that employees speak only English on the job may violate Title VII unless an employer shows that the requirement is necessary for conducting business. If the employer believes such a rule is necessary, employees must be informed when English is required and the consequences for violating the rule.

The Immigration Reform and Control Act (IRCA) of 1986 requires employers to assure that employees hired are legally authorized to work in the United States. However, an employer who requests employment verification only for individuals of a particular national origin, or individuals who appear to be or sound foreign, may violate both Title VII and the Immigration Reform and Control Act (IRCA); verification must be obtained from all applicants and employees. Employers who impose citizenship requirements or give preferences to U.S. citizens in hiring or employment opportunities also may violate IRCA.

Additional information about IRCA may be obtained from the Office of Special Counsel for Immigration-Related Unfair Employment Practices at 1-800-255-7688 (voice), 1-800-237-2515 (TTY for employees/applicants) or 1-800-362-2735 (TTY for employers) or at www.usdoj.gov/crt/osc.

Religious Accommodation

An employer is required to reasonably accommodate the religious belief of an employee or prospective employee, unless doing so would impose an undue hardship.

Sex Discrimination

Title VII's broad prohibitions against sex discrimination specifically cover the following:

- Sexual Harassment — This includes practices ranging from direct requests for sexual favors to workplace conditions that create a hostile environment for persons of either gender, including same sex harassment. (The "hostile environment" standard also applies to harassment on the bases of race, color, national origin, religion, age, and disability.)
- Pregnancy Based Discrimination — Pregnancy, childbirth, and related medical conditions must be treated in the same way as other temporary illnesses or conditions.

Additional rights are available to parents and others under the Family and Medical Leave Act (FMLA), which is enforced by the U.S. Department of Labor. For information on the FMLA, or to file an FMLA complaint, individuals should contact the nearest office of the Wage and Hour Division, Employment Standards Administration, U.S. Department of Labor. The Wage and Hour Division is listed in most telephone directories under U.S. Government, Department of Labor or at www.dol.gov/esa/whd_org.htm.

Age Discrimination in Employment Act

The ADEA's broad ban against age discrimination also specifically prohibits the following:

- Statements or specifications in job notices or advertisements of age preference and limitations. An age limit may only be specified in the rare circumstance where age has been proven to be a bona fide occupational qualification (BFOQ).
- Discrimination on the basis of age by apprenticeship programs, including joint labor–management apprenticeship programs.
- Denial of benefits to older employees. An employer may reduce benefits based on age only if the cost of providing the reduced benefits to older workers is the same as the cost of providing benefits to younger workers.

Equal Pay Act

The Equal Pay Act (EPA) prohibits discrimination on the basis of sex in the payment of wages or benefits, where men and women perform work of similar skill, effort, and responsibility for the same employer under similar working conditions.

Note the following:

- Employers may not reduce wages of either sex to equalize pay between men and women.
- A violation of the EPA may occur where a different wage was or is paid to a person who worked in the same job before or after an employee of the opposite sex.
- A violation may also occur where a labor union causes the employer to violate the law.

TITLES I AND V OF THE AMERICANS WITH DISABILITIES ACT

The ADA prohibits discrimination on the basis of disability in all employment practices. It is necessary to understand several important ADA definitions to know who is protected by the law and what constitutes illegal discrimination:

Individual with a Disability: An individual with a disability under the ADA is a person who has a physical or mental impairment that substantially limits one or more major life activities, has a record of such an impairment, or is regarded as having such an impairment. Major life activities are activities that an average person can perform with little or no difficulty such as walking, breathing, seeing, hearing, speaking, learning, and working.

Qualified Individual with a Disability: A qualified employee or applicant with a disability is someone who satisfies skill, experience, education, and other job-related requirements of the position held or desired, and who, with or without reasonable accommodation, can perform the essential functions of that position.

Reasonable Accommodation: Reasonable accommodation may include, but is not limited to, making existing facilities used by employees readily accessible to and usable by persons with disabilities; job restructuring; modification of work schedules; providing additional unpaid leave; reassignment to a vacant position; acquiring or modifying equipment or devices; adjusting or modifying examinations, training materials, or policies; and providing qualified readers or interpreters. Reasonable accommodation may be necessary to apply for a job, to perform job functions, or to enjoy the benefits and privileges of employment that are enjoyed by people without disabilities. An employer is not required to lower production standards to make an accommodation. An employer generally is not obligated to provide personal use items such as eyeglasses or hearing aids.

Undue Hardship: An employer is required to make a reasonable accommodation to a qualified individual with a disability unless doing so would impose an undue hardship on the operation of the employer's business. Undue hardship means an action that requires significant difficulty or expense when considered in relation to factors such as business size, financial resources, and the nature and structure of the business operation.

Prohibited Inquiries and Examinations: Before making an offer of employment, an employer may not ask job applicants about the existence, nature, or severity of a disability. Applicants may be asked about their ability to perform job functions. A job offer may be conditioned on the results of a medical examination, but only if the examination is required for all entering employees in the same job category. Medical examinations of employees must be job related and consistent with business necessity.

Drug and Alcohol Use: Employees and applicants currently engaging in the illegal use of drugs are not protected by the ADA when an employer acts on the basis of such use. Tests for illegal use of drugs are not considered medical examinations and, therefore, are not subject to the ADA's restrictions on medical examinations. Employers may hold individuals who are illegally using drugs and individuals with alcoholism to the same standards of performance as other employees.

THE CIVIL RIGHTS ACT OF 1991

The Civil Rights Act of 1991 made major changes in the federal laws against employment discrimination enforced by EEOC. Enacted in part to reverse several Supreme Court decisions that limited the rights of persons protected by these laws, the Act also provides additional protections. The Act authorizes compensatory and punitive damages in cases of intentional discrimination, and provides for obtaining attorneys' fees and the possibility of jury trials. It also directs the EEOC to expand its technical assistance and outreach activities.

Appendix E

SECRETARY OF LABOR, Complainant v. **SUMMITT CONTRACTORS, INC.,** Respondent	**(OSHRC Docket No. 03-1622)**

Appearances

Stephen D. Turow, Attorney; Ann Rosenthal, Counsel for Appellate Litigation; Daniel J. Mick, Counsel for Regional Trial Litigation; Joseph M. Woodward, Associate Solicitor; Howard M. Radzely, Solicitor; U.S. Department of Labor, Washington, DC

    for the Complainant

Robert E. Rader, Jr., Esq.; Rader & Campbell, Dallas, TX

    for the Respondent

Arthur G. Sapper, Esq.; Robert C. Gombar, Esq.; James A. Lastowka, Esq.; McDermott Will & Emery LLP, Washington, DC

    for Amici National Association of Home Builders; Contractors' Association of Greater New York; Texas Association of Builders; and Greater Houston Builders Association

Victoria L. Bor, Esq.; Sue D. Gunter, Esq.; Sherman, Dunn, Cohen, Leifer & Yellig, P.C., Washington, DC

    for Amicus Building and Construction Trades Department, AFL-CIO

DECISION

Before: RAILTON, Chairman; ROGERS and THOMPSON, Commissioners.

    BY RAILTON, Chairman:

    At issue before the Commission is a decision of Judge Ken S. Welsch affirming a citation issued to Summit Contractors, Inc. ("Summit") for an alleged scaffolding violation under 29 C.F.R. § 1926.451(g)(1)(vii).

[1] Section 1926.451(g) (1) (vii) states:

For all scaffolds not otherwise specified in paragraphs (g)(1)(i) through (g)(1)(vi) of this section, each
employee shall be protected by the use of personal fall arrest systems or guardrail systems meeting the
requirements of paragraph (g)(4) of this section.

Commissioner Thompson and I join in vacating the citation in its entirety.

[2] This case was consolidated solely for purposes of oral argument before the Commission with Docket
Number 05-0839, another case involving Summit.

BACKGROUND

Summit is a general building contractor with its corporate office located in Jacksonville, Florida. In June 2003, Summit was the prime contractor for the construction
of a college dormitory in Little Rock, Arkansas. Summit employed only a job superintendent and three assistant superintendents at the worksite. The superintendents
were responsible for coordinating the vendors, scheduling the work for the various subcontractors, and ensuring that the work of the subcontractors was performed
according to contract. Summit subcontracted the project's exterior brick masonry
work to All Phase Construction, Inc. ("All Phase"). All Phase workers used scaffolding to perform their work.

On June 18 and 19, 2003, an Occupational Safety and Health Administration
("OSHA") Compliance Safety and Health Officer ("CSHO") observed and photographed All Phase employees who were not protected from falls working from scaffolds at twelve to eighteen feet above the ground. The CSHO also observed other
employees working from a scaffold inside a building on June 19; these workers were
also not protected against falls. None of the exposed workers were employed by
Summit. Summit did not create the hazardous conditions observed by the CSHO.
Some of Summit's superintendents were present at the worksite on June 18 and 19,
and some of the instances were in plain view of Summit's trailer located on the worksite. Summit does not claim it lacked knowledge of the violative conditions observed
by the CSHO.

The CSHO did not perform a walk-around inspection, however, until June 24,
2003, when Summit's safety officer could be present. At the time of the walk-around
inspection, the scaffolding violations the CSHO observed on June 18 and 19 had
been corrected. According to Summit's project superintendent, Jimmy Guevara,
he had previously observed All Phase employees working on scaffolds that lacked
guardrails. Guevara had instructed All Phase to install guardrails two or three times
prior to the OSHA inspection. Each time, All Phase would address the violation but
then fall out of compliance when the scaffolding was moved to a different area.

Based on the CSHO's observations on June 18 and 19, OSHA issued Summit a citation for a violation of the construction safety standard set forth at
§ 1926.451(g)(1)(vii) as a "controlling" employer in accordance with the agency's
multi-employer worksite doctrine extant at the time. All Phase was also cited under
the doctrine as the employer who created the hazard and as the employer having
employees exposed to the hazard.

³ All Phase did not contest the citations and paid the penalties proposed by the Secretary.

Before the judge, Summit argued that the multi-employer worksite doctrine is invalid to a general contractor who neither created, nor had employees exposed to, the alleged and cited hazard. In other words, Summit challenged the Secretary's application of the doctrine to controlling contractors who have contractual authority over subcontractors. Summit argued before the judge, and also contends on review, that the doctrine as expressed in OSHA Directive CPL 2-0.124 (Multi-Employer Citation Policy) is not enforceable because it is contrary to 29 C.F.R. § 1910.12(a) which states as follows:

> *Standards.* The standards prescribed in Part 1926 of this chapter are adopted as occupational safety and health standards under section 6 of the Act and shall apply, according to the provisions thereof, to every employment and place of employment of every employee engaged in construction work. Each employer shall protect the employment and places of employment of each of his employees engaged in construction work by complying with the appropriate standards prescribed in this paragraph.

29 C.F.R. § 1910.12(A).

As the judge noted, Summit's argument focuses on the second sentence of this regulation. Specifically, Summit's position is that because it had no employees exposed to the hazard, and did not create the hazard; the regulation prohibits the issuance of a citation to Summit for the hazard created by the subcontractor, All Phase. The judge noted that the Commission has on numerous occasions applied the doctrine to controlling employers like Summit and, therefore, rejected the argument. Among others, he cited the Commission's decision in *Access Equipment Systems, Inc.*, 18 BNA OSHC 1718, 1999 CCH OSHD ¶ 31,821 (No. 95-1449, 1999), and *McDevitt Street Bovis, Inc.*, 19 BNA OSHC 1108, 2000 CCH OSHD ¶ 32,204 (No. 97-1918, 2000). As for the specific argument relating to § 1910.12(a), the judge simply noted his view that the regulation does not prohibit finding an employer responsible for the safety of employees of other employers.

DISCUSSION

In a decision rendered almost thirty-one years ago, the Commission stated that "the general contractor is well situated to obtain abatement of hazards either through its own resources or through its supervisory capacity." *Grossman Steel & Aluminum Corp.*, 4 BNA OSHC 1185, 1188, 1975-76 CCH OSHD ¶ 20,691, p. 24,791 (No. 12775, 1976). The Commission went on to say that "we will hold the general contractor responsible for violations it could reasonably have expected to prevent or abate by reason of its supervisory capacity." *Id.* This holding was characterized as "*dictum*" in a footnote. *Id.* at 1188-89 n.6, 1975-76 CCH OSHD at p. 24,791 n.6. Nevertheless, it took on a life of its own during ensuing years as the Commission and some circuit courts relied on these statements to find some general contractors in violation of construction safety standards simply by virtue of their "supervisory capacity."

[4] In effect, the Commission in 1976 was stating a policy decision. At that time *some* Commissioners believed they were charged under the Act to set policy. *See, for example, Cuyahoga Valley Ry. Co.*, 10 BNA OSHC 2156, 1982 CCH OSHD ¶ 26,296 (No. 76-1188, 1982), *aff'd*, 748 F.2d 340 (6th Cir. 1984), *rev'd*, 474 U.S. 3 (1985) (Supreme Court reversed Commission and circuit court's decision that Secretary cannot unilaterally withdraw citation without Commission approval); *Am. Cyanamid Co.*, 8 BNA OSHC 1346, 1980 CCH OSHD ¶ 24,424 (No. 77-3752, 1980), *rev'd*, 647 F.2d 383 (3d Cir. 1981) (circuit court reversed Commission decision holding that Commission had authority to determine whether abatement has occurred under a settlement agreement); *Sun Petroleum Products Co.*, 7 BNAOSHC 1306, 1979 CCH OSHD ¶ 23,502 (No. 76-3749, 1979), *aff'd on other grounds*, 622F.2d 1176 (3d Cir. 1980) (circuit court held that Commission's role in settlement process was limited where Commissioners split on whether Commission had authority to reject settlement agreement); *IMC Chem. Group*, 6 BNA OSHC 2075, 1978 CCH OSHD ¶ 23,149 (No. 76-4761, 1978), *rev'd*, 635 F.2d 544 (6th Cir. 1980) (circuit court reversed Commission's decision that, after notice of contest has been filed, Secretary may not withdraw citation without Commission approval).

See, for example, Universal Constr. Co. v. OSHRC, 182 F.3d 726 (10th Cir. 1999); *R.P. Carbone Constr. Co. v. OSHRC*, 166 F.3d 815 (6th Cir. 1998); *Brennan v. OSHRC (Underhill Constr. Corp.)*, 513 F.2d 1032 (2d Cir. 1975); *McDevitt Street Bovis, Inc.*, 19 BNA OSHC 1108, 2000 CCH OSHD ¶ 32,204 (No. 97-1918, 2000); *Blount Int'l Ltd.*, 15 BNA OSHC 1897, 1991-93 CCH OSHD ¶ 29,854 (No. 89-1394, 1992); *Gil Haugan*, 7 BNA OSHC 2004, 2006, 1979 CCH OSHD ¶ 24,105 (Nos. 76-1512 & 76-1513, 1979). Usually in these situations, the subcontractor responsible for the creation of the hazard and who had employees exposed to the hazard was also cited for the same violation.

The Commission, however, has been told in no uncertain terms by several courts that it is not a policy-setting agency. *See, for example, Donovan v. A. Amorello & Sons, Inc.*, 761 F.2d 61, 65 (1st Cir. 1985) (analyzing legislative history and determining that "Congress did not intend OSHRC to possess broad powers to set policy . . ."); *Marshall v. OSHRC (IMC Chem. Group)*, 635 F.2d 544, 547 (6th Cir. 1980) ("Whatever 'policies' the Commission establishes are indirect. Only those established by the Secretary are entitled to enforcement and defense in court." (quoting *Madden Constr. Inc. v. Hodgson*, 502 F.2d 278, 280 (9th Cir. 1974))). According to these decisions, that function belongs to the Secretary. *See Madden Constr.*, 502 F.2d at 280 ("[T]he Act imposes policy-making responsibility upon the Secretary, not the Commission."). The Secretary's citation policy on multi-employer construction worksites has a checkered history. Indeed, as the doctrine developed over the years, the Secretary's application and elucidation of her enforcement policy has been anything but consistent. *See IBP Inc. v. Herman (IBP)*, 144 F.3d 861, 865 n.3 (D.C. Cir. 1998) (detailing doctrine's "checkered history"). An analysis of the Secretary's own guidelines regarding the doctrine show the myriad changes in her interpretation as to how the doctrine should be applied. *Cf. Martin v. OSHRC (CF & I Steel Corp.)*, 499 U.S. 144, 158 (1991) (reviewing court may consult less formal means of interpreting regulations, such as the OSHA Field Operations Manual, to determine whether the Secretary has consistently applied her position, a factor in determining the reasonableness of Secretary's position (citing *Ehlert v. United States*, 402 U.S. 99, 105 (1971))).

In its first Field Operations Manual ("FOM") issued contemporaneously with § 1910.12(a), OSHA permitted the citation of employers who expose their own employees to hazards as well as employers who create a hazardous condition or supply hazardous equipment, *whether or not their own employees were exposed. See* OSHA FOM p. VII-6-8 para. 10 (May 20, 1971). The manual was revised six months later to remove the reference to employers who supply unsafe equipment. *See* OSHA Compliance Operations Manual ("COM") p. VII-7-8 para. 13 (Nov. 15, 1971). Approximately three years later, OSHA again narrowed its citation policy. In July 1974, OSHA amended the FOM, instructing compliance personnel to cite only an employer on a construction site who has *exposed his own employees* to an unsafe condition. OSHA FOM ¶ 4380.6 (July, 1974). In essence, OSHA eliminated any practice of making multiple employers, other than exposing employers, responsible for the abatement of the same hazard on construction sites. Indeed, OSHA instructed compliance personnel in this revised version of the FOM, as follows: "An employer will not be cited if his employees are not exposed or potentially exposed to an unsafe or unhealthful condition — even if that employer created the condition." *Id. See also* OSHA FOM ¶ 4380.6 (Jan. 1, 1979) (identical language).

Four years later, OSHA again changed its interpretation of the doctrine. In the revised 1983 version of the FOM, the Secretary announced that an employer on a multi-employer worksite could defend by showing that it did not create the hazard, could not correct the hazard, and had made an effort to persuade the controlling employer to correct the hazard, or had alerted employees to the dangers associated with the hazard. OSHA FOM ¶ 265 (Apr. 18, 1983). This version of the FOM specified that compliance personnel should cite the exposing employer(s), unless all exposing employers could establish the defense. In that case, compliance personnel should cite the employer in the best position to correct the hazard. *Id.* at ¶ 264-65. *See also* OSHA Instruction CPL 2.42B (June 15, 1989) (identical language).

Eleven years after that, OSHA again changed course and issued the multi-employer policy at issue in this case. In 1994, OSHA revised its compliance instructions and issued a new manual called the Field Inspection Reference Manual or "FIRM." There, OSHA stated that citations should be issued not only to exposing employers, but also to creating, controlling and correcting employers "whether or not their own employees are exposed ... " OSHA Field Inspection Reference Manual (FIRM) § V.C.6 (Sept. 26, 1994). *See also* OSHA Instruction CPL 2-0.124 (Dec. 10, 1999) (identical language; current multi-employer worksite doctrine).

In sum, OSHA issued § 1910.12(a) in May 1971, and almost simultaneously stated a policy for issuing citations on construction sites. The employer exposing its employees to hazards was to be cited, and employers who created or supplied hazardous equipment could also be cited. OSHA altered this policy six months later to eliminate citations to suppliers of faulty equipment. Citations to hazard-creating employers were eliminated next in 1974, and it was not until 1983 that such employers were returned to the mix, but only if every exposing employer had a defense. Then, in 1994, OSHA changed its policy significantly to allow citation of essentially every employer who might have some association with the hazard (i.e., the exposing employer, the creating employer, the controlling employer, and the correcting employer—the one who could abate the hazard). The Secretary never indicated the reasons behind her multiple changes in policy. *See Greater Boston Television Corp. v. FCC*, 444 F.2d 841, 852 (D.C. Cir. 1971) ("agency changing its course must supply

a reasoned analysis indicating that prior policies and standards were being deliberately changed, not casually ignored"). Furthermore, at no time throughout this
period of over twenty years did the Secretary ever note that § 1910.12(a) contains
language which on its face is in apparent conflict with the policy.

It is not as if this conflict has gone unnoticed by the courts or even the Commission. As early as 1995, the United States Court of Appeals for the District of
Columbia Circuit noted a "marked tension" between the language of § 1910.12(a)
and the Secretary's multi-employer policy. *Anthony Crane Rental, Inc. v. Reich*, 70
F.3d 1298, 1306-07 (D.C. Cir. 1995). The court went on to say: "Here, the relevant
regulation by its terms only applies to an employer's *own employees*, seemingly leaving little room for invocation of the [multi-employer] doctrine." *Id.* at 1307 (emphasis
in the original). The court, after noting that the issue had not been briefed and had
not been addressed by any other court, left "to a later date the critical decision of
whether to apply the multi-employer doctrine where an employer has been cited
under … [§ 1910.12]." *Id.* 1998, another panel of the same court similarly noted the
tension between the regulation and the policy. *IBP*, 144 F.3d at 865-66. It too determined that it was unnecessary to decide the issue.

---

[5] In its most recent multi-employer decision, the United States Court of Appeals for the Tenth Circuit
noted the D.C. Circuit's decision in *IBP* but did not address the issue of conflict between the multi-
employer policy and § 1910.12(a). *Universal Constr. Co. v. OSHRC*, 182 F.3d 726, 729-30 (10th Cir.
1999).

Id. at 866

In a like manner, the Commission in two recent cases noted the existence of the
problem but, like the D.C. Circuit, declined to address it for not having been briefed.
See Access Equip., 18 BNA OSHC at 1725 n.12, 1999 CCH OSHD at p. 46,780 n.12
(equipment supplier and installer was liable as such notwithstanding its defense that
it was not a contractor); *McDevitt Street Bovis, Inc.*, 19 BNA OSHC at 1112-13, 2000
CCH OSHD at p. 48,782-83 (general contractor was responsible for scaffold violation as a controlling employer). As the judge pointed out here, Summit has raised the
issue of the conflict or tension between § 1910.12(a) and the existing multi-employer
policy in this and a number of other cases. While I firmly believe that cases should
be disposed of on narrow grounds wherever possible, I do not see how the issue
raised by Summit can be avoided in this case.

The problem I see is the one recognized by the court in *Anthony Crane Rental,
Inc.*: that the limitation in § 1910.12(a) making the compliance obligation of employers for violations of standards applicable only to "*his employees*" precludes issuance
of a citation to a general contractor having none of its employees exposed to the hazard. *See Anthony Crane Rental*, 70 F.3d at 1306-07. It seems to me that the checkered
history of the multi-employer doctrine as expressed in the Secretary's ever-changing
compliance guidelines — be it the FOM, COM, CPL, or FIRM — taken in contrast
with a regulation which has not been amended since 1971, results in the latter trump-
ing whatever reliance the Commission can place on the varying nature of the policy.
Cf. Christensen v. Harris County, 529 U.S. 576, 587 (2000) (policy statements while
"entitled to respect" are not given *Chevron* deference like promulgated standards)

(citing *Chevron, USA, Inc. v. NRDC*, 467 U.S. 837 (1984)); *Union Tank Car Co.*, 18 BNA OSHC 1067, 1069, 1995 CCH OSHD ¶ 31,445, p. 44,470 (No. 96-0563, 1997) (in assessing reasonableness of Secretary's interpretation, Commission considers, *inter alia*, whether her interpretation "'sensibly conforms to the purpose and wording of the regulation[]', taking into account 'whether the Secretary has consistently applied the interpretation embodied in the citation.'" (quoting *CF & I Steel Corp.*, 499 U.S. at 150, 157-58)).

   I find unpersuasive the Secretary's argument in this litigation that the first sentence of the regulation permits or allows a broader class of employers, including those not having employees exposed to the cited hazard, to be cited under the policy.

⁶ Commissioner Rogers finds the regulation to be ambiguous. I do not agree that the regulation is ambiguous. It seems to me that both sentences are plain in their meaning. Here, I agree with the D.C. Circuit that the meaning of the regulation is "plain" and that the regulation "by its terms only applies to an employer's *own employees*." *See Anthony Crane*, 70 F.3d at 1303, 1307 (emphasis in the original). *See also Sec'y v. Simpson, Gumpertz & Heger, Inc.*, 3 F.3d 1, 4 (1st Cir. 1993) (holding that meaning of § 1910.12(a) is "plain"). The first sentence makes the construction safety standards applicable to "every employment and place of employment of every employee engaged in construction work." The second sentence makes each employer engaged in construction work responsible for "his employees." Were it otherwise, deference to the Secretary's interpretation is likely owed, as Commissioner Rogers states.

While I may be sympathetic to such an argument, it simply does not explain why the Secretary has sat on her hands for ten years after being alerted twice to the problem by the D.C. Circuit in *Anthony Crane* and *IBP*. She even issued a compliance instruction in 1999 and, while iterating her policy adopted in 1994, failed to address the significant issue and tension mentioned by the court. Beyond that, the Commission has alerted her to the issue in both *Access Equipment* and *McDevitt*, yet the Secretary still did not act.

   Moreover, to construe the first sentence of § 1910.12(a) as the Secretary argues in this litigation is to ignore or eliminate the language "each of his employees" used in the second sentence. *See United States v. Menasche*, 348 U.S. 528, 538-39 (1955) ("It is our duty 'to give effect, if possible, to every clause and word of a statute,' rather than to emasculate an entire section." (citations omitted)). In other words, the Secretary improperly suggests the meaning of the regulation would not change even if the words "his employees" were missing. *See AFL-CIO v. Chao*, 409 F.3d 377, 384 (D.C. Cir. 2005) ("the court is obligated not only to construe the statute as a whole but to give meaning to each word of the statute"). In my view, her interpretation is untenable. The Commission must give effect to the plain language of the regulation, especially in the face of the Secretary's inconsistent doctrine. *See Arcadian Corp.*, 17 BNA OSHC 1345, 1347, 1995-97 CCH OSHD ¶ 30,856 p. 42,917 (statutory analysis ends if language is plain), *aff'd*, 110 F.3d 1192 (5th Cir. 1997). *See also FDIC v. Philadelphia Gear Corp.*, 476 U.S. 426, 438-39 (1986) (affording deference to agency's contemporaneous understanding of ambiguous term where understanding had been fortified by agency's consistent behavior over the following decades).

[7] The Secretary's position is also weakened by the contrast between 29 C.F.R. § 1926.16 and § 1910.12(a). Section 1926.16, which was promulgated at nearly the same time as § 1910.12(a), applies to government jobs under the Contract Work Hours and Safety Standards Act. *See* 36 Fed. Reg. 1802 (Feb. 2, 1971); 36 Fed. Reg. 7340 (Apr. 17, 1971) (adopting Part 1926). Section 1926.16, in contrast to § 1910.12(a), contains language extending an employer's liability beyond his own employees. The fact that such language is absent from § 1910.12(a) is further evidence that § 1910.12(a) should be read as limiting an employer's liability to "his employees."

ORDER

For these reasons, I find the Secretary's reliance on her multi-employer worksite doctrine to cite Summit in this case to be impermissible given the contrary language of her regulation at § 1910.12(a). Accordingly, based on this analysis and that set forth in Commissioner Thompson's concurring opinion, we vacate the citation.

SO ORDERED.

/s/ _____

W. Scott Railton

Chairman

Dated: April 27, 2007

THOMPSON, Commissioner, concurring:

In this case, the Secretary seeks to enforce the duty of a "controlling employer" pursuant to her current multi-employer citation policy.

[1] The definition of a "controlling employer" is found in Section X.E.1 of OSHA's current multi-employer citation policy: "An employer who has general supervisory control over the worksite, including the power to correct safety violations itself or require others to correct them. Control can be established by contract or in the absence of ... contractual provisions, by the exercise of control in practice." OSHA Instruction CPL 2-0.124 at X.E.1 (Dec. 10, 1999). The controlling employer's "duty of reasonable care" is set forth in Sections X.E.3 and X.E.4. *Id.* at X.E.3-4.

The citation alleges a violation of a Part 1926 construction standard, 29 C.F.R. § 1926.451(g)(1)(vii), against Summit Contractors, Inc. ("Summit"), a general construction contractor who, the Secretary concedes, neither created the violative conditions nor exposed any of its own employees to these conditions. For the separate reasons I state below, I join Chairman Railton in vacating the citation because I conclude that 29 C.F.R. § 1910.12(a) prevents the Secretary from citing Summit in this case.

DISCUSSION

My colleague Commissioner Rogers notes that Commission precedent establishes that section 5 (a)(2) of the Occupational Safety and Health Act of 1970, 29 U.S.C. § 655 ("OSH Act"), grants the Secretary broad discretion to promulgate a multi-employer citation policy. *See Arcadian Corp.*, 17 BNA OSHC 1345, 1352, 1995-97

CCH OSHD ¶ 30,856, p. 42,918 (No. 93-3270, 1995), *aff'd*, 110 F.2d 1192 (5th Cir. 1997). In fact, more than thirty years ago, the Secretary published, but then withdrew, a *Federal Register* notice seeking comment on a proposed multi-employer citation policy. *See* 41 Fed. Reg. 17,639, 17,640 (Apr. 27, 1976).

However, having said that precedent grants the Secretary broad statutory discretion to adopt and enforce specific standards does not *a fortiori* define the limitations the Secretary voluntarily imposed on that discretion when she adopted a specific standard or set of standards.

---

² *See generally Anning-Johnson Co. v. OSHRC*, 516 F.2d 1081, 1091-92 (7th Cir. 1975) (Tone, J., concurring).

Thus, in this case, it remains to be resolved how § 1910.12(a) limits the discretion of the Secretary to issue citations for violations of 29 C.F.R. Part 1926 standards. Section 1910.12(a) states, in pertinent part:

> The standards prescribed in Part 1926 of this chapter are adopted as occupational safety and health standards ... and shall apply, according to the provisions thereof, to every employment and place of employment of every employee engaged in construction work. Each employer shall protect the employment and places of employment of each of his employees engaged in construction work by complying with the appropriate standards prescribed in this paragraph.

29 C.F.R. § 1910.12(A).

The Secretary issued § 1910.12(a) pursuant to section 6(a) of the OSH Act in order to adopt the Part 1926 standards originally enforced under the Contract Work Hours and Safety Standards Act, 40 U.S.C. § 333 ("Construction Safety Act" or "CSA").

---

³ Nat'l Consensus Standards and Established Fed. Standards, 36 Fed. Reg. 10,466, 10,469 (May 29, 1971). Former Part 1518 of Title 29, C.F.R. was subsequently redesignated as Part 1926. Redesignation, 36 Fed. Reg. 25,232 (Dec. 30, 1971).

See Coughlan Constr. Co., 3 BNA OSHC 1636, 1638, 1975-76 CCH OSHD ¶ 20,106, p. 23,923 (Nos. 5303 & 5304, 1975). The scope and application provisions of § 1910.12(a) define the "regulatory universe" to which those construction standards apply. *See Reich v. Simpson, Gumpertz & Heger, Inc.*, 3 F.3d 1, 4-5 (1st Cir. 1993). Neither a reviewing court nor the Commission has ever before sought to resolve the "marked tension" between the Secretary's multi-employer citation policy and § 1910.12(a). *See Anthony Crane Rental Inc. v. Reich*, 70 F. 3d 1298, 1307 (D.C. Cir. 1995) (recognizing "the marked tension" between the multi-employer citation policy and "the language of § 1910.12[(a)] . . . that '[e]ach employer shall protect the employment and places of employment of each of *his employees*,'" but failing to reach the issue).

---

⁴ On remand from the D.C. Circuit, the Commission found that Anthony Crane Rental was an exposing employer and so failed to reach any conclusion regarding the relationship between § 1910.12(a) and the multi-employer citation policy. *Anthony Crane*, 17 BNAOSHC 2107, n.1, 1995-97 CCH OSHD ¶ 31,251, p. 43,840 n.1 (No. 91-556, 1997).

I agree with my colleagues that the Commission should address this "tension" herein, which has been squarely presented, thoroughly briefed, and comprehensively analyzed during oral argument.

[5] In two prior cases, the Commission declined to rule on the issue of whether § 1910.12(a) is consistent with, or has any affect on, the multi-employer citation policy because the issue had not been briefed. However, in doing so, the Commission made no suggestion that the issue had been foreclosed or previously decided. *See McDevitt Street Bovis, Inc.*, 19 BNAOSHC 1108, 1112-13, 2000 CCH OSHD ¶ 32,204 p. 48,782-83 (No. 97-1918, 2000) (declining to address issue because Commission directed reconsideration on the "very narrow question of whether adverse circuit law precludes application of Commission precedent"; Commissioner Visscher dissented because he "share[d] the D.C. Circuit's concern as to the legal basis for multi-employer liability."); *Access Equip. Sys. Inc.*, 18 BNAOSHC 1718, 1725-26 n.12, 1999 CCH OSHD ¶ 31,821, p. 46,780 n.12 (No. 95-1449, 1999) (declining to address issue because it was neither argued nor briefed by the parties, nor ever considered by any court that adopted the multi-employer citation policy). *Cf., e.g., Underhill Constr. Corp. v. OSHRC*, 526 F. 2d 53, 54 n.3 (2d Cir. 1975) (interpretation of the "effective date" issue raised under § 1910.12(d) not precluded by prior decision of the instant court enforcing identical standard against same employer where the parties previously "neither briefed nor argued the [effective date] issue."). The reluctance of the courts and the Commission in the past to attempt to unravel the perplexing legal maelstrom surrounding this issue may suggest that resolution of the issue may ultimately depend on rulemaking by the Secretary.

Summit argues that the "his employees" phrase of the second sentence of § 1910.12(a) describes a construction employer's duty that is limited to his own employees. The Secretary argues that the first sentence describes a duty that is as broad as the working conditions of all employees on the construction site, effectively ignoring the "his employees" clause of the second sentence. To avoid the dilemma described in the Hindu parable of the blind observers disagreeing about the shape of an elephant after each grasped only his trunk, tusk or leg, I would not limit my perception of possible reasonable interpretations of the scope and application of § 1910.12(a) by focusing on only one clause or sentence. Read together, the two sentences of the regulation require an employer to "protect the employment and places of employment of each of his employees ... by complying with [Part 1926 standards]" applicable to "every employment and place of employment of every employee engaged in construction work." *See* 29 C.F.R. § 1910.12(a). Reading the provision in a manner consistent with the universal interpretation of the general duty clause, it is clear and unambiguous on the face of the regulation that the duty of a construction employer under § 1910.12(a) is owed to protect only "his employees," permitting only an employment-based enforcement scheme. What remains to be determined is whether a "controlling employer" theory of liability, defined by the Secretary as an enforcement scheme grounded in contract or quasi-contract, fits within the full scope and application of this "employment-based" duty under § 1910.12(a) of a construction employer to "protect ... his employees" by complying with the Part 1926 standards.

[6] The Secretary concedes that citation under the language of section 5(a) (1), semantically identical to the second sentence of § 1910.12(a), is limited to exposing employers. *See* OSHA Field Inspection Reference Manual § III.C.2.c. (2)(a) 2 (Sept. 26, 1994) stating: "The employees exposed to the Section 5(a) (1) hazard must be the employees of the cited employer." *See also* Letter to James H. Brown from OSHA Director of Construction Russell B. Swanson (July 25, 2003) (relying upon 1999 Multi-Employer Citation Policy: "[O]nly exposing employers can be cited for General Duty Clause violations."). *See, e.g., Access Equipment*, 18 BNA OSHC at 1724; 1999 CCH OSHD at p. 46,778.

It is clear and unambiguous on the face of the regulation that the duty of a construction employer under § 1910.12(a) is owed to protect only "his employees", permitting only an employment-based enforcement scheme. What remains to be determined is whether a "controlling employer" theory of liability, defined by the Secretary as an enforcement scheme grounded in contract or quasi-contract, fits within the full scope and application of this "employment-based" duty under § 1910.12(a) of a constrction employer to "protect ... his employees" by complying with the Part 1926 standards.

The full scope and application of the construction employer's § 1910.12(a) employment-based duty can be determined by analyzing the agency's original intent when it drafted and began enforcement of the regulation. *See Am. Waterways Operators, Inc. v. United States*, 386 F. Supp. 799, 803-04 (D.D.C. 1974) (construction of act by the agency charged with its administration is accorded great weight if reasonable, but "of higher significance" is the construction of the act by those who participated in the act's drafting and who directly made their views known to Congress), *aff'd*, 421 U.S. 1006 (1975). The first construction of a new act by the body charged with enforcing it is "entitled to more than usual deference accorded an agency's interpretation" of an act or regulation. *See Power Reactor Dev. Co. v. Int'l Union of Elec. Workers*, 367 U.S. 396, 408 (1961) (contemporaneous construction "by the men charged with the responsibility of setting its machinery in motion" is entitled to particular respect); *Nat'l Cable Television Ass'n v. Copyright Royalty Tribunal*, 689 F.2d 1077, 1081 (D.C. Cir. 1982) (affording "more than the usual deference due an agency's interpretation of its enabling act" to Copyright Royalty Tribunal's reading of the Copyright Act because it "was the first construction of a new act by the body charged with the responsibility for setting its machinery in motion."). The regulation's preamble says nothing about the Secretary's original intent. *See* 36 Fed. Reg. 10,466 (May 29, 1971). However, the Secretary did indicate her original intent to limit enforcement of Part 1926 standards, through promulgation of § 1910.12(a), against a class of employers similar to non-creating non-exposing "controlling employers" as defined in the Secretary's current multi-employer citation policy. Her intent is evident in two distinct actions: First, the Secretary excluded the Construction Safety Act duties of the prime (general) contractor, which are parallel to "controlling employer" duties, when she adopted the Construction Safety Act standards as OSH Act standards. Second, the Secretary precluded enforcement of any duties against the general contractor parallel to "controlling employer" duties when she issued the original enforcement guidelines directing citations at multi-employer construction sites.

The first demonstration of the Secretary's original intent is the striking contrast between the language of the second sentence of § 1910.12(a), which imposes an OSH Act duty on construction employers to protect their own employees through compliance with Part 1926 standards, and the language of § 1926.16, which imposed a Construction Safety Act duty on prime (general) contractors to protect the employees of subcontractors through assuring their compliance with the same standards. Indeed, contrary to the assertion of my colleague Commissioner Rogers at footnote 4, § 1910.12(a) was plainly intended as a limit. It was intended to limit the Secretary's discretion to impose under the OSH Act the duty under the CSA of prime (general) contractors at construction sites. The Secretary's intent to limit her discretion to enforce the adopted standards is clear from the dramatic distinction between what the Secretary had written as CSA regulations and standards, and the limited parts she adopted through § 1910.12(a). On May 29, 1971, in accordance with section 6(a) of the OSH Act, the Secretary promulgated § 1910.12. Section 1910.12 adopted as occupational safety and health standards those standards that had been issued under the Construction Safety Act in 29 C.F.R. Part 1518 (now 29 C.F.R. Part 1926).

[7] Former Part 1518 of Title 29, C.F.R. was subsequently redesignated as Part 1926. 36 Fed. Reg. 25,232 (1971). In addition, on February 17, 1972, the Secretary published a *Federal Register* notice clarifying "which regulations had been adopted under OSHA by the May 29, 1972 promulgation [.] [T]he Secretary added a new paragraph to the OSHA regulations entitled 'Construction Safety Act distinguished.'" *Underhill*, 526 F.2d at 56. Specifically, the Secretary added § 1910.12(c).

Through § 1910.12, the Secretary made "the standards (substantive rules)" published in Subpart C of Part 1926 applicable to construction employers in general, but left Subparts A and B of Part 1926 applicable only to federal contractors. Notably, § 1926.16 in Subpart B, expressly imposes liability on the prime (general) contractor for violations by subcontractors. The failure of the Secretary to adopt § 1926.16 through § 1910.12, or to use similar language when describing an employer's duties under the OSH Act in § 1910.12(a), is indication that she intended the duties of an employer (in this case, a prime (general) contractor) under the OSH Act to be more limited than the duties of a prime (general) contractor under the Construction Safety Act. Commissioner Rogers' footnote 4 is correct to the extent she concedes the text of § 1910.12(c) is a clear statement of the reason for the Secretary's failure to incorporate Subparts A and B of 1926, *i.e.*, the contractually-based enforcement scheme of the CSA was inconsistent with the Secretary's construction of an employment-based enforcement scheme under the OSH Act. This concedes the point that if the Secretary had originally intended to exercise discretion under section 5(a) (2) of the OSH Act to cite general contractors at multi-employer construction sites on a contractually-based "controlling employer" theory, she could have done so by adopting the enforcement scheme of § 1926.16, absent the federal contractor predicate, pursuant to section 6(a) of the OSH Act.

[8] For a period of two years after the effective date of the OSH Act, the Secretary had the authority to "promulgate as an occupational safety or health standard any national consensus standard, and any established Federal standard, unless he determines that the promulgation of such a standard would not result in improved safety or health … ." 29 U.S.C. § 655. The definition of "standard" and the phrase "established Federal standards" "make clear that the Secretary intended to adopt, indeed had the statutory authority to adopt, only those provisions in the CSA regulations which require 'conditions, or the adoption or use of one or more practices, means, methods, operations, or processes, reasonably necessary or appropriate to provide safe or healthful employment and places of employment.'" *See Underhill* 526 F.2d at 57.

[9] "Control [constituting an employer as a "controlling employer"] can be established by contract or, in the absence of explicit contractual provisions, by the exercise of control in-practice." OSHA Instruction CPL 2-0.124 at X.E.1. In short, control can be established by contract or quasi-contract.

The second demonstration of the Secretary's original intent is the exclusion of a "controlling employer" basis for citations from the Secretary's original multi-employer citation policy. Almost simultaneously with the promulgation of § 1910.12(a), the Secretary adopted her first Field Operations Manual ("FOM"), originally called the "Compliance Operations Manual."

[10] The FOM was published on May 20, 1971, and § 1910.12(a) appeared in the *Federal Register* nine days later on May 29, 1971.

The FOM published guidelines for OSHA's field officers charged with conducting workplace inspections to enforce, *inter alia,* Part 1926 standards. According to the original FOM, an employer may be cited at a multi-employer construction worksite for exposing its own employees to a hazard, even if it did not create the hazard, p.VII-7 ¶ 10c; or by creating a hazard, even if it did not expose its own employees to that hazard, p.VII-7 ¶ 10b. The simultaneous production by OSHA of two separate documents (the FOM and § 1910.12(a)), both limiting the Secretary's enforcement of Part 1926 standards, cannot be dismissed as a mere unrelated "coincidence." The May 1971 FOM is an indicator of the original intent of the drafters of § 1910.12(a) because (1) the Secretary contemporaneously drafted both § 1910.12(a) and the FOM guidelines for enforcement of the safety and health standards that were adopted by § 1910.12(a); (2) the enforcement guidelines in the FOM could not yet have been influenced by interpretations of the Secretary's citation authority by the newly-formed Occupational Safety and Health Review Commission; and (3) the FOM explicitly included guidelines for citations, *inter alia*, at multi-employer construction sites under the very standards adopted by § 1910.12(a). The original FOM, as well as the amendment to the FOM six months later, both set forth the two duties of an employer at a multi-employer construction worksite: (1) to not expose its employees to a hazard; and (2) to not create violative conditions.

I find it dispositive to a determination of the scope and application of the employer's duty to "protect ... his employees" under § 1910.12(a) that not only did the Secretary fail to adopt the "controlling employer" concept from the CSA when she adopted its body of standards, neither did she in her original enforcement guidelines direct field personnel to cite non-creating, non-exposing, controlling employers at a multi-employer construction worksite. In fact, during the next dozen years of enforcement of the OSH Act — one-third of OSHA's statutory life — official agency guidelines made it clear that the Secretary's power to cite an employer at a multi-employer worksite extended only to creating or exposing employers; controlling employers were never mentioned. *See* OSHA Compliance Operations Manual ("COM") p. VII-7-8 para. 13 (Nov. 15, 1971) (citation of creating or exposing employers); OSHA FOM ¶ 4380.6 (July, 1974) (citation of exposing employers only); OSHA FOM ¶ 4380.6 (Jan. 1, 1979) (same). It was not until 1983, twelve years after the Act's effective date, that OSHA for the first time directed its compliance officers to consider citation of a controlling employer. *See* OSHA FOM ¶ 265 (Apr. 18, 1983). That expansion then was limited to the narrow circumstances where a general contractor is informed of, but fails to abate, a hazard that cannot be abated by any exposing employer.

[11] Significantly, beginning in 1983 and for years after that, OSHA continued to proscribe citations against non-exposing employers except in those limited circumstances when "the exposing employer . . . did not create the hazard; . . . did not have the authority or ability to correct the hazard; made an effort to persuade the controlling employer to correct the hazard; [and] . . . has taken alternative means of protecting employees from the hazard ... ." *See* OSHA FOM ¶ 265 (April 18, 1983).

The Commission will normally defer to the Secretary's reasonable interpretation of a regulation. *See Martin* 499 U.S. at 150. I find the Secretary's original multi-employer citation policy, allowing citation of creating as well as exposing employers, is consistent with § 1910.12(a)'s requirement that an employer must "protect the employment and places of employment of his employees." It also comports with the purpose of the Act.

[12] That is, an employer owes a duty to "his employees" to refrain from creating hazardous working conditions and to prevent "his employees" from being exposed to hazardous conditions to assure that "every working man and woman in the Nation has safe and healthful working conditions." 29 U.S.C. § 651(b).

The creation of violative employment conditions puts all employees at risk. Here I agree with the statement made at oral argument by Summit's amicus that the Secretary recognized when she drafted

§ 1910.12(a) and the original FOM that reasonably predictable exposure generally runs with creation of a hazard.

[13] *See* Transcript of Oral Argument, Argument of Arthur G. Sapper, Esq., on behalf of National Association of Home Builders, Contractor's Association of Greater New York; Texas Association of Builders; and Greater Houston Builders Association, Amici at 17:1–3. Mr. Sapper is correct that while it is possible under current Commission case law for creating employer to create a hazard without exposing any of its own employees to the created hazard, as in the case of excavating an unshored trench, that would appear to be the rare case. *See Smoot Constructio*n, 21 BNA OSHC 1555, 1557, 2005 CCH OSHD ¶ 32,829, p. 52,723 (No. 05-0652, 2006); *Flint Engineering & Constr. Co.*, 15 BNA OSHC 2052,2055, 1993 CCH OSHD ¶ 29,923, p. 40,853 (No. 90-2873, 1992). Indeed, the Commission has affirmed citations against so-called "non-exposing" creating employers when, on closer review, the employees of the creating employer, originally found to have been unexposed, were in fact exposed to the hazard. *See, e.g., Anthony Crane Rental*, 17 BNA OSHC 2107, 1995-97 CCH OSHD ¶ 31,251 (No. 91-556, 1997).

On the other hand, as Chairman Railton adequately explains, deference to OSHA's "checkered history" of reinterpretation of the multi-employer citation policy after 1971 would yield an inconsistent, and therefore unreasonable interpretation of § 1910.12(a). Moreover, the Secretary cannot in this case simply ignore a regulatory limitation on her discretion, albeit that it was voluntarily imposed. As the regulation now exists, the agency has *ab initio* limited its discretion to expand the duties of employers beyond those duties originally intended when the Secretary adopted the Part 1926 standards. Unless and until the agency modifies or repeals the employment-based limitations imposed by the regulation, it may not by simple policy directive remove the substantive limitations on official discretion that now exist. In *Vitarelli v. Seaton*, the Supreme Court held that even agencies with broad discretion must adhere to internally promulgated regulations limiting the exercise of that discretion. *Vitarelli v. Seaton*, 359 U.S. 535, 539-40 (1959). *See also Graham vs. Ashcroft*, 358 F.3d 931, 932 (D.C. Cir 2004) ("It is well settled that an agency, even one that enjoys broad discretion, must adhere to voluntarily adopted, binding policies that limit its discretion." (citing *Padula vs. Webster*, 822 F.2d 97, 10 (D.C. Cir. 1987))).

[14] My colleague Commissioner Rogers suggests interpretation of § 1910.12(a) as requiring employment-related enforcement leads to numerous situations where *no one* on construction site will have *both* the practical ability and legal obligation to ensure safety compliance. This suggestion fails to explain why the exposing construction subcontractor cannot avail it of contractual remedies to ensure non-violative working conditions for its employees.

In this case, it is undisputed that Summit is a non-creating, non-exposing employer. In other words, the only basis for issuing the citation to Summit is that Summit is a "controlling employer" under the Secretary's current multi-employer citation policy. As explained above, however, I find § 1910.12(a) cannot be interpreted to permit citation for a violation of a Part 1926 standard of a controlling employer who neither created the violative conditions nor exposed his employees to the hazard.

Conclusion

For the foregoing reasons, I concur with the Chairman's conclusion that § 1910.12(a) prevents the Secretary from enforcing her current multi-employer citation policy to cite a non-exposing non-creating employer such as Summit, for violation of § 1926.451(g) (1) (vii). Therefore, I join Chairman Railton in vacating the citation.

/s/ _____

Horace A. Thompson, III

Commissioner

Dated: April 27, 2007

ROGERS, Commissioner, dissenting:

By their decision today, my colleagues have reversed over thirty years of Commission precedent that has had the effect of enhancing worker safety on construction worksites with multiple employers. In voting as they have to eliminate the Secretary's ability to cite general contractors under her multi-employer enforcement policy, my colleagues have deprived the Secretary of a very important tool to hold accountable those often in the best position to ensure safety on construction worksites.

The rejection of the multi-employer precedent has at least three additional undesirable results. First, it usurps for the Review Commission the Secretary's policy-making role under the Occupational Safety and Health Act ("the Act"). Second, it trivializes the Secretary's prosecutorial discretion and ability to develop and refine enforcement policies consistent with the Act. Finally, it de-stabilizes a body of law that, while not perfect or totally comprehensive, offers rationality and predictability.

I would uphold the long-standing precedent and continue to recognize the Secretary's authority to cite general contractors under her multi-employer enforcement policy.

Overview — The Multi-Employer Construction Worksite Doctrine

For over thirty years, the Commission has affirmed the validity of the multi-employer construction worksite doctrine. As described by the Commission, this doctrine, rooted in the Act, the principles of the common law, and the realities of the construction workplace, provides that:

> [A]n employer who either creates or controls the cited hazard has a duty under [section] 5(a)(2) of the Act, 29 U.S.C. § 666(a)(2), to protect not only its own employees, but those of other employers "engaged in the common undertaking." *Anning-Johnson Co.*, 4 BNA OSHC 1193, 1199, 1975-76 CCH OSHD ¶ 20,690, p. 24,784 (No. 3694, 1976); *Grossman Steel [& Aluminum Corp.]*, 4 BNA OSHC [1185], 1188, 1975-76 CCH OSHD [¶ 20,691], p. 24,791 [(No. 12775, 1976)]. Specifically, the Commission has concluded that an employer may be held responsible for the violations of other employers "where it could reasonably be expected to prevent or detect and abate the violations due to its supervisory authority and control over the worksite." *Centex-Rooney [Constr.*

Co.], 16 BNA OSHC [2127], 2130, 1993-95 CCH OSHD ¶ 30,621, p. 42,410
[(No. 92-0851, 1994)].

McDevitt Street Bovis, Inc., 19 BNA OSHC 1108, 1109, 2000 CCH OSHD ¶ 32,204,
p. 48,780 (No. 97-1918, 2000) (*McDevitt*).

A. The Act, Commission Precedent, and Circuit Court Precedent All Support the Secretary's Authority to Apply the Multi-Employer Worksite Doctrine

Respondent would have the Commission believe that there is simply no legal authority for the Secretary's use of the multi-employer doctrine and that it was invented
out of whole cloth. Notwithstanding Respondent's view of what the law should look
like, over the last thirty years, this Commission and most of the circuit courts that
have considered the doctrine have repeatedly affirmed the validity of the Secretary's authority and discretion to apply the multi-employer doctrine at construction
worksites. *See McDevitt*, 19 BNA OSHC at 1111-12, 2000 CCH OSHD at p. 48,782.
The Secretary's authority to apply the doctrine under the Act has been repeatedly
affirmed with respect to at least three classes of employers: exposing employers, *see,
e.g., Bratton Corp.*, 6 BNA OSHC 1327, 1978 CCH OSHD ¶ 22,504 (No. 12255,
1978), *aff'd*, 590 F.2d 273 (8th Cir. 1979); *Grossman Steel*, 4 BNA OSHC 1185, 1975-
76 CCH OSHD ¶ 20,691 (No. 12775, 1976); *Anning-Johnson*, 4 BNA OSHC 1193,
1975-76 CCH OSHD ¶ 20,690 (No. 3694, 1976) (consolidated); creating employers,
see, e.g., Beatty Equip. Leasing, Inc., 4 BNA OSHC 1211, 1975-76 CCH OSHD ¶
20,694 (No. 3901, 1976), *aff'd*, 577 F.2d 534 (9th Cir. 1978); and, at issue here, controlling employers (usually general contractors), *see Knutson Constr. Co.*, 4 BNA
OSHC 1759, 1976-77 CCH OSHD ¶ 21,185 (No. 765, 1976), *aff'd*, 566 F.2d 596 (8th
Cir. 1977).

[1] The Chairman states that the Commission and, by implication, the various Circuit Courts, were effectively creating "policy" in upholding the multi-employer doctrine as applied to general contractors.
While the discussion in *Grossman Steel* was characterized as *dictum*, the context was the adjudication
of the Secretary's citation where the Commission was explaining the contours of the Secretary's permissible authority to hold employers liable under the multi-employer doctrine. *Grossman Steel*, 4 BNA
OSHC at 1188-89 n.6, 1975-76CCH OSHD at p. 24,791 n.6. It is the Secretary — the policy maker
— who chooses whether to cite an employer under the doctrine, not the Commission.

The doctrine reflects a valid use of the Secretary's enforcement authority under
the Act. An employer's duties under the Act stem from section 5(a). *Anning-Johnson
Co. v. OSHRC*, 516 F.2d 1081, 1084 (7th Cir. 1975) (employer's duty flows from
section 5(a) (1) and (2)). In particular, section 5(a) (2) states broadly that an employer
"shall comply with … standards," thus indicating a duty to comply with specific
OSHA standards for the benefit of *all* employees on a worksite. *See* 29 U.S.C. §
654(a) (2); *United States v. Pitt-Des Moines, Inc.*, 168 F.3d 976, 982-83 (7th Cir.
1999).

In contrast, under section 5(a) (1), the general duty clause, an employer is required
to "furnish to each of *his employees* employment and a place of employment which

are free from recognized hazards that are causing or likely to cause death or serious physical harm to *his employees.*" 29 U.S.C. § 654(a) (1) (emphasis added). The use of the phrase "his employees" delineates that the general duty imposed by section 5(a) (1) is specifically limited to an employer's own employees. *See Pitt-Des Moines,* 168 F.3d at 982. *See also* S. Rep. No. 1282, 91st Cong., 2d Sess. 9 (1970), *reprinted in* Senate Comm. on Labor and Public Welfare, 92d Cong., 1st Sess., *Legislative History of the Occupational Safety and Health Act of 1970,* at 149; H.R. Rep. No. 1291, 91st Cong., 2d Sess. 21 (1970), Leg. Hist., at 851. *See also Pitt-Des Moines,* 168 F.3d at 983 ("Where Congress includes particular language in one section of a statute but omits it in another section of the same Act, it is generally presumed that Congress acts intentionally and purposefully in the disparate inclusion or exclusion." (quoting *Russello v. United States,* 464 U.S. 16, 23 (1983))); *Marshall v Knutson Constr. Co.,* 566 F.2d 596, 599 (8th Cir. 1977) (*per curiam*) (*Knutson*); *Teal v. E.I. DuPont de Nemours & Co.,* 728 F.2d 799, 804 (6th Cir. 1984) (*Teal*); *Beatty Equip. Leasing, Inc. v. Sec'y of Labor,* 577 F.2d 534, 536-37 (9th Cir. 1978) (*Beatty*).

Moreover, the Secretary's authority under the doctrine is supported by the Act's broad purpose, set forth at section 2(b) of the Act, 29 U.S.C. § 651(b), "to assure so far as possible *every* working man and woman in the Nation safe and healthful working conditions" (emphasis added). *See Pitt-Des Moines,* 168 F.3d at 983; *Knutson,* 566 F.2d at 600 n.7; *Teal,* 728 F.2d at 803; *Beatty,* 577 F.2d at 537; *Brennan v. OSHRC (Underhill Constr. Co.),* 513 F.2d 1032, 1038 (2d Cir. 1975) (*Underhill*). In addition, section 2(b)(1), 29 U.S.C. § 651(b)(1), states that an additional purpose of the Act is to encourage the reduction of hazards to employees "at their places of employment," indicating the Act's focus was on making places of employment safe from work related hazards. *See Pitt-Des Moines,* 168 F.3d at 983; *Underhill,* 513 F.2d at 1038. Thus, "once an employer is deemed responsible for complying with OSHA regulations, it is obligated to protect *every employee* who works in its workplace." *See Pitt-Des Moines,* 168 F.3d at 983 (quoting *Teal,* 728 F.2d at 805 (emphasis added)).

More specifically, both the Commission and the courts have upheld the Secretary's use of her authority under the Act to hold a general contractor liable under the doctrine "for violations it could reasonably have been expected to prevent or abate by reason of its supervisory capacity," because of the general contractor's unique position of control over the construction site and authority to obtain abatement. *See Grossman Steel,* 4 BNA OSHC at 1188, 1975-76 CCH OSHD at p. 24,791. Three circuits have specifically applied the doctrine to cases involving such controlling employers. *See Universal Constr. Co. v. OSHRC,* 182 F.3d 726, 727-32 (10th Cir. 1999) (*Universal*); *R.P. Carbone Constr. Co. v. OSHRC,* 166 F.3d 815, 817-19 (6th Cir. 1998) (*Carbone*); *Knutson,* 566 F.2d at 597-98 (8th Cir. 1977) (Commission's decision that general contractor had duty with respect to subcontractor's safety violations but that, in this case, general contractor lacked sufficient control to be held liable was "reasonable and … consistent with the purpose of the Act."). *See also Bratton Corp. v. OSHRC,* 590 F.2d 273, 276 (8th Cir. 1979) (discussing circuit's previous approval of application of multi-employer doctrine to general contractor in *Knutson*).

Indeed, it is the unique position of the general contractor — whose main function is to supervise the work of subcontractors — that gives it the control to ensure hazard abatement. *See Knutson*, 566 F.2d at 599 (general contractors have "the responsibility and the means to assure that other contractors fulfill their obligations with respect to employee safety where those obligations affect the construction worksite"); *Universal*, 182 F.3d at 730 (as practical matter, general contractor may be only on-site person with authority to compel OSHA compliance); *Carbone*, 166 F.3d at 818 (6th Cir. 1998) (it is presumed that general contractor has enough control over subcontractors to require that they comply with OSHA standards). *See also* Recent Case, *Administrative Law – Occupational Safety & Health Act – On Multiemployer Jobsite, When Employees of any Employer are Affected by Noncompliance with a Safety Standard, Employer in Control of Work Area Violates Act; Employer Not in Control of Work Area Does Not Violate Act, Even If His Own Employees Are Affected, Provided the Hazard is "Nonserious,"* 89 Harv. L. Rev. 793, 797 (1976) (person controlling work area in best position to prevent hazards). As noted in *Universal*, at times it is *only* the general contractor who can ensure that compliance takes place. *Universal*, 182 F.3d at 730. As such, the congressional command in section 5(a)(2) of the Act would be a dead letter unless it also ran to a general contractor with supervisory control over the worksite. It is important to emphasize, however, as I previously pointed out in *McDevitt*, that the general contractor's liability under the doctrine is not without limits. *See McDevitt*, 19 BNA OSHC at 1109 n.3, 2000 CCH OSHD at p. 48,779 n.3 (Rogers, Commissioner, noting that liability of general contractor is based on reasonableness standard and is "far from strict liability"). *See also Knutson*, 566 F.2d at 601 (general contractor's duty depends on what measures are commensurate with its degree of supervisory capacity).

[2] The multi-employer worksite doctrine is also consistent with the common law. The doctrine's focus on control is echoed in the rule set forth at § 414 of the Restatement (Second) of Torts (1965) which states that an employer is liable for the negligence of its contractor where the employer retains control of any part of the work performed by the contractor and fails to exercise that control with reasonable care. *See* Restatement (Second) of Torts § 414 cmt. a (1965).

This view also finds support in the cases, under which general contractors may be subject to various types of direct and vicarious liability. *See, e.g., Ghaffari v. Turner Constr. Co.* 699 N.W.2d 687, 694 (Mich. 2005) (as overall coordinator of construction activity, general contractor is "best situated to ensure workplace safety at the least cost"); *Shannon v. Howard S. Wright Constr. Co.*, 593 P.2d 438, 441-45 (Mont. 1979) (general contractor had duty to provide employees of subcontractors a safe place to work because it retained control over working conditions at site); *Kelley v. Howard S. Wright Constr. Co.*, 582 P.2d 500, 505-06 (Wash. 1978) (general contractor had duty, within scope of control over work, to provide safe place of work); *Funk v. Gen. Motors Corp.*, 220 N.W.2d 641, 646 (Mich. 1974) (holding general contractors liable for worksite safety makes it more likely that subcontractors or general contractor will implement safety precautions; often general contractor is only entity in position to provide expensive safety measures that will protect employees of multiple subcontractors, and subcontractors may be unable to rectify situations).

B. Section 1910.12(a) Does Not Limit the Secretary's Authority to Cite Controlling Employers Under the Act

Notwithstanding this long-standing precedent, my colleagues — like Respondent — now seek to turn back the clock and rewrite history more to their liking. Although the Commission has apparently never viewed it as such over the thirty years it has applied the doctrine, my colleagues now seem to separately suggest that 29 C.F.R. § 1910.12(a) should be viewed as a self-imposed limit on the Secretary's authority under section 5(a)(2) of the Act to utilize the multi-employer policy.

[3] Section 1910.12(a) provides:

> The standards prescribed in part 1926 of this chapter are adopted as occupational safety and health standards under section 6 of the Act and shall apply, according to the provisions thereof, to every employment and place of employment of every employee engaged in construction work. Each employer shall protect the employment and places of employment of each of his employees engaged in construction work by complying with the appropriate standards prescribed in this paragraph.

The rather sparse preamble gives no indication that § 1910.12(a) was at all intended to address multi-employer situations. *See* 36 Fed. Reg. 10,466 (May 29, 1971). Indeed, other than my colleagues' pure speculation, based on a coincidence in timing, there is no evidence that § 1910.12(a) was intended as a limit on an employer's duty to comply with construction standards, a duty which derives directly from section 5(a)(2) of the Act.

[4] My colleague Commissioner Thompson suggests that § 1910.12(c) supports his view that § 1910.12(a) was intended as a limit on an employer's duty to comply with construction standards. In that regard, he contends that is why the Secretary did not incorporate Subparts A and B of Part 1926 (including the provisions of § 1926.16 with respect to the responsibilities of a "prime contractor") as OSHA standards. Rather, my colleague seems to prefer his own speculative reason for the Secretary's action, instead of the reason the Secretary actually articulated in the text of § 1910.12(c) itself: "Subparts A and B have pertinence only to the application of section 107 of the Contract Work Hours and Safety Standards Act (the Construction Safety Act). ... [because certain Construction Safety Act terms and concepts, such as the interpretation of the statutory term 'subcontractor' in §1926.13 have] no significance in the application of the [Occupational Safety and Health] Act, which was enacted under the Commerce Clause and which establishes duties for 'employers' which are not dependent for their application upon any contractual relationship with the Federal Government or upon any form of Federal financial assistance."

My colleague also mischaracterizes the Secretary's explanation of the distinction between the two statutory schemes. Contrary to Commissioner Thompson, the Secretary was not foreswearing consideration of *private* contractual relationships between general contractors and subcontractors for Occupational Safety and Health Act enforcement purposes by the language of § 1910.12(c). Rather, she was indicating that the nexus of jurisdiction under the Occupational Safety and Health Act (unlike with the Construction Safety Act) was not predicated upon a contract *involving the Federal government.* Furthermore, as my colleague concedes by citing the language of CPL 2-0.124 in n.9 the Secretary does not rely solely on a contract to show control in multi-employer situations.

Thus the Secretary's failure to incorporate § 1926.16 as an Occupational Safety and Health Act standard is of no moment, contrary to the suggestions by both of my colleagues. As the Secretary explained, the unincorporated provisions were necessary to address terms and concepts from the Construction Safety Act in light of the fact that the jurisdictional predicate of the Construction Safety Act was a contractual relationship involving the Federal government, but they had no "pertinence" to the Occupational Safety and Health Act, which had a different jurisdictional predicate.

See Universal, 182 F.3d at 728-30. The Commission should not effectively reverse over thirty years of precedent and rewrite history based on rank speculation. Similarly, there is no indication that the multi-employer policy was intended as an interpretation of § 1910.12(a), as my colleagues separately seem to suggest. Rather, as the Commission and the courts have continuously held, the multi-employer policy represents the Secretary's expression of how she intends to exercise her permissible prosecutorial discretion within the parameters allowed by the Act itself. *Limbach Co.*, 6 BNA OSHC 1244, 1245, 1977-78 CCH OSHD ¶ 22,467, pp. 27,080-81 (No. 14302, 1977) (multi-employer policy represents general statement of policy for guidance of inspectors). *See Universal*, 182 F.3d at 730 (Secretary's interpretation of section 5(a) (2) consistent with Act).

Contrary to the suggestion by my colleagues, it is for this same reason that rule-making was not required here because the multi-employer worksite doctrine is not a substantive rule, but merely an interpretation of the OSH Act and recognition of the obligations already contained therein. *See Universal*, 182 F.3d at 728 n.2 (employer's position that rulemaking was required before applying multi-employer worksite doctrine "clearly is incorrect"); *Limbach Co.*, 6 BNA OSHC 1245, 1977-78 CCH OSHD at pp. 27,080-1 (multi-employer worksite doctrine is not substantive rule). Accordingly, given the case law, there was no need for the Secretary to initiate a rulemaking merely to respond to *dicta* in court and Commission decisions. Furthermore, Summit was on ample notice of its possible liability because the doctrine is well-established and has been in existence for many years. *See Universal*, 182 F.3d at 728 n.2 (noting doctrine's long history).

To the extent the Secretary has clarified the details of the policy over the years, those clarifications merely reflect adjustments in how the Secretary has chosen to exercise her permissible prosecutorial discretion, within the bounds of the Act and informed by her experiences in enforcing the Act. After all, such policy guidelines "'merely announce [the Secretary's] tentative intentions for the future, leaving himself free to exercise his informed discretion.'" *See Sec'y of Labor v. Twentymile Coal Co.*, 456 F.3d 151, 159 (D.C. Cir. 2006) (citation omitted) (*Twentymile*). As the Secretary correctly points out, to some extent, the Secretary has even altered the application of her policy in response to decisions of the Commission. Resp. Br. for the Sec'y of Labor, Summit Contractors, Inc., Docket No. 03-1622, at p. 26 n.14. *See also* 41 Fed. Reg. 17,639 (Apr. 27, 1976) (Secretary discusses evolution of multi-employer case law); Recent Case, *Administrative Law — Occupational Safety & Health Act — On Multiemployer Jobsite, When Employees of any Employer are Affected by Noncompliance with a Safety Standard, Employer in Control of Work Area Violates Act; Employer Not in Control of Work Area Does Not Violate Act, Even If His Own Employees Are Affected, Provided the Hazard is "Nonserious,"* 89 Harv. L. Rev. 793, 797 n.34 (1976) (discussing possible changes in Secretary's multi-employer policy in response to Commission and court decisions). It is highly ironic for my colleagues to use those Commission-driven changes against her.

In claiming that the Secretary has been inconsistent because her policy has evolved over the years, my colleagues have a fundamental misunderstanding of enforcement guidelines and seek to impose on the Secretary an inappropriate strait-jacket that would deprive her of the ability to make adjustments in her enforcement policies. My colleagues even suggest that the Secretary recognized she lacked the authority to cite controlling employers because she did not seek to cite them in her first enforcement policy. There is a significant difference between an agency not exercising the full scope of its statutory authority for reasons of enforcement discretion and an agency explicitly recognizing that it lacks statutory authority. I am not aware that the Secretary has ever taken the position that she lacked the authority to cite controlling employers. *See Kaspar Wire Works, Inc. v. Sec'y of Labor*, 268

F.3d 1123, 1131 (D.C. Cir. 2001) (Secretary had never taken position that she lacked authority to issue per-instance penalties).

In any event, courts have recognized the danger of "transmogrify[ing]" written guidelines that aid an agency's exercise of discretion into binding norms, as my colleagues inappropriately seek to do here. *See Comty. Nutrition Inst. v. Young*, 818 F.2d 943, 949 (D.C. Cir. 1987). Unfortunately, this "pernicious" practice of denying the Secretary her lawful prosecutorial discretion, second guessing her legitimate policy choices, and "substitut[ing] its views of enforcement policy for those of the Secretary" is becoming all too common.

[5] *See Beverly Healthcare-Hillview*, 21 BNA OSHC 1684, 1688, 2006 CCH OSHD ¶ 32,845, pp. 52,838-39 (No. 04-1091, 2006) (consolidated) (Rogers, Commissioner, partial concurrence and dissent), *appeal docketed*, No. 06-4810 (3d Cir. Nov. 17, 2006); *Cagle's Inc.*, 21 BNA OSHC 1738, 1746, 2006 CCH OSHD ¶ 32,846, p. 52,849 (No. 98-0485, 2006) (Rogers, Commissioner, partial concurrence and dissent), *appeal docketed*, No. 06-16172 (11th Cir. Nov. 28, 2006); *U.S. Postal Serv.*, 21 BNA OSHC 1767, 1776 (No. 04-0316,2006) (Rogers, Commissioner, dissent).

See Twentymile, 456 F.3d at 158. Indeed, were my colleagues to have their way, the Secretary would be required to embark on a series of never-ending rulemakings merely to maintain her statutory authority.

Even assuming, *arguendo*, that my colleagues are right about § 1910.12(a) as having some relevance to the Secretary's multi-employer citation authority under section 5(a)(2) of the Act, I would read paragraph (a), taken as a whole, as ambiguous. The first sentence makes clear that the construction standards apply to the "place of employment of every" construction employee and is similar in breadth to section 5(a)(2) of the Act. Thus, to the extent a general contractor exercises control over such a place of employment (and recognizing in some cases that *only* the general contractor can ensure safety compliance), it is reasonable to read the regulation as imposing on that controlling general contractor a duty to comply with the specific construction standards which apply to that place of employment. Read in this context, the second sentence merely emphasizes the primary responsibility of the direct employer to comply with the appropriate standards, but it is not drafted as a limitation, does not by its terms impose the duty exclusively on the direct employer (i.e., it does not say "[e]ach employer shall protect the employment and places of employment of *only* each of his employees …"), and is not inconsistent with the more generalized duty imposed by the first sentence and its statutory analog, section 5(a)(2) of the Act.

[6] My colleague, Commissioner Thompson, cites *Vitarelli v. Seaton*, 359 U.S. 535 (1959), to support his argument that the Secretary somehow violated a self-imposed limitation by applying her multi-employer policy to controlling contractors. The *Vitarelli* case is easily distinguishable. In *Vitarelli*, the Department of Interior specifically bound itself to certain procedural requirements for the dismissal of employees on security grounds and relied upon those requirements as authority in its actual dismissal notice of Vitarelli. *Id.* at 538-40. The Court, not surprisingly, found the agency was bound by its own internal procedural requirements for the dismissal of an employee on security grounds. *Id.* Here, in contrast, despite my colleague's speculation, there is no evidence that the Secretary intended §1910.12(a) to have relevance to the multi-employer question before us; even if § 1910.12(a) did have relevance, it does not operate as a limit on her authority under section 5(a)(2) of the Act to cite controlling contractors; and, in any event, in citing employers under the multi-employer policy, the Secretary does not rely on § 1910.12(a) as her authority.

My colleagues suggest that § 1910.12(a) should be interpreted in a manner similar to section 5(a) (1) of the Act, in light of the reference in the second sentence of the regulation to "each of his employees." But they appear to overlook the fact that section 5(a)(1) of the Act lacks the broad first sentence — similar in breadth to section 5(a)(2) — which appears in § 1910.12(a) of the regulation. Indeed, in that respect, § 1910.12(a) is more akin to sections 5(a) (1) and 5(a) (2) of the Act *combined*. Accordingly, the two sentences of § 1910.12(a) must be read together and in the context of the duty imposed by section 5(a) (2) of the Act.

Thus, to the extent that § 1910.12(a) might be viewed as having some relevance to the Secretary's multi-employer citation authority under the Act, I would defer to the Secretary's reasonable and longstanding interpretation of § 1910.12(a) as permitting her to cite controlling contractors under the multi-employer doctrine. *See Martin v. OSHRC*, 499 U.S. 144 (1991).

[7] To be sure, the D.C. Circuit has questioned the validity of the multi-employer worksite doctrine, although it has scrupulously avoided reaching the issue. *See IBP, Inc. v. Herman*, 144 F.3d 861, 865-66 (D.C. Cir. 1998) (non-construction case). That circuit has observed that the doctrine has a "checkered history," and that it "see[s] tension" between the doctrine and the language of the statute and regulations." *Id.* at 865 & n.3. *See also Anthony Crane Rental, Inc. v. Reich*, 70 F.3d 1298, 1306 (D.C. Cir. 1995) (noting that "it is not clear to us that the multi-employer doctrine is consistent with the Secretary's own construction industry regulation, 29 C.F.R. § 1910.12(a)" because "the language of § 1910.12, which says that '[e]ach employer shall protect the employment and places of employment of each of *his employees*' … is in marked tension with the multi-employer doctrine we are asked to apply here." (emphasis in original)). However, since the D.C. Circuit has never reached the validity of the doctrine, there is no legal basis to overturn over thirty years of our own precedent based on concerns expressed in *dicta* with respect to issues the court did not address. Furthermore, in *Universal*, the Tenth Circuit considered the concerns expressed by the D.C. Circuit in *Anthony Crane* and *IBP*, but did not view them as sufficient to cast aside the Secretary's interpretation. *Universal*, 182 F.3d at 731. Thus, it appears that the Tenth Circuit did not view § 1910.12(a) as a bar to the Secretary's exercise of the multi-employer policy.

The Chairman suggests that both the D.C. Circuit, in *Anthony Crane Rental*, and the First Circuit, in *Secretary v. Simpson, Gumpertz & Heger, Inc.*, 3 F.3d 1 (1st Cir. 1993), had found the meaning of § 1910.12 "plain" with respect to its effect on the Secretary's multi-employer policy. In fact, both circuits were addressing a *different issue* — the meaning of "places of employment" for the purpose of determining whether the worksites were "places of employment" for the respective respondents' employees which the respondents had a duty to protect based on their employees' presence on the site. *See Anthony Crane Rental*, 70F.3d at 1303, *Simpson, Gumpertz & Heger*, 3 F.3d at 5. Of course, despite the Chairman's implication, the D.C. Circuit did not address the validity of the multi-employer policy in *Anthony Crane Rental*.

The Fifth Circuit alone has seemingly rejected the theory of multi-employer liability, although it has not reviewed a relevant Commission decision since the Commission accepted the doctrine. *Melerine v. Avondale Shipyards, Inc.*, 659 F.2d 706, 712 (5th Cir. Unit A 1981) (tort case). *See also Southeast Contractors, Inc. v. Dunlop*, 512 F.2d 675 (1975) (percuriam); *McDevitt*, 19 BNA OSHC at 1110, 2000 CCH OSHD at p. 48,781 (No. 97-1918, 2000). But that is the clear minority view. *See Universal*, 182 F.3d at 731 (no Fifth Circuit case "persuasively explain[s] the basis for rejection of the [multi-employer] doctrine."); *United States v. Pitt-Des Moines, Inc.*, 168 F.3d 976, 983 (7th Cir. 1999) ("*Melerine* … doesn't persuade us that the doctrine is an inappropriate by-product of the Act's language or purpose.").

While the Fourth Circuit had affirmed an early Commission decision rejecting the Secretary's use of the multi-employer doctrine, *Brennan v. Gilles & Cotting, Inc.*, 504 F.2d1255 (4th Cir. 1974), it did so based on deference to the Commission. As the D.C. Circuit has pointed out, *IBP, Inc.*, 144 F.3d at 865-66 n.3, that decision has effectively been overruled by the Supreme Court's subsequent decision in *Martin v. OSHRC*, 499 U.S. 144(1991).

CONCLUSION — WE SHOULD UPHOLD LONG-STANDING PRECEDENT

As discussed above, and as recognized by many courts, it is often *only* the general contractor who can ensure safety and OSHA compliance at a construction site populated by an array of subcontractors, particularly in the context of a dispute among subcontractors. By freeing the general contractor of any safety compliance obligations as the controlling employer, my colleagues have ensured that there will be numerous situations where *no one* on a construction site will have *both* the practical ability and legal obligation to ensure safety compliance. With respect to those situations, they are reading section 5(a)(2) out of the Act and are creating a dangerous "no-man's land" of safety non-compliance.

For the reasons stated, I would not rewrite thirty years of history. I would maintain our long-standing precedent and continue to hold that the Secretary has the lawful authority to apply the multi-employer doctrine to general contractors at construction worksites. In the context of over thirty years of precedent, I cannot join those who would reverse that precedent and further straitjacket the Secretary's lawful exercise of prosecutorial discretion in protecting worker safety.

I respectfully dissent.

/s/ _____

Thomasina V. Rogers

Commissioner

Dated: April 27, 2007

SECRETARY OF LABOR, Complainant v. SUMMITT CONTRACTORS, INC., Respondent	(OSHRC Docket No. 03-1622)

Appearances:

Robert C. Beal, Esquire, Office of the Solicitor, U.S. Department of Labor, Dallas, Texas

   for Complainant

Robert E. Rader, Esquire, Rader & Campbell, Dallas, Texas

   for Respondent

Before: Administrative Law Judge Ken S. Welsch

DECISION AND ORDER

Summit Contractors, Inc. (Summit), contests a serious citation for violation of 29 C.F.R. § 1926.451(g) (1) (vii) issued on August 25, 2003, by the Occupational Safety and Health Administration (OSHA). The citation alleges that Summit, as the general contractor for the construction of a college dormitory in Little Rock, Arkansas, failed to ensure that employees of a masonry subcontractor were utilizing fall protection while working on scaffolds in excess of 12 feet above the ground. The citation proposes a penalty of $4000.

The hearing was held in Little Rock, Arkansas, on January 27, 2004, and the record remained opened until March 1, 2004, for the inclusion of two trial depositions. Jurisdiction and coverage are stipulated (Tr. 4).

Summit does not dispute the existence of the violative conditions as described in the citation. Summit asserts, however, that as general contractor who neither created nor had employees exposed to the fall hazard, it cannot be cited for the violation. Summit argues that the multi-employer worksite doctrine is invalid, and that it lacked sufficient control of the jobsite as general contractor to prevent or abate the violation.

For the reasons discussed, Summit's arguments are rejected. The serious citation is affirmed and a penalty of $2000 is assessed.

Background

Summit is in business as a general contractor overseeing construction projects. Its corporate office is located in Jacksonville, Florida. Summit employs approximately 180 employees. It employs no construction trade employees (Tr. 258-259).

In December 2002, Summit and Collegiate Development Services, LP, the developer for the property owner, entered into a construction contract to build new student housing for Philander Smith College in Little Rock, Arkansas. Summit contracted to serve as general contractor and assumed the construction responsibilities for the project (Exhs. C-7, JT -2, pp. 5–6). The proposed dormitory consisted of a three-story building with 134 units comprising approximately 90,000 square feet (Exh. Jt-1, p. 42; Tr. 142, 192).

To perform the construction work, Summit contracted approximately fifteen subcontractors and nine vendors (Exh. C-6; Tr. 104). Summit's project superintendent, Jimmy Guevara, and three assistant project superintendents worked at the project coordinating the vendors, scheduling the work of the various subcontractors, and ensuring that a subcontractor's work was performed in accordance with the subcontract agreement (Tr. 101-102, 110-111). Summit's project manager Jon Lee visited the site twice a month to check on the progress and schedules (Exh. Jt-1, p. 5; Tr. 102-103).

Superintendent Guevara testified that thirty to forty subcontractors worked on the project (Tr. 104). However, Summit's list of subcontractors shows only fifteen subcontractors and nine vendors (Exh. C-6).

The site clearing and foundation preparation work began in January 2003. The framing work commenced on April 28, 2003 (Exh. Jt-1, p. 19; Tr. 103, 193). Summit's contract with the developer required Summit to complete the project in 150 days. Otherwise, Summit was subject to paying liquidated damages (Exh. Jt-1, p. 19). The dormitory was completed on schedule on August 15, 2003 (Tr. 193).

Summit subcontracted All Phase Construction, Inc. (All Phase), to complete the exterior brick masonry work for the new building (Exh. Jt-1, p. 20, C-8; Tr. 104). All Phase started the brick work on May 23, 2003 (Exh. C-9). To access the building's exterior, All Phase leased scaffolds which it installed and moved as its brick work progressed around the building (Tr. 202).

Summit's project superintendent Guevara testified that prior to OSHA's inspection, he had observed All Phase employees on the scaffold without using personal fall protection. The scaffold also lacked guardrails. Guevara stated that he told the All Phase superintendent of the lack of fall protection and advised them to correct it (Tr. 116, 119-120). According to Guevara, All Phase would implement fall protection until the scaffold was moved to another location when employees again would work without fall protection. Guevara explained that this occurred two or three times prior to the OSHA inspection (Tr. 129).

On June 18, 2003, OSHA Compliance Officer (CO) Richard Watson, while driving to another inspection site, observed and photographed employees on a scaffold at the student housing project laying bricks approximately twelve feet above the ground without fall protection (Exh. C-1; Tr. 33). After receiving permission from his office to conduct an inspection, CO Watson returned to the project on June 19, 2003. He again observed and photographed employees on a scaffold laying bricks without fall protection (Exhs. C-2, C-3; Tr. 37). Upon entering the project, CO Watson was informed by project superintendent Guevara that the masonry contractor was All Phase (Exh. C-9; Tr. 169). However, Summit would not permit CO Watson to conduct a walk-around inspection until its safety officer who lived in Jacksonville, Florida, was present (Tr. 38, 40). OSHA agreed to wait, and the walk-around inspection was performed on June 24, 2003. However, All Phase was not on site (Exh. C-9; Tr. 38-40, 233-234).

As a result of CO Watson's observations on June 18 through 19, 2003, Summit received a serious citation for violation of 29 C.F.R. § 1926.451(g) (1) (vii). All Phase also received a citation which included an alleged violation of § 1926.451(g) (1) (vii) and a proposed penalty of $2500 (Tr. 79, 81).

DISCUSSION

It is undisputed that Summit did not create, nor were its employees exposed to, the lack of fall protection on the scaffold (Tr. 27). The scaffold was leased and erected by All Phase. There is no evidence that Summit or other subcontractors ever used the scaffold or that their employees were exposed to a fall hazard. The exposed employees were employed by All Phase, a subcontractor hired by Summit (Tr. 79, 83, 202-203). CO Watson observed All Phase employees on the scaffold without fall protection on two successive days (June 18 and 19, 2003).

Summit does not dispute that the cited standard, § 1926.451(g) (1) (vii), applies to the scaffolding conditions existing at the construction site or that All Phase's employees were exposed to a fall hazard of 12 feet and 18 feet without personal fall protection or a guardrail system on the scaffold (Tr. 26, 37). Section 1926.450(g) (1) (vii) applies to all scaffolds used in workplaces covered by the construction industry standards. See 29 C.F.R. § 1926.450(a).

During the hearing, Summit speculated that the exposed employees were independent contractors hired by All Phase (Tr. 26, 79). However, this was not established or argued in its post-hearing brief. Therefore, the exposed employees are considered employed by All Phase. This was CO Watson's understanding during his inspection (Tr.47). Also, the subcontract agreement requires Summit to approve in writing the hiring of contractors by subcontractors (Exh.C-8, Art. 8). Summit offered no written approvals.

Summit stipulates that it was aware that All Phase's employees were not utilizing personal fall protection and that the scaffold lacked guardrails (Tr. 24, 48, 116). The lack of fall protection was open and obvious and in plain view from the street and Summit's jobsite trailer (EHz. C-1, C-2; Tr. 33, 36-37, 46). Summit's superintendent inspected the jobsite once or twice each day, and his three assistants were on site overseeing the subcontractors' work. They were on the jobsite on June 18 and 19, 2003, at the time of the alleged violations (Tr. 137-138, 140, 200). On June 19, the superintendent had walked the jobsite prior to CO Watson's arrival (Tr. 140).

Summit's superintendent had observed the same violations several times earlier by All Phase and had asked All Phase to correct the violations (Tr. 119-120, 122).

Also, there is no dispute that the superintendent knew the scaffolding fall protection requirements since he had previously received OSHA training (Tr. 123).

Based on these undisputed stipulations, if Summit is found to have sufficient authority and control to prevent or abate the scaffold violation under the multi-employer worksite doctrine, a serious violation of § 1926.451(g) (1) (viii) is supported by the record.

In its answer, Summit asserts an infeasibility defense. Other than its lack of control argument, Summit doesn't argue, and the record does not support, that fall protection for employees on the scaffold was technically or economically infeasible. Also, there is no showing that the effect of implementing measures against All Phase to assure compliance would adversely affect Summit's financial condition. Summit's affirmative defense of infeasibility is rejected.

MULTI-EMPLOYER WORKSITE DOCTRINE

Under the multi-employer worksite doctrine, an employer, including a general contractor who controls or creates a worksite safety hazard, may be liable for violations of the Occupational Safety and Health Act (Act) even if the employees exposed to the hazard are solely employees of another employer. A general contractor may be held responsible on a construction site to ensure a subcontractor's compliance with safety standards, such as fall protection requirements, if it can be shown that the general contractor could reasonably be expected to prevent or detect and abate the violative condition by reason of its supervisory capacity and control over the worksite. *Centex-Rooney Construction Co.*, 16 BNA OSHC 2127, 2129-2130 (No. 92-0851, 1994).

As it has argued in earlier cases,

Summit has a history of cases before Commission judges on the issue of the multi-employer worksite doctrine. Summit's alleged violations for the most part were vacated on the basis of lack of knowledge and, in one case, lack of control. *See Summit Contractors, Inc.*, 20 BNA OSHC 1118 (No. 01-1891, 2003) (ALJ Spies) (after rejecting arguments that the multi-employer worksite doctrine contravenes the Act and that Summit lacked sufficient control, the citation was vacated because of Summit's lack of knowledge of the unsafe condition); *Summit Contractors, Inc.*, 19 BNA OSHC 2089 (No. 01-1614, 2002) (ALJ Schoenfeld) (vacated citation based on Summit's lack of sufficient authority to control the manner a subcontractor complied with the safety requirements and finding that the authority to terminate a subcontract is not a sufficient basis to hold the general contractor responsible for the subcontractor's violations); *Summit Contractors, Inc.*, 19 BNA OSHC 1270 (No. 00-0838, 2000) (ALJ Spies) (as general contractor and controlling employer who had two employees exposed, Summit had the responsibility to comply with the fire extinguishing standard); *Summit Contractors, Inc.*, 18 BNA OSHC 1861 (No. 98-1015, 1999) (ALJ Spies) (the citation was affirmed because Summit retained a safety consultant to advise it of potential safety hazards and issued fines to subcontractors for safety violations); and *Summit Contractors, Inc.*, 17 BNA OSHC 1854 (No. 96-55, 1996) (ALJ Welsch) (the citation was vacated because Summit as general contractor lacked knowledge of the hazard).

Summit challenges the multi-employer worksite doctrine. In this case, Summit has moved for declaratory relief asserting that there is no basis in the Act and regulations for the multi-employer worksite doctrine. However, since the doctrine is based on Review Commission precedent, it is not appropriate for a Commission judge to engage in such declaratory relief. Also, the Commission has already rejected many of the arguments raised by Summit and discussed the basis for the doctrine. *Access Equipment Systems, Inc.*, 18 BNA OSHC 1718, 1723-1724 (No. 95-1449, 1999).

The multi-employer worksite doctrine, as applied by the Review Commission, has been accepted in one form or another in at least six circuits and rejected outright in only one. See *U.S. v. Pitt-Des Moines, Inc.*, 168 F.3d 976 (7th Cir. 1999); *R.P. Carbone Constr. Co. v. OSHRC*, 166 F.3d 815 (6th Cir. 1998); *Beatty Equip. Leasing, Inc. v. Secretary of Labor*, 577 F.2d 534 (9th Cir. 1978); *Marshall v. Knutson Constr. Co.*, 566 F.2d 596 (8th Cir. 1977); *Brennan v. OSHRC* 513 F.2d 1032 (2d Cir. 1975); and *Universal Construction Company Inc. v. OSHRC*, 182 F.3d 726 (10th Cir. 1999). But *see Southeast Contractors, Inc. v. Dunlop*, 512 F.2d 675 (5th Cir. 1975).

In this case, Summit's office is located in Jacksonville, Florida, and the worksite at issue was in Arkansas. These states are located in the Eleventh and Eighth Circuits where this case could be appealed.

"Where it is highly probable that a Commission decision would be appealed to a particular circuit, the Commission has generally applied the precedent of that circuit in deciding the case-even though it may differ from the Commission's precedent." *Kerns Brothers Tree Service,* 18 BNA OSHC 2064, 2067 (No. 96-1719, 2000).

The Eighth Circuit has accepted the doctrine. *Marshall v. Knutson Constr. Co.*, 566 F.2d 596 (8th Cir. 1977) (an employer who has control over an entire worksite must take whatever measures are "commensurate with its degree of supervisory capacity"). The Eleventh Circuit has not had an opportunity to rule on the doctrine. Although several employers have argued that the Eleventh Circuit has rejected the multi-employer worksite doctrine based on earlier Fifth Circuit case law, the Review Commission has ruled otherwise. *McDevitt Street Bovis, Inc.*, 19 BNA OSHC 1108, 1111-1112 (No. 97-1918, 2000) (case law decided by the former Fifth Circuit rejecting the multi-employer worksite doctrine does not preclude application of the Review Commission's precedent regarding the doctrine in the Eleventh Circuit). Additionally, Summit could appeal to the D.C. Circuit. Although the D.C. Circuit has questioned the doctrine's validity in a manufacturing plant in *IBP, Inc. v. Herman*, 144 F.3d 861 (D.C. Cir. 1998), it did not specifically reject the doctrine. The multi-employer worksite doctrine is still viable before the Review Commission. *McDevitt Street Bovis, Inc., Id.*

In *IBP, Inc.*, 17 BNA OSHC 2073 (No. 93-3059, 1997), the Review Commission held the owner of a plant responsible for LOTO violations of an independent contractor while cleaning meat processing machinery. The Commission found that the plant owner had supervisory authority over the worksite; it had contractual authority to bar entry to the independent contractor and, although its employees were not exposed, it owned the machinery which gave it responsibility to do what was reasonably expected to abate violations. The D.C. Court of Appeals reversed the Commission finding that the Secretary had not shown sufficient control. A contract provision allowing the owner to terminate the contract was not sufficient to show control. Control and preventability are the keys to the applicability of the doctrine, not whether the employer is a general contractor.

In this case, Summit asserts that OSHA's Directive CPL 2-0.124 ("Multi-Employer Citation Policy") issued by the Secretary on December 10, 1999, is not enforceable because it is contrary to the OSHA's published regulation at 29 C.F.R. §1910.12.

Section 1910.12(a) provides in part that "[e]ach employer shall protect the employment and places of employment of each of *his employees* engaged in construction work by complying with the appropriate standards prescribed in this paragraph" (emphasis added). Summit argues that because § 1910.12(a) places safety responsibility on the employer for its own employees engaged in construction work, OSHA's multi-employer worksite citation policy in OSHA Directive CPL 2-0.124 ("Multi-Employer Citation Policy") which permits citing a non-exposing and non-creating employer, is unenforceable.

Summit's argument regarding OSHA's multi-employer citation policy is rejected. The citation at issue alleges Summit violated § 1926.451(g) (1) (vii). In deciding this case, it is the applicable Review Commission precedent which determines if Summit, as a general contractor, is responsible for the alleged scaffold violation and not an internal guideline used by an OSHA compliance officer. The Review Commission does not consider an OSHA CPL or other internal directives as binding on the Commission, and may only look to them as an aid in resolving interpretations under the Act. The CPL does not confer procedural or substantive rights on employers and does not have the force and effect of law. *Drexel Chemical Company*, 17 BNA OSHC 1908, 1910, n. 3 (No. 94-1460, 1997). Also, Summit's reading of § 1910.12 is too narrow. The standard does not prohibit application of an employer's safety responsibility to employees of other employers.

SUMMIT'S CONTROL OF THE WORKSITE

Summit concedes that it knew of All Phase's repeated failure to provide fall protection or require employees to utilize personal fall protection while on a scaffold more than ten feet above the ground (Tr. 24, 26, 48, 116). The Secretary concedes that Summit was not a creating or exposing employer (Tr. 27, 79).

The issue in dispute is whether Summit had sufficient supervisory authority and control of the student housing worksite to prevent and abate the violative condition which exposed All Phase's employees to a fall hazard.

Summit's argument that Judge Shoenfeld's decision involving Summit, 19 BNA OSHC 2089 (No. 01-1614, 2002), finding a lack of control is res judicata is rejected. Judge Shoenfeld's decision is a non-binding, un-reviewed decision of a Commission judge based on the facts in his case. Judge Shoenfeld's decision is not a final adjudication on all issues. Leone Construction Co., 3 BNA OSHC 1979 (No. 4090, 1976). Also, since res judicata was not pled until Summit's post-hearing brief, the affirmative defense was waived. 29 C.F.R. 2200.34(b) (4).

To determine whether a general contractor such as Summit is a controlling employer for purposes of multi-employer responsibility, the general contractor must be in a position to prevent or correct a violation or to require another employer to prevent or correct the violation. Such control may be in the form of an explicit or implicit contract right to require another employer to adhere to safety requirements and to correct violations the controlling employer discovers.

Summit maintains that it is company policy not to be responsible for the safety of a subcontractor's employees or for any OSHA requirements placed on subcontractors (Exh. Jt-2, pp. 4-5; Tr. 42). This policy is reflected in Summit's subcontract agreements and its safety manual. Summit's subcontract with All Phase, as well as with its other subcontractors, provides that:

All parties hereby agree that SUBCONTRACTOR has sole responsibility for compliance with all of the requirements of the Occupational Safety and

Health Act of 1970 and agrees to indemnify and hold harmless CONTRAC-TOR against any legal liability or loss including personal injuries which CON-TRACTOR may incur due to SUBCONTRACTOR's failure to comply with the above referenced act. In the event any fines or legal costs are assessed against CONTRACTOR by any governmental agency due to noncompliance of safety codes or regulations by SUBCONTRACTOR, such cost will be deducted, by change order, from SUBCONTRACTOR's Subcontract amount (Exh. C-8, Attach A, section 4).

Summit's safety manual provides that:

[b]ecause the subcontractors and sub-subcontractors are each separate employ-ers, Summit is not legally responsible for their compliance with OSHA. Nor would it be feasible or reasonable for Summit to assume responsibility for any subcontractor's compliance with OSHA because Summit has no control over a subcontractor's hiring, training or disciplinary practices (Exh. C-5, p. DOL 18).

If a subcontractor's safety violations are observed, the safety manual provides that:

If, during the normal course of operations, an open and obvious hazard is observed, Summit will contact the appropriate trade supervisor/foreman and ask that they correct the hazard. Summit encourages all trades to emphasize safety while they are on the project. In cases where questions arise regarding some safety or health issue, Summit's Director of Safety will, if asked, act as a resource in an attempt to assist a subcontractor with their question by pro-viding copies of relevant standards or other helpful information (Exh. C-5, p. DOL 19).

Regardless of its stated company policy,

It is noted that an employer cannot contract away its responsibilities under the Act. *Pride Oil Well Service*, 15BNA OSHC 1809 (No. 87-692, 1992).

Summit, pursuant to the contract with the owner's representative in this case, agreed to be responsible for the safety of subcontractors' employees. In its contract with Collegiate Development, the owner's representative, Summit, agreed to "indemnify and hold harmless the Design-Builder, the Owner and their respective agents, servants and employees from and against claims, dam-ages, losses and expenses, including but not limited to, attorneys' fees arising out of or resulting from performance of the Work, provided that such claim, damage, loss or expense is attributable to ...

(e) liability imposed upon any Indemnified party directly or indirectly by Contractor's failure or the failure of any of Contractor's or a Subcontractor's employees to comply with any Occupational Safety and Health Administration (or related statutes) violations and any penalties including enhancements, resulting in whole or in part from Contractor's acts or omissions . ..." (Exh. C-7, Section XII).

Summit also accepted responsibility "for initiating, maintaining and supervising all safety precautions and programs in connection with the performance of the Contract" and to "take reasonable precautions for safety of, and shall provide reasonable protection to prevent damage, injury or loss to: a. employees on the Work and other persons who may be affected thereby" (Exh. C-7, Section XIII, para. A.1 and para. B.1). Summit agreed to "comply with applicable laws, ordinances, rules, regulations and lawful orders of public authorities bearing on the safety of persons or property or their protection from damage, injury or loss (Exh. C-7, Section XIII, para. B.2). Summit acknowledged responsibility "for all general conditions work such as, but not by way of limitation, hoists, safety equipment, and portable toilets" (Exh. C-7, Section I, para. F).

Based on these contractual obligations with the property owner's representative, Summit explicitly agreed to protect the safety of subcontractors' employees. The plain, unambiguous language of the agreement provides that Summit had to protect all "employees on the Work" and "other persons who may be affected thereby." Summit's former vice-president defined the term "Work" to include any type of construction work performed by any worker of a subcontractor (Exh. Jt-2, p. 16).

Despite Summit's former vice-president's testimony that he did not see anything in the contract regarding Summit's responsibility for subcontractor safety and, if he had, he would have tried to negotiate changes; the clear language of the contract provides otherwise (Exh. Jt-2, p. 7).

If Summit failed to comply with these safety obligations, the owner had the right to terminate the agreement or to pursue other remedies (Exh. C-7, Section XXVII, para. A and C).

When Summit attempts to avoid responsibility for the safety of subcontractors' employees on a given construction project, it attaches an addendum to the general contract with the owner. Summit normally uses a standard American Institute of Architects (AIA) contract form which holds the general contractor responsible. To the general contract, Summit attaches an addendum that expressly negates the responsibility for the safety of a subcontractor's employees (Tr. 247, 266-267).

However, for the Philander Smith College construction project, Summit did not avoid such responsibility and no addendum was attached to the general contract. Summit admits that the agreement with the owner's representative was different from Summit's typical agreement (Tr. 247). The agreement for this project was based on a form provided by the owner's representative. Summit did not write the contract or negotiate any changes (Exh. Jt-2, p. 5-6, 13).

Based on its agreement with the owner's representative, Summit contracted with various subcontractors, including All Phase, to perform the actual construction work for the new student housing. Summit used its standard subcontract agreement form which it required all subcontractors to sign (Exh. Jt-1, pp. 49-50).

The subcontract with All Phase also establishes Summit's requisite control over the safety of All Phase's workers. In Article 6 of the subcontract with All Phase, the

> SUBCONTRACTOR agrees to be bound to CONTRACTOR by the terms and conditions of the General Contract between CONTRACTOR and OWNER as well as this Subcontract Agreement and hereby assumes towards the CON-TRACTOR all of the duties, obligations and responsibilities applicable to SUBCONTRACTOR's work which the CONTRACTOR owes towards the Owner under the General Contract. (Exh. C-8)

Summit also required the subcontractor to "comply with all laws, ordinances, rules, regulations and orders of any public authority bearing on the performance of the Work" (Exh. C-8, Art. 9). The subcontract required All Phase to warrant and guarantee that all of its work would be "in compliance with all federal, state and local codes and requirements" (Exh. C-8, Art. 15). Although the subcontract attempts to place responsibility for compliance with the Occupational Safety and Health Act (Act) on the subcontractor, the subcontractor is required to hold Summit harmless against any liability, including the assessment of OSHA fines and legal costs. Summit is reimbursed for fines assessed and legal costs incurred as a result of the subcontractor's failure to comply with safety requirements. Summit retained the authority to deduct the OSHA fines and legal costs from the subcontract amount by change order.

Additionally, other provisions of the subcontract show Summit's control over the safety of All Phase's employees. All Phase's subcontract provided that the subcontractor could not subcontract without the prior written consent of Summit, and Summit had sole discretion on whether to approve a subcontractor's subcontractor. Also, subcontractors were required to keep their work areas clean and orderly subject to Summit's approval. The subcontract required All Phase to have on site at all times a "competent superintendent and necessary assistants all approved by" Summit, one of which had to be able to speak English (Exh. C-8, Attachment A, para. 17, 33, 45). All Phase agreed that "any scaffolding installed by SUBCONTRACTOR to install this scope of work shall be OSHA approved" — meaning that it would comply with OSHA regulations (Exh. C-8, Attachment B, para. 20). Summit required that All Phase comply with all governing laws imposed by all Federal governing authorities, including the Occupational Safety and Health Act (Exh. C-8, Attachment A, para. 42 and Attachment B, preamble).

Moreover, Summit's control over All Phase's worksite is addressed in paragraph 5 of Attachment A to the subcontract (Exh. C-8) which provides that:

> All parties hereby agree that control of the Work Schedule, use of the site and coordination of all on-site personnel will be performed under the com-plete direction of CONTRACTOR's supervisory staff. CONTRACTOR may enforce upon SUBCONTRACTOR any of the following actions in order to expedite or coordinate the work. However, CONTRACTOR does not assume any liability for delays to SUBCONTRACTOR or third parties in connection

with coordination of on-site personnel. These actions include, but are not limited to, the following:

A) Designated storage, designated unloading and parking areas.
B) Require unacceptable materials, equipment or vehicles to be removed from the project.
C) Limit the use of the site by SUBCONTRACTOR's equipment, vehicles, personnel or stored materials.
D) Temporarily or permanently bar specific personnel from the site. Listed below is a partial list of reasons to deny a person access to the project.
   1) Drug or alcohol use
   2) Fighting, possession of weapons
   3) Theft
   4) Harassment of anyone on or off the project
   5) Personal use of the areas near the project limits for parking, eating, sleeping, and so forth
   6) Failure to cooperate with CONTRACTOR's supervisory personnel or comply with project documents.

Summit's authority explicitly granted by a combination of contract provisions is broad enough to necessarily involve subcontractor employees' safety. Summit held authority over the subcontractor's actions, as well as authority over conditions affecting general safety on the worksite. The authority granted Summit mirrored how Summit actually controlled the project. In addition to accepting responsibility for compliance with OSHA's safety requirements in its contract with the owner, and by requiring its subcontractors to hold Summit harmless for a failure to comply, Summit held sufficient authority and control over the worksite and the safety of the employees.

The Review Commission considers supervisory authority and control sufficient where the general contractor has specific authority to demand a subcontractor's compliance with safety requirements, stop a contractor's work for failure to observe safety precautions, and remove a contractor from the worksite. *McDevitt Street Bovis, Inc.*, *supra*. Summit held this control over All Phase.

Thirty-eight employees of four subcontractors including All Phase were working on the student housing project on June 18 and 19, 2003 (Exh. C-9). Summit's project superintendent and his three assistants were also present on site. The assistants were assigned to particular locations in the building where the subcontractors performed their jobs, and the project superintendent inspected the site twice daily to ensure progress and quality of work. Summit kept track of the subcontractor's activities on the worksite. Guevara, as project superintendent, prepared a project diary and daily report at the end of the day which detailed the activities performed by the subcontractors and the occurrence of any problems (Exh. C-9).

Respondent held the power to hire and fire subcontractors (Tr. 104, 109, 149-150). Summit controlled the sequencing of work, telling subcontractors when to start and finish their work (Tr. 109, 144). Summit controlled the quality of work, ensuring through inspections that subcontractors performed their work in accordance with the contract specifications and blueprints (Tr. 109-111). Summit had authority to correct deficiencies in the work of the subcontractors (Tr. 144). Summit conducted injury investigations for employees of subcontractors who were injured at the work-site (Exh. C-9, entry April 16, 2003; Tr. 163-167). At the preconstruction meeting, Summit conducted a safety presentation which included fall protection and invited subcontractors to attend (Tr. 213-214).

If, during the normal course of his activities, Summit's superintendent observed an obvious safety concern, the superintendent requested the subcontractor to rectify the hazard immediately (Tr. 222, 244). This is what superintendent Guevara advised several times prior to the OSHA inspection. He mentioned to All Phase at least twice that its employees were not using fall protection while laying bricks from the scaf-fold (Tr. 120, 129, 205). As recognized by Superintendent Guevara, subcontractors generally complied with his safety warnings (Tr. 129).

Summit's claim that it has only a limited ability to require a subcontractor to correct safety violations is disingenuous. The subcontract which Summit drafted and required subcontractors to sign in order to work on the student housing project retained Summit's authority to terminate, suspend, or withhold contract payments from any subcontractor who failed to abide by its directions. Summit, not the sub-contractors, dictated the terms of the subcontract and what occurred on the worksite. Guevara testified that subcontractors never refused any of its requests concerning safety (Tr. 113, 129). This shows recognition by the subcontractors of Summit's con-trol and authority over the worksite.

As a general contractor, Summit held a unique position on the construction project. The subcontract agreement provided Summit multiple methods to enforce All Phase's compliance with OSHA requirements. Summit chose the subcontractors for the work, controlled the scheduling of their work, and could enforce penalties or ultimately ter-minate the subcontract if the subcontractor failed to meet its schedule. Summit had the right to terminate All Phase for convenience or for cause if the subcontractor failed to "perform the Work in Accordance with the Contract Documents," disregarded "Laws, Codes or Regulations of any public body having jurisdiction," or "otherwise violates in any way provisions of the Contract Documents" (Exh. C-8, Art. 14(b)). This right included the power to fire a subcontractor for the violation of OSHA regulations (Exh. Jt-1, pp. 14-15). Although termination of a subcontractor could cause serious problems with the scheduling; nevertheless, Summit has exercised that ultimate control when necessary.

Summit also had the right to exclude All Phase from the jobsite and to take pos-session of the Work (Exh. C-8, Art. 14). Summit could temporarily or permanently bar specific personnel of All Phase from the jobsite for failure to cooperate with Summit's supervisors (Exh. C-8, Attachment A, para 5). In fact, Summit's safety and health manual contemplated that a partial or total work stoppage might be required until corrective action is taken (Exh. C-5, p. 72). In Article 14 of the subcontract agreement, Summit retained the authority to suspend the subcontractor for not more than 90 days without cause.

If termination or suspension were too harsh a remedy, the subcontract provided other methods by which to enforce All Phase's compliance with OSHA. Summit had the right to retain 10 percent of the contract amount until All Phase satisfied all of its contractual obligations (Exh. C-8, Art. 3(d); Tr. 114-115). Summit's safety policy also provides that the project superintendent could solicit assistance from Summit's safety director or the project manager (Tr. 222, 229-230). In fact, the project superintendent testified that when he had encountered a problem with a roofing subcontractor during a rain, he took the problem to his project manager who corrected it by dealing directly with the subcontractor's officers (Tr. 169-171). This was not done, however, when All Phase repeatedly failed to require fall protection for employees (Tr. 231).

Within its control and authority over the safety of All Phase employees, Summit failed to exercise reasonable care. Summit had observed scaffolding violations several times by All Phase prior to the OSHA inspection. On each occasion, Summit did no more than ask All Phase to correct the violation. Despite having knowledge of the June 18 and 19 violations at issue, Summit did not request All Phase to correct the violations (Tr. 121-122). Instead Summit rescheduled the OSHA inspection to June 24, 2003, when All Phase was not onsite. There is no showing that Summit took any corrective action such as inspecting All Phase for fall protection requirements, conducting worksite safety meetings or training, and enforcing compliance with a graduated system of enforcement.

Summit's violation of § 1926.451(g) (1) (vii) is established.

SERIOUS CLASSIFICATION

In order to establish that a violation is "serious" under § 17(k) of the Act, the Secretary must establish that there is a substantial probability of death or serious physical harm that could result from the cited condition and the employer knew or should have known of the violation. Showing the likelihood of an accident is not required. *Spancrete Northeast, Inc.,* 15 BNA OSHC 1020, 1024 (No. 86-521, 1991).

Summit's violation of § 1926.451(g) (1) (vii) is properly classified as serious. Summit stipulates that it knew of the lack of fall protection by All Phase employees. Its employees were exposed to falls of twelve feet or eighteen feet from a scaffold to the ground. Such a fall could cause serious physical harm or possibly death.

PENALTY CONSIDERATION

In determining an appropriate penalty, consideration of the size of the employer's business, history of the employer's previous violations, the employer's good faith, and the gravity of the violation is required. Gravity is the principal factor.

Having 148 employees and a history of past serious citations, Summit is not entitled to credit for size or history. However, Summit is entitled to credit for good faith. There is no showing that Summit's safety program is inadequate in protecting its employees. Although its company's policy is to avoid safety responsibilities for subcontractors' employees, Summit does attempt to advise subcontractors of known safety hazards.

A penalty of $2000 is reasonable for Summit's violation of § 1926.451(g) (1) (vii). Summit was the general contractor and had no employees exposed to the lack of fall protection. Summit did not create the unsafe condition. The subcontractor who caused the violation and had employees exposed received a $2500 penalty from OSHA.

FINDINGS OF FACT AND CONCLUSIONS OF LAW

The foregoing decision constitutes the findings of fact and conclusions of law in accordance with Rule 52(a) of the Federal Rules of Civil Procedure.

ORDER

Based upon the foregoing decision, it is ORDERED:

Serious violation of § 1926.451(g) (1) (vii), is affirmed and penalty of $2000 is assessed.

/s/ _____

Ken S. Welsch

Judge

Date: June 14, 2004

U.S. SUPREME COURT

MARSHALL V. BARLOW'S, INC., 436 U.S. 307 (1978)

436 U.S. 307

MARSHALL, SECRETARY OF LABOR, ET AL. v. BARLOW'S, INC.
APPEAL FROM THE UNITED STATES DISTRICT COURT FOR THE DISTRICT OF IDAHO

No. 76-1143

Argued January 9, 1978
Decided May 23, 1978

Appellee brought this action to obtain injunctive relief against a warrantless inspection of its business premises pursuant to 8 (a) of the Occupational Safety and Health Act of 1970 (OSHA), which empowers agents of the Secretary of Labor to search the work area of any employment facility within OSHA's jurisdiction for safety hazards and violations of OSHA regulations. A three-judge District Court ruled in appellee's favor, concluding, in reliance on *Camara v. Municipal Court,* 387 U.S. 523, 528-529, and *See v. Seattle,* 387 U.S. 541, 543, that the Fourth Amendment required a warrant for the type of search involved and that the statutory authorization for warrantless inspections was unconstitutional. Held: The inspection without a warrant or its equivalent pursuant to 8 (a) of OSHA violated the Fourth Amendment, Pp. 311-325.

(a) The rule that warrantless searches are generally unreasonable applies to commercial premises as well as homes. *Camara v. Municipal Court,* supra, and *See v. Seattle,* supra. Pp. 311-313.

(b) Though an exception to the search warrant requirement has been recognized for "closely regulated" industries "long subject to close supervision and inspection," *Colonnade Catering Corp. v. United States,* 397 U.S. 72, 74, 77, that exception does not apply simply because the business is in interstate commerce. Pp. 313-314.

(c) Nor does an employer's necessary utilization of employees in his operation mean that he has opened areas where the employees alone are permitted to, the warrantless scrutiny of Government agents. Pp. 314-315.

(d) Insofar as experience to date indicates, requiring warrants to make OSHA inspections will impose no serious burdens on the inspection system or the courts. The advantages of surprise through the opportunity of inspecting without prior notice will not be lost if, after entry to an inspector is refused, an ex parte warrant can be obtained, facilitating an inspector's reappearance at the premises without further notice; and appellant Secretary's entitlement to a warrant will not depend on his demonstrating probable cause to believe that conditions on the premises violate OSHA but merely that reasonable legislative or administrative standards for conducting an inspection are satisfied with respect to a particular establishment. Pp. 315-321.

(e) Requiring a warrant for OSHA inspections does not mean that, as a practical matter, warrantless-search provisions in other regulatory statutes are unconstitutional, as the reasonableness of those provisions depends upon the specific enforcement needs and privacy guarantees of each statute. Pp. 321-322.

424 F. Supp. 437, affirmed.

WHITE, J., delivered the opinion of the Court, in which BURGER, C. J., and STEWART, MARSHALL, and POWELL, JJ., joined. STEVENS, J., filed a dissenting opinion, in which BLACKMUN and REHNQUIST, JJ., joined, post, p. 325. BRENNAN, J., took no part in the consideration or decision of the case.

Solicitor General McCree argued the cause for appellants. With him on the briefs were Deputy Solicitor General Wallace, Stuart A. Smith, and Michael H. Levin.

John L. Runft argued the cause for appellee. With him on the brief was Iver J. Longeteig.*

MR. JUSTICE WHITE delivered the opinion of the Court.

Section 8 (a) of the Occupational Safety and Health Act of 1970 (OSHA or Act)[1] empowers agents of the Secretary of Labor (Secretary) to search the work area of any employment facility within the Act's jurisdiction. The purpose of the search is

---

* Warren Spannaus, Attorney General of Minnesota, Richard B. Allyn, Solicitor General, and Steven M. Gunn and Richard A. Lockridge, Special Assistant Attorneys General, filed a brief for eleven States as amici curiae urging reversal, joined by the Attorneys General for their respective States as follows: Frank J. Kelley of Michigan, William F. Hyland of New Jersey, Toney Anaya of New Mexico, Rufus Edmisten of North Carolina, Robert P. Kane of Pennsylvania, Daniel R. McLeod of South Carolina, M. Jerome Diamond of Vermont, Anthony F. Troy of Virginia, and V. Frank Mendicino of Wyoming. Briefs of amici curiae urging reversal were filed by J. Albert Woll and Laurence Gold for the American Federation of Labor and Congress of Industrial Organizations; and by Michael R. Sherwood for the Sierra Club et al.

Briefs of amici curiae urging affirmance were filed by Wayne L. Kidwell, Attorney General of Idaho, and Guy G. Hurlbutt, Chief Deputy Attorney General, Robert B. Hansen, Attorney General of Utah, and Michael L. Deamer, Deputy Attorney General, for the States of Idaho and Utah; by Allen A. Lauterbach for the American Farm Bureau Federation; by Robert T. Thompson, Lawrence Kraus, and Stanley T. Kaleczyc for the Chamber of Commerce of the United States; by Anthony J. Obadal, Steven R. Semler, Stephen C. Yohay, Leonard J. Theberge, Edward H. Dowd, and James Watt for the Mountain States Legal Foundation; by James D. McKevitt for the National Federation of Independent Business; and by Ronald A. Zumbrun, John H. Findley, Albert Ferri, Jr., and W. Hugh O'Riordan for the Pacific Legal Foundation.

Briefs of amici curiae were filed by Robert E. Rader, Jr., for the American Conservative Union; and by David Goldberger, Barbara O'Toole, McNeill Stokes, Ira J. Smotherman, Jr., and David Rudenstine for the Roger Baldwin Foundation, Inc., of the American Civil Liberties Union, Illinois Division.

to inspect for safety hazards and violations of OSHA regulations. No search warrant or other process is expressly required under the Act.

On the morning of September 11, 1975, an OSHA inspector entered the customer service area of Barlow's, Inc., an electrical and plumbing installation business located in Pocatello, Idaho. The president and general manager, Ferrol G. "Bill" Barlow, was on hand; and the OSHA inspector, after showing his credentials,[2] informed Mr. Barlow that he wished to conduct a search of the working areas of the business. Mr. Barlow inquired whether any complaint had been received about his company. The inspector answered no, but that Barlow's, Inc., had simply turned up in the agency's selection process. The inspector again asked to enter the nonpublic area of the business; Mr. Barlow's response was to inquire whether the inspector had a search warrant. The inspector had none. Thereupon, Mr. Barlow refused the inspector admission to the employee area of his business. He said he was relying on his rights as guaranteed by the Fourth Amendment of the United States Constitution.

Three months later, the Secretary petitioned the United States District Court for the District of Idaho to issue an order compelling Mr. Barlow to admit the inspector.[3] The requested order was issued on December 30, 1975, and was presented to Mr. Barlow on January 5, 1976. Mr. Barlow again refused admission, and he sought his own injunctive relief against the warrantless searches assertedly permitted by OSHA. A three-judge court was convened. On December 30, 1976, it ruled in Mr. Barlow's favor. 424 F. Supp. 437. Concluding that *Camara v. Municipal Court,* 387 U.S. 523, 528-529 (1967), and *See v. Seattle,* 387 U.S. 541, 543 (1967), controlled this case, the court held that the Fourth Amendment required a warrant for the type of search involved here[4] and that the statutory authorization for warrantless inspections was unconstitutional. An injunction against searches or inspections pursuant to 8 (a) was entered. The Secretary appealed, challenging the judgment, and we noted probable jurisdiction. 430 U.S. 964.

I

The Secretary urges that warrantless inspections to enforce OSHA are reasonable within the meaning of the Fourth Amendment. Among other things, he relies on 8 (a) of the Act, 29 U.S.C. 657 (a), which authorizes inspection of business premises without a warrant and which the Secretary urges represents a congressional construction of the Fourth Amendment that the courts should not reject. Regrettably, we are unable to agree.

The Warrant Clause of the Fourth Amendment protects commercial buildings as well as private homes. To hold otherwise would belie the origin of that Amendment, and the American colonial experience. An important forerunner of the first ten Amendments to the United States Constitution, the Virginia Bill of Rights, specifically opposed "general warrants, whereby an officer or messenger may be commanded to search suspected places without evidence of a fact committed."[5] The general warrant was a recurring point of contention in the Colonies immediately preceding the Revolution.[6] The particular offensiveness it engendered was acutely felt by the merchants and businessmen whose premises and products were inspected for compliance with the several parliamentary revenue measures that most irritated the colonists.[7] "[T]he Fourth Amendment's commands grew in large measure out of the colonists' experience with the writs of assistance ... [that] granted sweeping power to customs officials and other agents of the King to search at large for smuggled goods." *United States v. Chadwick*, 433 U.S. 1, 7-8 (1977).

See also *G.M. Leasing Corp. v. United States*, 429 U.S. 338, 355 (1977). Against this background, it is untenable that the ban on warrantless searches was not intended to shield places of business as well as of residence.

This Court has already held that warrantless searches are generally unreasonable, and that this rule applies to commercial premises as well as homes. In *Camara v. Municipal Court*, supra, at 528-529, we held:

> "[E]xcept in certain carefully defined classes of cases, a search of private property without proper consent is 'unreasonable' unless it has been authorized by a valid search warrant."

On the same day, we also ruled:

> "As we explained in Camara, a search of private houses is presumptively unreasonable if conducted without a warrant. The businessman, like the occupant of a residence, has a constitutional right to go about his business free from unreasonable official entries upon his private commercial property. The businessman, too, has that right placed in jeopardy if the decision to enter and inspect for violation of regulatory laws can be made and enforced by the inspector in the field without official authority evidenced by a warrant." *See v. Seattle*, supra, at 543.

These same cases also held that the Fourth Amendment prohibition against unreasonable searches protects against warrantless intrusions during civil as well as criminal investigations. Ibid. The reason is found in the "basic purpose of this Amendment ... [which] is to safeguard the privacy and security of individuals against arbitrary invasions by governmental officials." Camara, supra, at 528. If the government intrudes on a person's property, the privacy interest suffers whether the government's motivation is to investigate violations of criminal laws or breaches of other statutory or regulatory standards. It therefore appears that unless some recognized exception to the warrant requirement applies, *See v. Seattle* would require a warrant to conduct the inspection sought in this case.

The Secretary urges that an exception from the search warrant requirement has been recognized for "pervasively regulated business[es]," *United States v. Biswell*, 406 U.S. 311, 316 (1972), and for "closely regulated" industries "long subject to close supervision and inspection." *Colonnade Catering Corp. v. United States*, 397 U.S. 72, 74, 77 (1970). These cases are indeed exceptions, but they represent responses to relatively unique circumstances. Certain industries have such a history of government oversight that no reasonable expectation of privacy, see *Katz v. United States*, 389 U.S. 347, 351-352 (1967), could exist for a proprietor over the stock of such an enterprise. Liquor (Colonnade) and firearms (Biswell) are industries of this type;

when an entrepreneur embarks upon such a business, he has voluntarily chosen to subject himself to a full arsenal of governmental regulation.

Industries such as these fall within the "certain carefully defined classes of cases," referenced in Camara, 387 U.S., at 528. The element that distinguishes these enterprises from ordinary businesses is a long tradition of close government supervision, of which any person who chooses to enter such a business must already be aware. "A central difference between those cases [Colonnade and Biswell] and this one is that businessmen engaged in such federally licensed and regulated enterprises accept the burdens as well as the benefits of their trade, whereas the petitioner here was not engaged in any regulated or licensed business. The businessman in a regulated industry in effect consents to the restrictions placed upon him." *Almeida-Sanchez v. United States,* 413 U.S. 266, 271 (1973).

The clear import of our cases is that the closely regulated industry of the type involved in Colonnade and Biswell is the exception. The Secretary would make it the rule. Invoking the Walsh-Healey Act of 1936, 41 U.S.C. 35 et seq., the Secretary attempts to support a conclusion that all businesses involved in interstate commerce have long been subjected to close supervision of employee safety and health conditions. But the degree of federal involvement in employee working circumstances has never been of the order of specificity and pervasiveness that OSHA mandates. It is quite unconvincing to argue that the imposition of minimum wages and maximum hours on employers who contracted with the Government under the Walsh-Healey Act prepared the entirety of American interstate commerce for regulation of working conditions to the minutest detail. Nor can any but the most fictional sense of voluntary consent to later searches be found in the single fact that one conducts a business affecting interstate commerce; under current practice and law, few businesses can be conducted without having some effect on interstate commerce.

The Secretary also attempts to derive support for a Colonnade-Biswell-type exception by drawing analogies from the field of labor law. In *Republic Aviation Corp. v. NLRB,* 324 U.S. 793 (1945), this Court upheld the rights of employees to solicit for a union during nonworking time where efficiency was not compromised. By opening up his property to employees, the employer had yielded so much of his private property rights as to allow those employees to exercise seven rights under the National Labor Relations Act. But this Court also held that the private property rights of an owner prevailed over the intrusion of non-employee organizers, even in nonworking areas of the plant and during nonworking hours. *NLRB v. Babcock & Wilcox Co.,* 351 U.S. 105 (1956).

The critical fact in this case is that entry over Mr. Barlow's objection is being sought by a Government agent.[8] Employees are not being prohibited from reporting OSHA violations. What they observe in their daily functions is undoubtedly beyond the employer's reasonable expectation of privacy. The Government inspector, however, is not an employee. Without a warrant he stands in no better position than a member of the public. What is observable by the public is observable, without a warrant, by the Government inspector as well.[9] The owner of a business has not, by the necessary utilization of employees in his operation, thrown open the areas where employees alone are permitted to the warrantless scrutiny of Government agents. That an employee is free to report, and the Government is free to use, any evidence

of noncompliance with OSHA that the employee observes furnishes no justification
for federal agents to enter a place of business from which the public is restricted and
to conduct their own warrantless search.[10]

II

The Secretary nevertheless stoutly argues that the enforcement scheme of the Act
requires warrantless searches, and that the restrictions on search discretion contained in the Act and its regulations already protect as much privacy as a warrant
would. The Secretary thereby asserts the actual reasonableness of OSHA searches,
whatever the general rule against warrantless searches might be. Because "reasonableness is still the ultimate standard," *Camara v. Municipal Court,* 387 U.S., at 539,
the Secretary suggests that the Court decide whether a warrant is needed by arriving
at a sensible balance between the administrative necessities of OSHA inspections
and the incremental protection of privacy of business owners a warrant would afford.
He suggests that only a decision exempting OSHA inspections from the Warrant
Clause would give "full recognition to the competing public and private interests
here at stake." Ibid.

The Secretary submits that warrantless inspections are essential to the proper
enforcement of OSHA because they afford the opportunity to inspect without prior
notice and hence to preserve the advantages of surprise. While the dangerous conditions outlawed by the Act include structural defects that cannot be quickly hidden
or remedied, the Act also regulates a myriad of safety details that may be ame-
nable to speedy alteration or disguise. The risk is that during the interval between an
inspector's initial request to search a plant and his procuring a warrant following the
owner's refusal of permission, violations of this latter type could be corrected and
thus escape the inspector's notice. To the suggestion that warrants may be issued ex
parte and executed without delay and without prior notice, thereby preserving the
element of surprise, the Secretary expresses concern for the administrative strain
that would be experienced by the inspection system, and by the courts, should ex
parte warrants issued in advance become standard practice.

We are unconvinced, however, that requiring warrants to inspect will impose
serious burdens on the inspection system or the courts, will prevent inspections necessary to enforce the statute, or will make them less effective. In the first place,
the great majority of businessmen can be expected in normal course to consent to
inspection without warrant; the Secretary has not brought to this Court's attention
any widespread pattern of refusal.[11] In those cases where an owner does insist on
a warrant, the Secretary argues that inspection efficiency will be impeded by the
advance notice and delay. The Act's penalty provisions for giving advance notice
of a search, 29 U.S.C. 666 (f), and the Secretary's own regulations, 29 CFR 1903.6
(1977), indicate that surprise searches are indeed contemplated. However, the Secretary has also promulgated a regulation providing that upon refusal to permit an
inspector to enter the property or to complete his inspection, the inspector shall
attempt to ascertain the reasons for the refusal and report to his superior, who shall
"promptly take appropriate action, including compulsory process, if necessary." 29
CFR 1903.4 (1977).[12] The regulation represents a choice to proceed by process where

entry is refused; and on the basis of evidence available from present practice, the Act's effectiveness has not been crippled by providing those owners who wish to refuse an initial requested entry with a time lapse while the inspector obtains the necessary process.[13] Indeed, the kind of process sought in this case and apparently anticipated by the regulation provides notice to the business operator.[14]

If this safeguard endangers the efficient administration of OSHA, the Secretary should never have adopted it, particularly when the Act does not require it. Nor is it immediately apparent why the advantages of surprise would be lost if, after being refused entry, procedures were available for the Secretary to seek an ex parte warrant and to reappear at the premises without further notice to the establishment being inspected.[15]

Whether the Secretary proceeds to secure a warrant or other process, with or without prior notice, his entitlement to inspect will not depend on his demonstrating probable cause to believe that conditions in violation of OSHA exist on the premises. Probable cause in the criminal law sense is not required. For purposes of an administrative search such as this, probable cause justifying the issuance of a warrant may be based not only on specific evidence of an existing violation[16] but also on a showing that "reasonable legislative or administrative standards for conducting an ... inspection are satisfied with respect to a particular [establishment]." *Camara v. Municipal Court,* 387 U.S., at 538. A warrant showing that a specific business has been chosen for an OSHA search on the basis of a general administrative plan for the enforcement of the Act derived from neutral sources such as, for example, dispersion of employees in various types of industries across a given area, and the desired frequency of searches in any of the lesser divisions of the area, would protect an employer's Fourth Amendment rights.[17] We doubt that the consumption of enforcement energies in the obtaining of such warrants will exceed manageable proportions.

Finally, the Secretary urges that requiring a warrant for OSHA inspectors will mean that, as a practical matter, warrantless-search provisions in other regulatory statutes are also constitutionally infirm. The reasonableness of a warrantless search, however, will depend upon the specific enforcement needs and privacy guarantees of each statute. Some of the statutes cited apply only to a single industry, where regulations might already be so pervasive that a Colonnade-Biswell exception to the warrant requirement could apply. Some statutes already envision resort to federal-court enforcement when entry is refused, employing specific language in some cases[18] and general language in others.[19] In short, we base today's opinion on the facts and law concerned with OSHA and do not retreat from a holding appropriate to that statute because of its real or imagined effect on other, different administrative schemes.

Nor do we agree that the incremental protections afforded the employer's privacy by a warrant are so marginal that they fail to justify the administrative burdens that may be entailed. The authority to make warrantless searches devolves almost unbridled discretion upon executive and administrative officers, particularly those in the field, as to when to search and whom to search. A warrant, by contrast, would provide assurances from a neutral officer that the inspection is reasonable under the Constitution, is authorized by statute, and is pursuant to an administrative plan containing specific neutral criteria.[20] Also, a warrant would then and there advise the owner of the scope and objects of the search, beyond which limits the inspector is not expected to proceed.[21] These are important functions for a warrant to perform, functions which underlie the Court's prior decisions that the Warrant Clause applies to inspections for compliance with regulatory statutes.[22] *Camara v. Municipal Court*, 387 U.S. 523 (1967); *See v. Seattle*, 387 U.S. 541 (1967). We conclude that the concerns expressed by the Secretary do not suffice to justify warrantless inspections under OSHA or vitiate the general constitutional requirement that for a search to be reasonable a warrant must be obtained.

III

We hold that Barlow's was entitled to a declaratory judgment that the Act is unconstitutional insofar as it purports to authorize inspections without warrant or its equivalent and to an injunction enjoining the Act's enforcement to that extent.[23] The judgment of the District Court is therefore affirmed.

So ordered.

MR. JUSTICE BRENNAN took no part in the consideration or decision of this case.

FOOTNOTES:

[1] "In order to carry out the purposes of this chapter, the Secretary, upon presenting appropriate credentials to the owner, operator, or agent in charge, is authorized

(1) to enter without delay and at reasonable times any factory, plant, establishment, construction site, or other area, workplace or environment where work is performed by an employee of an employer; and
(2) to inspect and investigate during regular working hours and at other reasonable times, and within reasonable limits and in a reasonable manner, any such place of employment and all pertinent conditions, structures, machines, apparatus, devices, equipment, and materials therein, and to question privately any such employer, owner, operator, agent, or employee." 84 Stat. 1598, 29 U.S.C. 657 (a).

[2] This is required by the Act. See n. 1, supra.

[3] A regulation of the Secretary, 29 CFR 1903.4 (1977), requires an inspector to seek compulsory process if an employer refuses a requested search. See infra, at 317, and n. 12.

[4] No res judicata bar arose against Mr. Barlow from the December 30, 1975, order authorizing a search, because the earlier decision reserved the constitutional issue. See 424 F. Supp. 437.

[5] H. Commager, Documents of American History 104 (8th ed. 1968).

[6] See, for example, Dickerson, Writs of Assistance as a Cause of the Revolution in The Era of the American Revolution 40 (R. Morris ed. 1939).

[7] The Stamp Act of 1765, the Townshend Revenue Act of 1767, and the tea tax of 1773 are notable examples. See Commager, supra, n. 5, at 53, 63. For commentary, see 1 S. Morison, H. Commager, and W. Leuchtenburg, The Growth of the American Republic 143, 149, 159 (1969).

[8] The Government has asked that Mr. Barlow be ordered to show cause why he should not be held in contempt for refusing to honor the inspection order, and its position is that the OSHA inspector is now entitled to enter at once, over Mr. Barlow's objection.

[9] Cf. *Air Pollution Variance Bd. v. Western Alfalfa Corp.*, 416 U.S. 861 (1974).

[10] The automobile-search cases cited by the Secretary are even less helpful to his position than the labor cases. The fact that automobiles occupy a special category in Fourth Amendment case law is by now beyond doubt due, among other factors, to the quick mobility of a car, the registration requirements of both the car and the driver, and the more available opportunity for plain-view observations of a car's contents. Cady v. Dombrowski, 413 U.S. 433, 441-442 (1973); see also Chambers v. Maroney, 399 U.S. 42, 48-51 (1970). Even so, probable cause has not been abandoned as a requirement for stopping and searching an automobile.

[11] We recognize that today's holding itself might have an impact on whether owners choose to resist requested searches; we can only await the development of evidence not present on this record to determine how serious an impediment to effective enforcement this might be.

[12] It is true, as the Secretary asserts, that 8 (a) of the Act, 29 U.S.C. 657 (a), purports to authorize inspections without warrant; but it is also true that it does not forbid the Secretary from proceeding to inspect only by warrant or other process. The Secretary has broad authority to prescribe such rules and regulations as he may deem necessary to carry out his responsibilities under this chapter, "including rules and regulations dealing with the inspection of an employer's establishment." 8 (g) (2), 29 U.S.C. 657 (g) (2). The regulations with respect to inspections are contained in 29 CFR Part 1903 (1977). Section 1903.4, referred to in the text, provides as follows:

> Upon a refusal to permit a Compliance Safety and Health Officer, in the exercise of his official duties, to enter without delay and at reasonable times any place of employment or any place therein, to inspect, to review records, or to question any employer, owner, operator, agent, or employee, in accordance with 1903.3, or to permit a representative of employees to accompany the Compliance Safety and Health Officer during the physical inspection of any workplace in accordance with 1903.8, the Compliance Safety and Health

Officer shall terminate the inspection or confine the inspection to other areas, conditions, structures, machines, apparatus, devices, equipment, materials, records, or interviews concerning which no objection is raised. The Compliance Safety and Health Officer shall endeavor to ascertain the reason for such refusal, and he shall immediately report the refusal and the reason therefore to the Area Director. The Area Director shall immediately consult with the Assistant Regional Director and the Regional Solicitor, who shall promptly take appropriate action, including compulsory process, if necessary.

When his representative was refused admission by Mr. Barlow, the Secretary proceeded in federal court to enforce his right to enter and inspect, as conferred by 29 U.S.C. 657.

[13] A change in the language of the Compliance Operations Manual for OSHA inspectors supports the inference that, whatever the Act's administrators might have thought at the start, it was eventually concluded that enforcement efficiency would not be jeopardized by permitting employers to refuse entry, at least until the inspector obtained compulsory process. The 1972 Manual included a section specifically directed to obtaining "warrants," and one provision of that section dealt with ex parte warrants:

In cases where a refusal of entry is to be expected from the past performance of the employer, or where the employer has given some indication prior to the commencement of the investigation of his intention to bar entry or limit or interfere with the investigation, a warrant should be obtained before the inspection is attempted. Cases of this nature should also be referred through the Area Director to the appropriate Regional solicitor and the Regional Administrator alerted. (Dept. of Labor, OSHA Compliance Operations Manual V-7, Jan. 1972)

The latest available manual, incorporating changes as of November 1977, deletes this provision, leaving only the details for obtaining "compulsory process" after an employer has refused entry (Dept. of Labor, OSHA Field Operations Manual, Vol. V. pp. V-4-V-5). In its present form, the Secretary's regulation appears to permit establishment owners to insist on "process"; and hence their refusal to permit entry would fall short of criminal conduct within the meaning of 18 U.S.C. 111 and 1114 (1976 ed.), which make it a crime forcibly to impede, intimidate, or interfere with federal officials, including OSHA inspectors, while engaged in or on account of the performance of their official duties.

[14] The proceeding was instituted by filing an "Application for Affirmative Order to Grant Entry and for an Order to show cause why such affirmative order should not issue." The District Court issued the order to show cause, the matter was argued,

and an order then issued authorizing the inspection and enjoining interference by Barlow's. The following is the order issued by the District Court:

IT IS HEREBY ORDERED, ADJUDGED AND DECREED that the United States of America, United States Department of Labor, Occupational Safety and Health Administration, through its duly designated representative or representatives, are entitled to entry upon the premises known as Barlow's Inc., 225 West Pine, Pocatello, Idaho, and may go upon said business premises to conduct an inspection and investigation as provided for in Section 8 of the Occupational Safety and Health Act of 1970 (29 U.S.C. 651, et seq.), as part of an inspection program designed to assure compliance with that Act; that the inspection and investigation shall be conducted during regular working hours or at other reasonable times, within reasonable limits and in a reasonable manner, all as set forth in the regulations pertaining to such inspections promulgated by the Secretary of Labor, at 29 C. F. R., Part 1903; that appropriate credentials as representatives of the Occupational Safety and Health Administration, United States Department of Labor, shall be presented to the Barlow's Inc. representative upon said premises and the inspection and investigation shall be commenced as soon as practicable after the issuance of this Order and shall be completed within reasonable promptness; that the inspection and investigation shall extend to the establishment or other area, workplace, or environment where work is performed by employees of the employer, Barlow's Inc., and to all pertinent conditions, structures, machines, apparatus, devices, equipment, materials, and all other things therein (including but not limited to records, files, papers, processes, controls, and facilities) bearing upon whether Barlow's Inc. is furnishing to its employees employment and a place of employment that are free from recognized hazards that are causing or are likely to cause death or serious physical harm to its employees, and whether Barlow's Inc. is complying with the Occupational Safety and Health Standards promulgated under the Occupational Safety and Health Act and the rules, regulations, and orders issued pursuant to that Act; that representatives of the Occupational Safety and Health Administration may, at the option of Barlow's Inc., be accompanied by one or more employees of Barlow's Inc., pursuant to Section 8 (e) of that Act; that Barlow's Inc., its agents, representatives, officers, and employees are hereby enjoined and restrained from in anyway whatsoever interfering with the inspection and investigation authorized by this Order and, further, Barlow's Inc. is hereby ordered and directed to, within five working days from the date of this Order, furnish a copy of this Order to its officers and managers, and, in addition, to post a copy of this Order at its employee's bulletin board located upon the business premises; and Barlow's Inc. is hereby ordered and directed to comply in all respects with this order and allow the inspection and investigation to take place without delay and forthwith.

[15] Insofar as the Secretary's statutory authority is concerned, a regulation expressly providing that the Secretary could proceed ex parte to seek a warrant or its equivalent

would appear to be as much within the Secretary's power as the regulation currently in force and calling for "compulsory process."

[16] Section 8 (f) (1), 29 U.S.C. 657 (f) (1), provides that employees or their representatives may give written notice to the Secretary of what they believe to be violations of safety or health standards and may request an inspection. If the Secretary then determines that "there are reasonable grounds to believe that such violation or danger exists, he shall make a special inspection in accordance with the provisions of this section as soon as practicable." The statute thus purports to authorize a warrantless inspection in these circumstances.

[17] The Secretary, Brief for Petitioner 9 n. 7, states that the Barlow inspection was not based on an employee complaint but was a "general schedule" investigation. "Such general inspections," he explains, "now called Regional Programmed Inspections, are carried out in accordance with criteria based upon accident experience and the number of employees exposed in particular industries. U.S. Department of Labor, Occupational Safety and Health Administration, Field Operations Manual, supra, 1 CCH Employment Safety and Health Guide 4327.2 (1976)."

[18] The Federal Metal and Nonmetallic Mine Safety Act provides:

> Whenever an operator ... refuses to permit the inspection or investigation of any mine which is subject to this chapter ... a civil action for preventive relief, including an application for a permanent or temporary injunction, restraining order, or other order, may be instituted by the Secretary in the district court of the United States for the district." 30 U.S.C. 733 (a). "The Secretary may institute a civil action for relief, including a permanent or temporary injunction, restraining order, or any other appropriate order in the district court ... whenever such operator or his agent ... refuses to permit the inspection of the mine. ... Each court shall have jurisdiction to provide such relief as may be appropriate." 30 U.S.C. 818. Another example is the Clean Air Act, which grants federal district courts jurisdiction "to require compliance" with the Administrator of the Environmental Protection Agency's attempt to inspect under 42 U.S.C. 7414 (1976 ed., Supp. I), when the Administrator has commenced "a civil action" for injunctive relief or to recover a penalty. 42 U.S.C. 7413 (b) (4) (1976 ed., Supp. I).

[19] Exemplary language is contained in the Animal Welfare Act of 1970 which provides for inspections by the Secretary of Agriculture; federal district courts are vested with jurisdiction "specifically to enforce, and to prevent and restrain violations of this chapter, and shall have jurisdiction in all other kinds of cases arising under this chapter." 7 U.S.C. 2146 (c) (1976 ed.). Similar provisions are included in other agricultural inspection Acts; see, for example, 21 U.S.C. 674 (meat product inspection); 21 U.S.C. 1050 (egg product inspection). The Internal Revenue Code, whose excise tax provisions requiring inspections of businesses are cited by the Secretary,

provides: "The district courts ... shall have such jurisdiction to make and issue in civil actions, writs and orders of injunction ... and such other orders and processes, and to render such ... decrees as may be necessary or appropriate for the enforcement of the internal revenue laws." 26 U.S.C. 7402 (a). For gasoline inspections, federal district courts are granted jurisdiction to restrain violations and enforce standards (one of which, 49 U.S.C. 1677, requires gas transporters to permit entry or inspection). The owner is to be afforded the opportunity for notice and response in most cases, but "failure to give such notice and afford such opportunity shall not preclude the granting of appropriate relief [by the district court]." 49 U.S.C. 1679 (a).

[20] The application for the inspection order filed by the Secretary in this case represented that "the desired inspection and investigation are contemplated as a part of an inspection program designed to assure compliance with the Act and are authorized by Section 8 (a) of the Act." The program was not described, however, or any facts presented that would indicate why an inspection of Barlow's establishment was within the program. The order that issued concluded generally that the inspection authorized was "part of an inspection program designed to assure compliance with the Act."

[21] Section 8 (a) of the Act, as set forth in 29 U.S.C. 657 (a), provides that "[i]n order to carry out the purposes of this chapter" the Secretary may enter any establishment, area, work place or environment "where work is performed by an employee of an employer" and "inspect and investigate" any such place of employment and all "pertinent conditions, structures, machines, apparatus, devices, equipment, and materials therein, and ... question privately any such employer, owner, operator, agent, or employee." Inspections are to be carried out "during regular working hours and at other reasonable times, and within reasonable limits and in a reasonable manner." The Secretary's regulations echo the statutory language in these respects. 29 CFR 1903.3 (1977). They also provide that inspectors are to explain the nature and purpose of the inspection and to "indicate generally the scope of the inspection." 29 CFR 1903.7 (a) (1977). Environmental samples and photographs are authorized, 29 CFR 1903.7 (b) (1977), and inspections are to be performed so as "to preclude unreasonable disruption of the operations of the employer's establishment." 29 CFR 1903.7 (d) (1977). The order that issued in this case reflected much of the foregoing statutory and regulatory language.

[22] Delineating the scope of a search with some care is particularly important where documents are involved. Section 8 (c) of the Act, 29 U.S.C. 657 (c), provides that an employer must "make, keep and preserve, and make available to the Secretary [of Labor] or to the Secretary of Health, Education and Welfare" such records regarding his activities relating to OSHA as the Secretary of Labor may prescribe by regulation as necessary or appropriate for enforcement of the statute or for developing information regarding the causes and prevention of occupational accidents and illnesses. Regulations requiring employers to maintain records of and to make periodic reports on "work-related deaths, injuries and illnesses" are also contemplated, as are

rules requiring accurate records of employee exposures to potential toxic materials and harmful physical agents.

In describing the scope of the warrantless inspection authorized by the statute, 8 (a) does not expressly include any records among those items or things that may be examined, and 8 (c) merely provides that the employer is to "make available" his pertinent records and to make periodic reports.

The Secretary's regulation, 29 CFR 1903.3 (1977), however, expressly includes among the inspector's powers the authority "to review records required by the Act and regulations published in this chapter, and other records which are directly related to the purpose of the inspection." Further, 1903.7 requires inspectors to indicate generally "the records specified in 1903.3 which they wish to review" but "such designations of records shall not preclude access to additional records specified in 1903.3." It is the Secretary's position, which we reject, that an inspection of documents of this scope may be effected without a warrant.

The order that issued in this case included among the objects and things to be inspected "all other things therein (including but not limited to records, files, papers, processes, controls and facilities) bearing upon whether Barlow's, Inc. is furnishing to its employees employment and a place of employment that are free from recognized hazards that are causing or are likely to cause death or serious physical harm to its employees, and whether Barlow's, Inc. is complying with ..." the OSHA regulations.

[23] The injunction entered by the District Court, however, should not be understood to forbid the Secretary from exercising the inspection authority conferred by eight pursuant to regulations and judicial process that satisfy the Fourth Amendment. The District Court did not address the issue whether the order for inspection that was issued in this case was the functional equivalent of a warrant, and the Secretary has limited his submission in this case to the constitutionality of a warrantless search of the Barlow establishment authorized by 8 (a). He has expressly declined to rely on 29 CFR 1903.4 (1977) and upon the order obtained in this case. Tr. of Oral Arg. 19. Of course, if the process obtained here, or obtained in other cases under revised regulations, would satisfy the Fourth Amendment, there would be no occasion for enjoining the inspections authorized by 8 (a).

MR. JUSTICE STEVENS, with whom MR. JUSTICE BLACKMUN and MR. JUSTICE REHNQUIST join, dissenting.

Congress enacted the Occupational Safety and Health Act to safeguard employees against hazards in the work areas of businesses subject to the Act. To ensure compliance, Congress authorized the Secretary of Labor to conduct routine, nonconsensual inspections. Today the Court holds that the Fourth Amendment prohibits such inspections without a warrant. The Court also holds that the constitutionally required warrant may be issued without any showing of probable cause. I disagree with both of these holdings.

The Fourth Amendment contains two separate Clauses, each flatly prohibiting a category of governmental conduct. The first Clause states that the right to be free from unreasonable searches "shall not be violated"[1]; the second unequivocally prohibits the issuance of warrants except "upon probable cause."[2] In this case the ultimate question is whether the category of warrantless searches authorized by the statute is "unreasonable" within the meaning of the first Clause.

In cases involving the investigation of criminal activity, the Court has held that the reasonableness of a search generally depends upon whether it was conducted pursuant to a valid warrant. See, for example, *Coolidge v. New Hampshire,* 403 U.S. 443. There is, however, also a category of searches which are reasonable within the meaning of the first Clause even though the probable-cause requirement of the Warrant Clause cannot be satisfied. See *United States v. Martinez-Fuerte,* 428 U.S. 543; *Terry v. Ohio,* 392 U.S. 1; *South Dakota v. Opperman,* 428 U.S. 364; *United States v. Biswell,* 406 U.S. 311. The regulatory inspection program challenged in this case, in my judgment, falls within this category.

I

The warrant requirement is linked "textually ... to the probable-cause concept" in the Warrant Clause. *South Dakota v. Opperman,* supra, at 370 n. 5. The routine OSHA inspections are, by definition, not based on cause to believe there is a violation on the premises to be inspected. Hence, if the inspections were measured against the requirements of the Warrant Clause, they would be automatically and unequivocally unreasonable. Because of the acknowledged importance and reasonableness of routine inspections in the enforcement of federal regulatory statutes such as OSHA, the Court recognizes that requiring full compliance with the Warrant Clause would invalidate all such inspection programs. Yet, rather than simply analyzing such programs under the "Reasonableness" Clause of the Fourth Amendment, the Court holds the OSHA program invalid under the Warrant Clause and then avoids a blanket prohibition on all routine, regulatory inspections by relying on the notion that the "probable cause" requirement in the Warrant Clause may be relaxed whenever the Court believes that the governmental need to conduct a category of "searches" outweighs the intrusion on interests protected by the Fourth Amendment.

The Court's approach disregards the plain language of the Warrant Clause and is unfaithful to the balance struck by the Framers of the Fourth Amendment — "the one procedural safeguard in the Constitution that grew directly out of the events which immediately preceded the revolutionary struggle with England."[3] This pre-constitutional history includes the controversy in England over the issuance of general warrants to aid enforcement of the seditious libel laws and the colonial experience with writs of assistance issued to facilitate collection of the various import duties imposed by Parliament. The Framers' familiarity with the abuses attending the issuance of such general warrants provided the principal stimulus for the restraints on arbitrary governmental intrusions embodied in the Fourth Amendment.

[O]ur constitutional fathers were not concerned about warrantless searches, but about overreaching warrants. It is perhaps too much to say that they feared the warrant more than the search, but it is plain enough that the warrant was the prime object of their concern. Far from looking at the warrant as a protection against unreasonable searches, they saw it as an authority for unreasonable and oppressive searches.[4]

Since the general warrant, not the warrantless search, was the immediate evil at which the Fourth Amendment was directed, it is not surprising that the Framers placed precise limits on its issuance. The requirement that a warrant only issue on a showing of particularized probable cause was the means adopted to circumscribe the warrant power. While the subsequent course of Fourth Amendment jurisprudence in this Court emphasizes the dangers posed by warrantless searches conducted without probable cause, it is the general reasonableness standard in the first Clause, not the Warrant Clause, that the Framers adopted to limit this category of searches. It is, of course, true that the existence of a valid warrant normally satisfies the reasonableness requirement under the Fourth Amendment. But we should not dilute the requirements of the Warrant Clause in an effort to force every kind of governmental intrusion which satisfies the Fourth Amendment definition of a "search" into a judicially developed, warrant-preference scheme.

Fidelity to the original understanding of the Fourth Amendment, therefore, leads to the conclusion that the Warrant Clause has no application to routine, regulatory inspections of commercial premises. If such inspections are valid, it is because they comport with the ultimate reasonableness standard of the Fourth Amendment. If the Court were correct in its view that such inspections, if undertaken without a warrant, are unreasonable in the constitutional sense, the issuance of a "new-fangled warrant" — to use Mr. Justice Clark's characteristically expressive term — without any true showing of particularized probable cause would not be sufficient to validate them.[5]

II

Even if a warrant issued without probable cause were faithful to the Warrant Clause, I could not accept the Court's holding that the Government's inspection program is constitutionally unreasonable because it fails to require such a warrant procedure. In determining whether a warrant is a necessary safeguard in a given class of cases, "the Court has weighed the public interest against the Fourth Amendment interest of the individual," *United States v. Martinez-Fuerte,* 428 U.S., at 555. Several considerations persuade me that this balance should be struck in favor of the routine inspections authorized by Congress.

Congress has determined that regulation and supervision of safety in the workplace furthers an important public interest and that the power to conduct warrantless searches is necessary to accomplish the safety goals of the legislation. In assessing the public interest side of the Fourth Amendment balance, however, the Court today substitutes its judgment for that of Congress on the question of what inspection authority is needed to effectuate the purposes of the Act. The Court states that if surprise is truly an important ingredient of an effective, representative inspection program, it can be retained by obtaining ex parte warrants in advance. The Court assures the Secretary that this will not unduly burden enforcement resources because most employers will consent to inspection.

The Court's analysis does not persuade me that Congress' determination that the warrantless-inspection power as a necessary adjunct of the exercise of the regulatory power is unreasonable. It was surely not unreasonable to conclude that the rate at which employers deny entry to inspectors would increase if covered businesses, which may have safety violations on their premises, have a right to deny warrantless entry to a compliance inspector. The Court is correct that this problem could be avoided by requiring inspectors to obtain a warrant prior to every inspection visit. But the adoption of such a practice undercuts the Court's explanation of why a warrant requirement would not create undue enforcement problems. For, even if it were true that many employers would not exercise their right to demand a warrant, it would provide little solace to those charged with administration of OSHA; faced with an increase in the rate of refusals and the added costs generated by futile trips to inspection sites where entry is denied, officials may be compelled to adopt a general practice of obtaining warrants in advance. While the Court's prediction of the effect a warrant requirement would have on the behavior of covered employers may turn out to be accurate, its judgment is essentially empirical. On such an issue, I would defer to Congress' judgment regarding the importance of a warrantless-search power to the OSHA enforcement scheme.

The Court also appears uncomfortable with the notion of second-guessing Congress and the Secretary on the question of how the substantive goals of OSHA can best be achieved. Thus, the Court offers an alternative explanation for its refusal to accept the legislative judgment. We are told that, in any event, the Secretary, who is charged with enforcement of the Act, has indicated that inspections without delay are not essential to the enforcement scheme. The Court bases this conclusion on a regulation prescribing the administrative response when a compliance inspector is denied entry. It provides: "The Area Director shall immediately consult with the Assistant Regional Director and the Regional Solicitor, who shall promptly take appropriate action, including compulsory process, if necessary." 29 CFR 1903.4 (1977). The Court views this regulation as an admission by the Secretary that no enforcement problem is generated by permitting employers to deny entry and delaying the inspection until a warrant has been obtained. I disagree. The regulation was promulgated against the background of a statutory right to immediate entry, of which covered employers are presumably aware and which Congress and the Secretary obviously thought would keep denials of entry to a minimum. In these circumstances, it was surely not unreasonable for the Secretary to adopt an orderly procedure for dealing with what he believed would be the occasional denial of entry. The regulation does not imply a judgment by the Secretary that delay caused by numerous denials of entry would be administratively acceptable.

Even if a warrant requirement does not "frustrate" the legislative purpose, the Court has no authority to impose an additional burden on the Secretary unless that burden is required to protect the employer's Fourth Amendment interests.[6] The essential function of the traditional warrant requirement is the interposition of a neutral magistrate between the citizen and the presumably zealous law enforcement officer so that there might be an objective determination of probable cause. But this purpose is not served by the newfangled inspection warrant. As the Court acknowledges, the inspector's "entitlement to inspect will not depend on his demonstrating probable cause to believe that conditions in violation of OSHA exist on the premises. … For purposes of an administrative search such as this, probable cause justifying the issuance of a warrant may be based … on a showing that 'reasonable legislative or administrative standards for conducting an … inspection are satisfied with respect to a particular [establishment].'" Ante, at 320. To obtain a warrant, the inspector need only show that "a specific business has been chosen for an OSHA search on the basis of a general administrative plan for the enforcement of the Act derived from neutral sources." Ante, at 321. Thus, the only question for the magistrate's consideration is whether the contemplated inspection deviates from an inspection schedule drawn up by higher level agency officials.

Unlike the traditional warrant, the inspection warrant provides no protection against the search itself for employers who the Government has no reason to suspect are violating OSHA regulations. The Court plainly accepts the proposition that random health and safety inspections are reasonable. It does not question Congress' determination that the public interest in workplaces free from health and safety hazards outweighs the employer's desire to conduct his business only in the presence of permittees, except in those rare instances when the Government has probable cause to suspect that the premises harbor a violation of the law.

What purposes, then, are served by the administrative warrant procedure? The inspection warrant purports to serve three functions: to inform the employer that the inspection is authorized by the statute, to advise him of the lawful limits of the inspection, and to assure him that the person demanding entry is an authorized inspector. *Camara v. Municipal Court,* 387 U.S. 523, 532. An examination of these functions in the OSHA context reveals that the inspection warrant adds little to the protections already afforded by the statute and pertinent regulations, and the slight additional benefit it might provide is insufficient to identify a constitutional violation or to justify overriding Congress' judgment that the power to conduct warrantless inspections is essential.

The inspection warrant is supposed to assure the employer that the inspection is in fact routine, and that the inspector has not improperly departed from the program of representative inspections established by responsible officials. But to the extent that harassment inspections would be reduced by the necessity of obtaining a warrant, the Secretary's present enforcement scheme would have precisely the same effect. The representative inspections are conducted "in accordance with criteria based upon accident experience and the number of employees exposed in particular industries." Ante, at 321 n. 17. If, under the present scheme, entry to covered premises is denied, the inspector can gain entry only by informing his administrative superiors of the refusal and seeking a court order requiring the employer to submit to the inspection. The inspector who would like to conduct a non-routine search is just as likely to be deterred by the prospect of informing his superiors of his intention and of making false representations to the court when he seeks compulsory process as by the prospect of having to make bad-faith representations in an ex parte warrant proceeding.

The other two asserted purposes of the administrative warrant are also adequately achieved under the existing scheme. If the employer has doubts about the official status of the inspector, he is given adequate opportunity to reassure himself in this regard before permitting entry. The OSHA inspector's statutory right to enter the premises is conditioned upon the presentation of appropriate credentials. 29 U.S.C. 657 (a) (1). These credentials state the inspector's name, identify him as an OSHA compliance officer, and contain his photograph and signature. If the employer still has doubts, he may make a toll-free call to verify the inspector's authority, *Usury v. Godfrey Brake & Supply Service, Inc.*, 545 F.2d 52, 54 (CA8 1976), or simply deny entry and await the presentation of a court order.

The warrant is not needed to inform the employer of the lawful limits of an OSHA inspection. The statute expressly provides that the inspector may enter all areas in a covered business "where work is performed by an employee of an employer," 29 U.S.C. 657 (a) (1), "to inspect and investigate during regular working hours and at other reasonable times, and within reasonable limits and in a reasonable manner ... all pertinent conditions, structures, machines, apparatus, devices, equipment, and materials therein." 29 U.S.C. 657 (a) (2). See also 29 CFR 1903 (1977). While it is true that the inspection power granted by Congress is broad, the warrant procedure required by the Court does not purport to restrict this power but simply to ensure that the employer is apprised of its scope. Since both the statute and the pertinent regulations perform this informational function, a warrant is superfluous.

Requiring the inspection warrant, therefore, adds little in the way of protection to that already provided under the existing enforcement scheme. In these circumstances, the warrant is essentially a formality. In view of the obviously enormous cost of enforcing a health and safety scheme of the dimensions of OSHA, this Court should not, in the guise of construing the Fourth Amendment, require formalities which merely place an additional strain on already overtaxed federal resources.

Congress, like this Court, has an obligation to obey the mandate of the Fourth Amendment. In the past the Court "has been particularly sensitive to the Amendment's broad standard of 'reasonableness' where ... authorizing statutes permitted the challenged searches." *Almeida-Sanchez v. United States,* 413 U.S. 266, 290 (WHITE, J., dissenting). In *United States v. Martinez-Fuerte,* 428 U.S. 543, for example, respondents challenged the routine stopping of vehicles to check for aliens at permanent checkpoints located away from the border. The checkpoints were established pursuant to statutory authority and their location and operation were governed by administrative criteria. The Court rejected respondents' argument that the constitutional reasonableness of the location and operation of the fixed checkpoints should be reviewed in a Camara warrant proceeding. The Court observed that the reassuring purposes of the inspection warrant were adequately served by the visible manifestations of authority exhibited at the fixed checkpoints.

Moreover, although the location and method of operation of the fixed checkpoints were deemed critical to the constitutional reasonableness of the challenged stops, the Court did not require Border Patrol officials to obtain a warrant based on a showing that the checkpoints were located and operated in accordance with administrative standards. Indeed, the Court observed that "[t]he choice of checkpoint locations must be left largely to the discretion of Border Patrol officials, to be exercised in accordance with statutes and regulations that may be applicable ... [and] [m]any incidents of checkpoint operation also must be committed to the discretion of such officials." 428 U.S., at 559-560, n. 13. The Court had no difficulty assuming that those officials responsible for allocating limited enforcement resources would be "unlikely to locate a checkpoint where it bears arbitrarily or oppressively on motorists as a class." Id., at 559.

The Court's recognition of Congress' role in balancing the public interest advanced by various regulatory statutes and the private interest in being free from arbitrary governmental intrusion has not been limited to situations in which, for example, Congress is exercising its special power to exclude aliens. Until today, we have not rejected a congressional judgment concerning the reasonableness of a category of regulatory inspections of commercial premises.[7] While businesses are unquestionably entitled to Fourth Amendment protection, we have "recognized that a business, by its special nature and voluntary existence, may open itself to intrusions that would not be permissible in a purely private context."

G.M. Leasing Corp. v. United States, 429 U.S. 338, 353. Thus, in *Colonnade Catering Corp. v. United States,* 397 U.S. 72, the Court recognized the reasonableness of a statutory authorization to inspect the premises of a caterer dealing in alcoholic beverages, noting that "Congress has broad power to design such powers of inspection under the liquor laws as it deems necessary to meet the evils at hand." Id., at 76. And in *United States v. Biswell,* 406 U.S. 311, the Court sustained the authority to conduct warrantless searches of firearm dealers under the Gun Control Act of 1968 primarily on the basis of the reasonableness of the congressional evaluation of the interests at stake.[8]

The Court, however, concludes that the deference accorded Congress in Biswell and Colonnade should be limited to situations where the evils addressed by the regulatory statute are peculiar to a specific industry and that industry is one which has long been subject to Government regulation. The Court reasons that only in those situations can it be said that a person who engages in business will be aware of and consent to routine, regulatory inspections. I cannot agree that the respect due the congressional judgment should be so narrowly confined.

In the first place, the longevity of a regulatory program does not, in my judgment, have any bearing on the reasonableness of routine inspections necessary to achieve adequate enforcement of that program. Congress' conception of what constitute urgent federal interests need not remain static. The recent vintage of public and congressional awareness of the dangers posed by health and safety hazards in the workplace is not a basis for according less respect to the considered judgment of Congress. Indeed, in Biswell the Court upheld an inspection program authorized by a regulatory statute enacted in 1968. The Court there noted that "[f]ederal regulation of the interstate traffic in firearms is not as deeply rooted in history as is governmental control of the liquor industry, but close scrutiny of this traffic is undeniably" an urgent federal interest. 406 U.S., at 315. Thus, the critical fact is the congressional determination that federal regulation would further significant public interests, not the date that determination was made.

In the second place, I see no basis for the Court's conclusion that a congressional determination that a category of regulatory inspections is reasonable need only be respected when Congress is legislating on an industry-by-industry basis. The pertinent inquiry is not whether the inspection program is authorized by a regulatory statute directed at a single industry, but whether Congress has limited the exercise of the inspection power to those commercial premises where the evils at which the statute is directed are to be found. Thus, in Biswell, if Congress had authorized inspections of all commercial premises as a means of restricting the illegal traffic in firearms, the Court would have found the inspection program unreasonable; the power to inspect was upheld because it was tailored to the subject matter of Congress' proper exercise of regulatory power. Similarly, OSHA is directed at health and safety hazards in the workplace, and the inspection power granted the Secretary extends only to those areas where such hazards are likely to be found.

Finally, the Court would distinguish the respect accorded Congress' judgment in Colonnade and Biswell on the ground that businesses engaged in the liquor and firearms industry "accept the burdens as well as the benefits of their trade."

Ante, at 313. In the Court's view, such businesses consent to the restrictions placed upon them, while it would be fiction to conclude that a businessman subject to OSHA consented to routine safety inspections. In fact, however, consent is fictional in both contexts. Here, as well as in Biswell, businesses are required to be aware of and comply with regulations governing their business activities. In both situations, the validity of the regulations depends not upon the consent of those regulated, but on the existence of a federal statute embodying a congressional determination that the public interest in the health of the Nation's workforce or the limitation of illegal firearms traffic outweighs the businessman's interest in preventing a Government inspector from viewing those areas of his premises which relate to the subject matter of the regulation.

The case before us involves an attempt to conduct a warrantless search of the working area of an electrical and plumbing contractor. The statute authorizes such an inspection during reasonable hours. The inspection is limited to those areas over which Congress has exercised its proper legislative authority.[9] The area is also one to which employees have regular access without any suggestion that the work performed or the equipment used has any special claim to confidentiality.[10] Congress has determined that industrial safety is an urgent federal interest requiring regulation and supervision, and further, that warrantless inspections are necessary to accomplish the safety goals of the legislation. While one may question the wisdom of pervasive governmental oversight of industrial life, I decline to question Congress' judgment that the inspection power is a necessary enforcement device in achieving the goals of a valid exercise of regulatory power.[11]

I respectfully dissent.

FOOTNOTES

[1] "The right of the people to be secure in their persons, houses, papers, and effects, against unreasonable searches and seizures, shall not be violated."

[2] "[A]nd no Warrants shall issue, but upon probable cause, supported by Oath or affirmation, and particularly describing the place to be searched, and the persons or things to be seized."

[3] J. Kandinsky, Search and Seizure and the Supreme Court 19 (1966).

[4] T. Taylor, Two Studies in Constitutional Interpretation 41 (1969).

[5] See v. Seattle, 387 U.S. 541, 547 (Clark, J., dissenting).

[6] When it passed OSHA, Congress was cognizant of the fact that in light of the enormity of the enforcement task "the number of inspections which it would be desirable to have made will undoubtedly for an unforeseeable period, exceed the capacity of the inspection force." Senate Committee on Labor and Public Welfare, Legislative History of the Occupational Safety and Health Act of 1970, 92d Cong., 1st Sess., 152 (Comm. Print 1971).

[7] The Court's rejection of a legislative judgment regarding the reasonableness of the OSHA inspection program is especially puzzling in light of recent decisions finding law enforcement practices constitutionally reasonable, even though those practices involved significantly more individual discretion than the OSHA program. See, for example, Terry v. Ohio, 392 U.S. 1; Adams v. Williams, 407 U.S. 143; Cady v. Dombrowski, 413 U.S. 433; South Dakota v. Opperman, 428 U.S. 364.

[8] The Court held:

> "In the context of a regulatory inspection system of business premises that is carefully limited in time, place, and scope, the legality of the search depends ... on the authority of a valid statute.
>
> We have little difficulty in concluding that where, as here, regulatory inspections further urgent federal interest, and the possibilities of abuse and the threat to privacy are not of impressive dimensions, the inspection may proceed without a warrant where specifically authorized by statute." 406 U.S., at 315, 317.

[9] What the Court actually decided in Camara v. Municipal Court, 387 U.S. 523, and See v. Seattle, 387 U.S. 541, does not require the result it reaches today. Camara involved a residence, rather than a business establishment; although the Fourth Amendment extends its protection to commercial buildings, the central importance of protecting residential privacy is manifest. The building involved in See was, of course, a commercial establishment, but a holding that a locked warehouse may not be entered pursuant to a general authorization to "enter all buildings and premises, except the interior of dwellings, as often as may be necessary," 387 U.S., at 541, need not be extended to cover more carefully delineated grants of authority. My view that

the See holding should be narrowly confined is influenced by my favorable opinion of the dissent written by Mr. Justice Clark and joined by Justices Harlan and STEW-ART. As Colonnade and Biswell demonstrate, however, the doctrine of stare decisis does not compel the Court to extend those cases to govern today's holding.

[10] The Act and pertinent regulation provide protection for any trade secrets of the employer. 29 U.S.C. 664-665; 29 CFR 1903.9 (1977).

[11] The decision today renders presumptively invalid numerous inspection provisions in federal regulatory statutes. For example, 30 U.S.C. 813 (Federal Coal Mine Health and Safety Act of 1969); 30 U.S.C. 723, 724 (Federal Metal and Nonmetallic Mine Safety Act); 21 U.S.C. 603 (inspection of meat and food products). That some of these provisions apply only to a single industry, as noted above, does not alter this fact. And the fact that some "envision resort to federal-court enforcement when entry is refused" is also irrelevant since the OSHA inspection program invalidated here requires compulsory process when a compliance inspector has been denied entry. Ante, at 321.

2005 WL 2697262 (O.S.H.R.C.)

OCCUPATIONAL SAFETY HEALTH REVIEW COMMISSION

*1 Secretary of Labor, Complainant
V. Saw Pipes USA, Inc., and Its Successors,
Respondent (OSHRC Docket No. 01-0422)
Final Order Date: September 28, 2005

Appearances

For the Complainant: Raquel Tamez, Esq., Madeleine Le, Esq., Office of the Solicitor, U.S. Department of Labor, Dallas, Texas
For the Respondent: Thomas H. Wilson, Esq., Julianne Merten, Esq., Michael J. Muskat, Esq., Vinson & Elkins, LLP, Houston, Texas

Before: Administrative Law Judge: James H. Barkley

Decision and Order

This proceeding arises under the Occupational Safety and Health Act of 1970 (29 U.S.C. Section 651 et seq.; hereafter called the "Act").

Respondent, Saw Pipes USA, Inc., and its successors (Saw), at all times relevant to this action maintained a place of business in Baytown, Texas, where it was engaged in manufacturing seamless welded line pipe (Tr. 272, 286). Respondent admits it is an employer engaged in a business affecting commerce and is subject to the requirements of the Act.

On July 27, 2000 the Occupational Safety and Health Administration (OSHA) began an inspection of Saw's Baytown, Texas, worksite. As a result of that inspection, Saw was issued a Willful citation alleging 67 violations of § 1904.2(a). The citation alleges sixty-six instances in which Saw failed to list recordable injuries or illnesses on its OSHA 200 form. Item 67 alleges that sixteen injuries or illnesses were incorrectly recorded. Penalties of $8000 per instance were proposed. By filing a timely notice of contest Saw brought this proceeding before the Occupational Safety and Health Review Commission (Commission). On January 22–23, 2002, a hearing was held in Houston, Texas. Pursuant to a joint statement of the parties filed at the hearing, the Secretary withdrew items 27 and 67 (subpart 16), and Saw withdrew its contest to the recordability of all but 18 of the 66 failures to record and all but 5 of the 16 items allegedly reported incorrectly. Saw continued to contest the classification of the citation as willful, and the proposed penalties. During the hearing, the Secretary also withdrew item 17 of Willful citation 1 (Tr. 157), and Saw withdrew its contest as to the recordability of items 29, 34, 38 and 49 and 63 (Tr. 146-148). Following the hearing, Complainant withdrew three of those items, numbers 29, 34, and 63 (Complainant's post-hearing brief, p. 37, 38,

47). Finally, in its post-hearing brief, Saw withdrew its notice of contest to items 11 and 20 based on the testimony presented at the hearing (Respondent's post-hearing brief, p. 2). The parties have submitted briefs on the issues remaining in dispute and this matter is ready for disposition.

FACTS

Robert Murphey, a certified industrial safety technician, has been Saw's safety manager since May 1998 (Tr. 64, 216; Exh. C-245). Murphey testified that, in addition to filling out the OSHA 200 logs, he is responsible for responding to and investigating all accidents and injuries that occur at the Saw facility (Tr. 65, 80-81). Plant managers and supervisors are instructed to inform him of injury accidents (Tr. 66-68). If Murphey is not in the plant, security is to be notified (Tr. 67). Security is instructed to notify Murphey, via pager or by way of a patrolman's report, of all accidents and injuries which occur outside of his shift (Tr. 69-71). Murphey testified that Saw's accident policy requires that all injuries be reported (Tr. 71, 113). Murphey testified that the reports are not always submitted in a timely manner, and that he has complained about the problem to his supervisor, Mr. Jones, to the plant manager, Mr. Turnipseed, and to the chief executive officer, Mr. Bhargava (Tr. 115).

*2 Carroll Caudill, Saw's maintenance manager, and Kurt Brodd, Saw's plant manager, both testified that when a Saw employee is injured, he notifies his supervisor. If required, medical treatment is immediately sought for the employee (Tr. 46, 160). Brodd testified that Robert Murphey is then contacted (Tr. 160). Where a reportable injury occurs, the supervisor fills out an accident or injury report which is submitted either to Robert Murphy, or to Murphey's supervisor, Gary Jones (Tr. 45-48, 58, 161). Brodd testified that it is Saw's policy to fill out an accident report in all cases (Tr. 162). Caudill, however, testified that reports are not always filled out for minor injuries (Tr. 58). Both Brodd and Caudill admitted that there are times when they have to be reminded to get a report to Murphey (Tr. 48, 56-57, 163, 168). However, Brodd stated that he was not aware of any accidents that were never reported to Mr. Murphey (Tr. 164). Caudill testified, if an employee's injury is serious enough to merit a trip to the hospital, the employee must obtain a doctor's release before returning to work (Tr. 53). According to Caudill, the release is returned to the personnel department (Tr. 53). Caudill was not aware of any instances where an employee returned to work without going through personnel (Tr. 54). Dr. Jerry McShane, owner of the San Augustine Industrial Clinic, where Saw employees have been treated since early 1999, testified that accident reports, which include Return to Work (RTW) restrictions, are faxed to Saw in virtually every case (Tr. 452-53, 468). Dr. Carl Davis, a physician associated with BayCoast, which provided patient care for Saw's employers in 1998, testified that it was the practice to fax a copy of any work restrictions to Saw (Tr. 409-10). Murphey testified that because it is his job to coordinate any work restrictions an employee might have as a result of his or her injury, any paperwork documenting physician imposed work restrictions comes to him (Tr. 75, 124). In addition to incident reports from supervisory personnel and plant security, and RTW forms from treating physicians, Murphey receives copies of the logs from the on-site clinic on a monthly basis (Tr. 75, Exh. C-248, C-249, C-250). If an employee visits the on-site clinic as a result of his or her injury, the employee is required to sign in on the clinic log (Tr. 73). Murphey testified that he uses the nurse's sign-in logs and any incident reports to assure that the OSHA 200 log is correct (Tr. 112). Finally, Murphey testified that he and Gary Jones review medical bills for employees who are referred to the off-site clinic. It is his job to verify any medical expenses billed before they are paid (Tr. 72, 121).

ALLEGED VIOLATIONS

Knowledge. Initially, this judge notes that all of the cited violations were discovered during a review of records obtained, considerably after the fact, in the course of the OSHA investigation. The records reviewed were in Saw's control or were contained in medical records controlled by Saw's agents. Such documents were available for the review of Saw's safety management at any time (Tr. 407, 469). Thus I find that even if Saw's safety managers lacked specific knowledge of any of the contested injuries, they could, with the exercise of reasonable diligence, have known of any medical treatment provided or work restriction imposed on any of the employees named in the citations. For this reason, Saw's constructive knowledge of the individual violations will not be further discussed.

*3 Willful citation 1, item 5 alleges that on March 2, 1998, Edward Giering sprained his left knee at Saw's facility. Complainant alleges that the sprain resulted in work restrictions and should have been, but was not recorded (Joint Statement, p. 1, Exh. C-12).

After spraining his knee Giering visited the Bay Coast Medical Center, where the knee was splinted in a brace supported with metal rods, which he was to wear for seven days (Giering Depo., p. 25). Giering returned to work with instructions to restrict the use of his leg, and to refrain from climbing, bending or stooping (Exh. C-12). In a deposition given on February 26, 2002, Giering, a tack welder, testified that he was instructed at the clinic to remain seated, and to keep his leg elevated (Giering Depo., p. 12). Giering stated that he tried doing his job sitting the first day, but was unable to perform his regular duties in that position (Giering Depo., p. 12-13). Giering testified that he was told that he "needed to be up there," and, believing his job was on the line, he resumed his duties despite his prescribed work restrictions.

The U.S. Department of Labor, Bureau of Labor Statistics publication, Recordkeeping Guidelines for Occupational Injuries and Illnesses, the "Blue Book" states that: Lost workday cases involving days of restricted work activity are those cases where, because of injury or illness, (1) The employee was assigned to another job on a temporary basis, or (2) the employee worked at a permanent job less than full time, or (3) the employee worked at his or her permanently assigned job but could not perform all the duties normally connected with it.

Restricted work activity occurs when the employee, because of the job-related injury or illness, is physically or mentally unable to perform all or any part of his or her normal assignment during all or part of the normal workday or shift. The emphasis is on the employee's inability to perform normal job duties over a normal work shift. (Exh. C-278, p. 48).

Saw argues that Giering was not unable to perform his job, because, at least after his first day back, he was able to stand on his braced left leg throughout his shift. Giering admitted that, because he believed that he would lose his job if he wasn't up on the line, he chose to ignore the physician's recommended work restrictions after the first day. This judge believes it would set a poor precedent to hold that the recordability of an injury may be based solely on an employee's behavior, where such employee acts under pressure from his employer to return to his normal duties against the advice of his physician. See, item 32, in which on February 2, 1999, the same employee reported to OccuCare Industrial Medicine Clinic, with "persistent" left ankle sprain, after ignoring the physician's work restrictions and returning to full duty after a January 15, 1999, sprain, again fearing the loss of his job (Giering Depo. p. 24; Tr. Exh. C-96).

*4 Because Mr. Giering was unable to perform his normal job duties without disregarding the advice of his physician, this injury was recordable.[1] Item 5 is affirmed.

Willful citation 1, item 21 alleges that on September 24, 1998, Craig Brodd suffered a contusion to his forearm at Saw's facility. Complainant alleges that the contusion required medical treatment and should have been, but was not recorded (Joint Statement, p. 1, Exh. C-57, C-58).

Medical Records from the BayCoast Clinic indicate that Brodd reported to the clinic in the early hours of September 25, where ice and Tylenol #3 were prescribed (Exh. C-58). Brodd visited Saw's on-site clinic later in that day, at which time Helen Stipe, the nurse on duty, examined him and made a note, Tylenol #3 TPD Q4H PRN, under the heading "Medications" (Exh. C-57). Complainant introduced no evidence from which this judge might ascertain the meaning of this notation.

The Blue Book states that the administration of a single dose of prescription medication on a first visit for a minor injury or discomfort is generally considered first aid treatment, and need not be recorded if the injury does not involve a loss of consciousness, or restriction of work or motion (Exh. C-278, p. 43). In the absence of any evidence that more than one dose of Tylenol #3 was prescribed, this judge cannot find that the cited injury was recordable. This item is vacated.

Willful citation 1, item 23 alleges that on September 29, 1998, Alberto Arredondo reported to the BayCoast Clinic complaining that a piece of metal had struck his right eye while he was working at Saw's facility. Complainant alleges that the resulting corneal abrasion required medical treatment and resulted in work restrictions which should have been, but were not recorded (Joint Statement, p. 2, Exh. C-63).

BayCoast's records indicate that Arredondo's vision was unaffected by the incident; however, Neosporin ophthalmic solution and an eye patch were applied to Arredondo's eye (Exh. C-63). The physician on duty, Richard Trifiro, then released Arredondo for work with instructions to refrain from driving any equipment (Exh. C-63). Arredondo returned for observation the following day, at which time Dr. Trifiro, who was unable to visualize a foreign body in Arredondo's eye, referred him to an ophthalmologist.

Complainant failed to introduce any evidence indicating that Mr. Arredondo was ever diagnosed with an injury requiring treatment beyond a single dose of Neosporin ointment, or that the eye patch Arredondo was required to wear interfered with the performance of his duties. Because the evidence does not establish that Mr. Arredondo suffered a recordable injury, item 23 is vacated.

Willful citation 1, item 31 alleges that on December 22, 1998, Edward Giering suffered an injury to his groin while working at Saw's facility. Complainant alleges that the groin injury resulted in work restrictions and should have been, but was not recorded (Joint Statement, p. 2, Exh. C-93).

*5 In his February 26, 2002, deposition, Giering testified that Gary Jones sent him to the on-site nurse, and subsequently had Ray Snell, Robert Murphey's assistant; take him to the San Augustine Industrial Clinic (Giering Depo., p. 15-16). Giering returned to work with instructions to refrain from repetitive lifting until January 5, 1999 (Exh. C-93). Prior to his injury Giering was employed in the quality assurance and compliance department. Part of his job, which occupied approximately an hour and a half each day, was to move 20 to 25 pound samples of cut steel piping used in expansion rate testing (Giering Depo., p. 14-18). After the injury, Giering could no longer lift the samples, and was strictly limited to performing testing (Tr. 17-19).

As noted above, where the employee can still work at his or her permanently assigned job but cannot perform all the duties normally connected with it, that injury involves restricted work activity. In this case, Mr. Giering, because of his job-related injury was physically unable to perform a part of his normal assignment during part of the normal workday or shift. The injury was recordable, and this item is affirmed.

Willful citation 1, item 32 alleges that on January 15, 1999, Edward Giering sprained his left ankle while working at Saw's facility. Complainant alleges that the ankle injury resulted in work restrictions and should have been, but was not recorded (Joint Statement, p. 2, Exh. C-95).

At his February 26, 2002, deposition, Giering testified that he was taken to the San Augustine Clinic on Gary Jones' instructions (Giering Depo., p. 21). Giering returned to work the same day with instructions to avoid excessive bearing of weight on his left foot, squatting, and climbing (Exh. C-95). Giering testified that his job required him "[t]o climb up to the scope and to the ID and OD welders" (Giering Depo. p. 24). Giering testified that his supervisor, Hank Gosnell, implied that Giering would lose his job if he could not perform the duties associated with it; Giering stated that he ignored the medical restrictions (Giering Depo., p. 24). On February 2, 1999, Giering visited the OccuCare Industrial Medicine Clinic, where he was diagnosed with "persistent" left ankle sprain. Giering was provided with an ankle brace and instructed to avoid excessive weight bearing on his left foot until further notice (Exh. C-96).

The record establishes that Giering could not perform his normal work activities without disregarding the restrictions imposed by his physician. He ignored those restrictions because he feared for his job, not because he felt the restrictions were unnecessary and, in fact, his injury persisted. For the reasons set forth in item 5, this judge finds that item 32 was recordable. This item is affirmed.

Willful citation 1, item 41 alleges that on March 23, 1999, Richard Lee strained his lower back while working at Saw's facility. Complainant alleges that the injury resulted in work restrictions and should have been, but was not recorded (Joint Statement, p. 2, Exh. C-120).

*6 At the hearing, Lee testified that Robert Murphy took him to the doctor, where he was provided with samples of 850 mg. Motrin and pain killers. X-rays were taken and Lee was given instructions for physical therapy (Tr. 251-52; Exh. C-120). Lee was allowed to return to work with restrictions; he was not to engage in excessive climbing, squatting or lifting of over 25 pounds (Tr. 252; Exh. C-120). After an April 7, 1999, follow-up appointment, Lee received instructions to remain on light duty until April 12, 1999 (Exh. C-120). Lee testified that his regular duties included helping out in other departments as needed. Lee testified that he helped the grinders at the expander on a daily basis (Tr. 255). Lee stated that, while helping out, he was required to squat, climb and lift over 25 pounds (Tr. 255). It was while he was helping out that he hurt his back (Tr. 254). Lee stated that his actual position was hydro tester/operator. That position does not require climbing, squatting, or lifting, and Lee was able to continue working as a hydro tester after his injury (Tr. 253-54).

Mr. Lee worked regularly in other departments, helping out on a daily basis. Because he was regularly assigned to assist the grinders at the expander, that duty was part of his normal job. Lee's inability to continue assisting in the expander department until his release from light duty on April 12, 1999, constitutes a recordable work restriction. This item is affirmed.

Willful citation 1, item 44 alleges that on April 8, 1999, Bobby Sindel suffered smoke inhalation while fighting a fire at Saw (Sindel Depo., p. 5-6; Exh. C-125, C-126). Complainant states that smoke inhalation is a recordable occupational illness which should have been, but was not recorded.

Sindel reported to Saw's in-house clinic on April 8, complaining of chest congestion and coughing. The nurse on duty reported that Sindel's throat was raw, but that his chest sounds were clear (Exh. C-125). Sindel was transported to OccuCare Clinic, where he was examined, and released for work without further treatment (Exh. C-125).

The Blue Book defines "occupational illness" as:

Any abnormal condition or disorder ... caused by exposure to environmental factors associated with employment. It includes acute and chronic illnesses or diseases which may be caused by inhalation, absorption, ingestion or direct contact.

Illnesses are to be listed in specific columns, by category. The relevant category in this case is: 7c. Respiratory conditions due to toxic agents. Examples: Pneumonitis, pharyngitis, rhinitis or acute congestion due to chemicals, dusts, gases or fumes.

The record does not establish that Sindel contracted an acute or chronic abnormal condition or disorder from the single exposure to smoke. Neither the in-house nurse, nor the physician at OccuCare reported finding any chest congestion (pneumonitis), or swelling of the throat (pharyngitis) or nasal passages (rhinitis). Because the Secretary failed to prove that Mr. Sindel was diagnosed with a recordable occupational illness, this item is vacated.

*7 Willful citation 1, item 47 alleges that on June 7, 1999, Saw employee Alejandro Dominguez suffered a corneal abrasion when debris flew into his left eye while he was grinding (Tr. 151; Exh. C-131, C-132). Complainant maintains that Dominguez received medical treatment, making the injury recordable.

Dominguez testified that after his injury, the on-site nurse flushed his eye (Tr. 151). The nurse noted that she provided Dominguez with Cortisporin ophthalmic solution and an eye patch (Exh. C-131). Robert Murphey then drove Dominguez to OccuCare (Tr. 152). The attending physician's notes indicate that he anesthetized the eye, removed a foreign body, and dosed Dominguez's eye with Blephamide, listed in the Physician's Desk Reference as a prescription medication. Dominguez was provided with an eye patch and released with a monocular vision restriction (Tr. 132; Exh. C-132). Dominguez testified that when the discomfort did not subside the following day, Mr. Murphey took him back to the clinic, at which time the doctor checked his eye again, and again dosed it with Blephamide, before sending him to an ophthalmologist (Tr. 153; Exh. C-132).

The Blue Book states that the removal of foreign bodies embedded in the eye is almost always a recordable injury (Exh. C-278, p. 43). In this case, it is more likely than not that the debris was embedded in Dominguez's eye, as the on-site nurse was unable to remove it with flushing. Moreover Dominguez was treated on multiple visits with the prescription medication, Blephamide. Injuries requiring multiple doses of prescription medications are recordable (Exh. C-278, p. 43). Item 47 is affirmed.

Willful Citation 1, item 48 alleges that on June 25, 1999, James Jones suffered a sprained left ankle (Exh. C-133, 134). Jones' injury required a brace, which resulted in work restrictions including instructions to avoid excessive standing, weight bearing on the left foot, and climbing (Exh. C-134). This judge takes note that in a February 6, 1998, Compliance Letter, the Secretary interpreted medical treatment to include the use of casts, splints and/or orthopedic devices designed to immobilize a body part. The Secretary's interpretation is reasonable, and this item is affirmed.

Willful Citation 1, item 67(5) alleges that on June 25, 1998 Marco Contresas strained his back while lifting ground wires (Exh. C-199, C-200). Contresas was treated at Bay Coast Clinic and was released for work with restrictions consisting of no climbing, bending, stooping, or lifting of any weight in excess of ten pounds for five to seven days (Exh. C-201).

The injury report that Robert Murphey filed with Texas' workers' compensation states that Contresas, a welder, was performing his regular job when injured (Exh. C-198). This judge finds it more likely than not that Contresas' restrictions would prevent him from stooping or bending to lift ground wires, a part of his normal duties, and the very activity which caused his injury. Item 67(5) is affirmed.

*8 Willful Citation 1, item 67(8) alleges that on March 11, 1999, James Hedden, a welder, strained his right wrist and hand while using a wrench. Hedden was treated, his wrist splinted and he was allowed to return to work with restrictions (Exh. C-213,

C-214, C-215, C-216). Hedden was instructed to avoid excessive use of his right hand and repetitive lifting of twenty pounds or more (Exh. C-216). Hedden did not testify, and in the absence of any evidence as to his normal duties, this judge cannot find that he was unable to perform his job while observing the listed restrictions. Item 67(8) is vacated.

Willful Citation 1, item 67(11) alleges that on March 9, 2000, Gary Cooper, a millwright, cut his right thumb (Exh. C-226). After receiving stitches and a tetanus shot at the clinic, Cooper returned to work with restrictions including instructions to avoid excessive pressure and/or use of his right hand, and to keep the wound dry and clean (Tr. 394-95; Exh. C-226). At the hearing, Cooper testified that his job requires that he use power and hand tools, and that he is right-handed (Tr. 395). Cooper testified that he could perform most of his job duties using his left hand; however, he could not operate a torch, which required him to use his thumb (Tr. 397).

Using a torch was part of Cooper's job, and the restriction which prevented him from full use of his right hand was recordable. This item is affirmed.

Willful Citation 1, item 67(13) alleges that on April 18, 2000, Michael Ellison, a welder, suffered a laceration to his right palm (Exh. C-230). After seeing the on-site nurse, Ellison went to the San Augustine clinic, where he was given stitches, and allowed to return to work with restrictions (Exh. C- 231). Ellison was instructed to avoid excessive use of his right hand and to keep the wound dry and clean (Exh. C-232). On April 19, and again on April 28, Ellison returned to San Augustine at which time the attending physician noted "no welding" on his Return to Work (RTW) form (Exh. C-232). Ellison was returned to full duty on May 2, 2000.

This judge finds it more likely than not that Ellison, a welder, would be unable to perform his normal duties with a medical restriction barring him from welding. Item 67(13) is affirmed.

Willful Citation 1, item 67(15) alleges that on Don Schwartz suffered a foreign body to his right eye. After the foreign body was removed with an algar brush at the San Augustine clinic, Schwartz returned to work with instructions to refrain from excessive monocular vision (Exh. C-235). Schwartz did not testify at the hearing, and in the absence of any evidence as to his normal duties, this judge cannot find that he was unable to perform his job with restricted vision. Item 67(15) is vacated.

WILLFULNESS

The Secretary maintains that Saw engaged in a deliberate pattern of under-reporting workplace injuries in order to avoid being targeted for OSHA inspection. Complainant argues that Saw's practice of under-reporting is demonstrated by a drop in its lost workday injury and illness (LWDII) rate from 9.6 in 1996 to reported rates of 2.7, 3.1 and 4.2 in the following years, 1997, 1998, and 1999. Complainant points out that Saw's witnesses can recall no significant changes that would have impacted safety at the Saw plant during 1997 and the following years, and cannot account for the dramatic drop in the LWDII during that time. Complainant's witnesses, on the other hand, testified that they believed the drop in the rate was the result of deliberate under-reporting of minor injuries pursuant to policy developed by Gary Jones. Complainant argues that its theory explains Jones' deliberate decision not to record injuries suffered by temporary laborers, and his decision to change Saw's Standard Industrial Classification (SIC) code in 1998.

*9 Saw maintains that the drop in the LWDII, the failure to record minor reportable injuries and injuries to temporary workers, and the change in Saw's SIC code were all attributable to unrelated misinterpretations of OSHA requirements. Saw argues that any and all misinterpretations of OSHA requirements were made in good faith. At no time did Saw intend to circumvent an OSHA inspection.

FACTS

Charles Kutan is currently the production planning manager for Saw (Tr. 257). Kutan was originally hired in June 1995 to serve as Saw's human resources and safety director (Tr. 257-58). Kutan testified that, in that position, it was his responsibility, in addition to dealing with personnel issues, to investigate accidents and to maintain the OSHA 200 logs (Tr. 261). Kutan stated that he maintained the logs until mid-1997 (Tr. 261). According to Kutan, his experience prior to 1995 was in shipping (Tr. 267). In order to fulfill his new safety duties at Saw, he had to familiarize himself with OSHA regulations regarding recordkeeping by conferring with Saw's Workman's Compensation carrier, and by reading OSHA publications, including the 200 form and the Blue Book (Tr. 267-68; Exh. C-278). Kutan testified that it was his practice to list recordable injuries on the OSHA 200 as they occurred (Tr. 266). Robert Whitmore, chief of OSHA's division of recordkeeping requirements, testified that in 1996 Charles Kutan submitted an LWDII rate of 9.6 on the 1996 OSHA 196 Summary form, based on information contained in 1996 OSHA 200, the only complete year for which Kutan maintained the 200 form (Tr. 305).

Prior to joining Saw, Gary Jones, who is an attorney, was employed for nineteen years as the employee relations representative at Brown & Root. As part of his job duties at Brown & Root, Jones dealt with a number of governmental agencies, including OSHA. He attended safety seminars, met with OSHA investigators, and responded to OSHA complaints (Tr. 182). During his tenure with Brown & Root, Jones occasionally reviewed the OSHA 200 logs (Tr. 200). He was familiar with the Blue Book (Tr. 200-01).

Jones took over as director of human relations at Saw in July 1997 (Tr. 181). Jones also acted as Saw's safety director, and his duties included ensuring compliance with OSHA reporting requirements (Tr. 206-07). In August 1997, Jones hired Ronnie Johnson to act as safety manager for Saw (Tr. 208). Jones testified that he chose Johnson because he was a medic, and was familiar with OSHA regulations (Tr. 208). Jones testified that he considered the OSHA 200 log to be Johnson's responsibility; however, at some point Jones became aware that Johnson was not keeping the log. Jones testified that, together, he and Johnson brought the log up to date (Tr. 209, 213). Jones testified that he never again consulted with Johnson about the recordability of any particular injuries (Tr. 215). According to Jones, Johnson completed and certified the 1997 OSHA 200 logs (Tr. 214; Exh. R-1). Jones, however, completed and certified the OSHA Form 196 Summary for 1997 (Tr. 207; Exh. C-276, p. 3). The LWDII rate Jones submitted for 1997 was 2.7 (Tr. 305).

***10** Robert Murphey, who replaced Johnson as safety manager in 1998, testified that he was familiar with OSHA regulations and standards, including the Blue Book (Tr. 65). As Saw's safety manager, he was responsible for responding to accidents and injuries and for maintaining the OSHA 200 logs (Tr. 65, 116). Though Murphey reported directly to Jones, he testified that Jones did not instruct him in the correct means of completing the logs (Tr. 64, 111, 216). As noted above, Murphey used the nurse's sign-in logs and accident and incident reports to assure the accuracy of the OSHA 200 log (Tr. 112). Murphey stated that he made every attempt to follow up on accidents and incidents at the Saw facility in order to ascertain whether all recordable injuries had been reported (Tr. 139-40). Murphey maintained that if he failed to record an injury, it was either because the nurse failed to turn over forms specifying medical treatment or work restrictions, or because supervisors failed to turn in accident reports (Tr. 124-126). Murphey completed and certified the OSHA 200 logs for 1998 and 1999, and kept the logs for the first part of 2000 (Tr. 98, 106-107, 111). Mr. Murphey was aware that LWDII rates were used to target OSHA inspections, though he did not know exactly what rates would trigger an inspection (Tr. 131). The LWDII rates for 1998 and 1999 were 3.1 and 4.2, respectively (Tr. 305). During the first half of 2000, Saw's LWDII rate was 0 (Tr. 305).

Craig Wetherington was hired in May 1998 to act as the safety director for Jindal Steel, a plate steel manufacturing facility located on the same fifty-nine-acre tract which houses Saw; both were part of an older USX steel mill facility (See, Secretary v. Jindal United Steel Corp. (Jindal), Docket No. 00-2231 [petition for discretionary review pending before the Commission] Tr. 583).[1] Jindal produces the plate steel which Saw uses for producing pipe. The two companies share some common ownership; Saw Pipes holds stock in Jindal (Jindal, Exh. A, pp. 141, 595). Wetherington was told when he was hired that Jones was the administrator of the safety program for both Jindal and Saw pipes (Jindal, Exh. A, pp. 66, 126), and that he had dual responsibility to both Jindal's production manager and to Gary Jones (Jindal, Exh. A, pp. 49, 66-67, 76, 98-99, 113). Wetherington stated that Jones instructed him not to record any injuries which were not also reported to Texas Workers' Compensation (Jindal, Exh. A, pp. 65, 68, 77-80). Jones also specifically told Wetherington that injuries suffered by temporary laborers would be reported by the agency through which the laborer was employed, and that Wetherington was not to report those injuries (Jindal, Exh. A, p. 74). Wetherington testified that when he told Jones this did not conform to OSHA reporting guidelines, Jones told him he interpreted the guidelines differently (Jindal, Exh. A, pp. 74, 93). As a result, Wetherington did not report any injuries to temporary laborers; he did not do any follow up on temporary employees who had been injured (Jindal, Exh. A, p. 81). Wetherington testified that he filled out the OSHA 200 forms in accordance with instructions provided him by Gary Jones (Jindal, Exh. A, p. 68), but told Jones he did not believe the 200 log was being filled out correctly. Jones told Wetherington that he interpreted OSHA regulations differently and that if Wetherington didn't want to record injuries according to his interpretation, Wetherington could find another job (Jindal, Exh. A, pp. 68-69, 555). Wetherington believed that Jones' intent was to misrepresent the number of accidents at the plant in order to evade inspections by OSHA by reporting a low LWDII (Jindal, Exh. A, pp. 109-10, 115-116).

*11 Ronnie Johnson testified that Gary Jones maintained the OSHA 200 logs while he was employed at Saw (Jindal, Exh. E, p. 401). Johnson stated that he was aware that otherwise recordable injuries were not being recorded on the OSHA 200 logs in an attempt to keep the numbers down (Jindal, Exh. E, pp. 396, 402-03). According to Johnson, Jones would not record an injury if Saw paid the injured employee's medical costs in house rather than submitting a Workman's Compensation claim (Jindal, Exh. E, pp. 395-96). In addition, Johnson testified, he knew that Jones was not recording otherwise recordable injuries if those injuries were suffered by temporary employees (Jindal, Exh. E, p. 397).

Carl Davis, the Director of Occupational Medicine for the Jacinto Medical Group in Baytown, Texas (Tr. 399), testified that in October 1998 he became aware that the Jacinto Medical Group supervised the Licensed Vocational Nurse (LVN) from BayCoast Medical Center who ran Saw's onsite medical facility (Tr. 400-01). On or about October 28, 1998, Davis visited the Saw facility, where he met with the LVN, Helen Stipe, Gary Jones and Robert Murphey (Tr. 402). During that visit, Davis testified, Jones took him aside and informed him that Saw "did not have recordable injuries" (Tr. 403). Davis testified that he subsequently spoke with Jones two or three times a week (Tr. 406). During those conversations, Jones often criticized Davis' decisions to restrict employees' work activities (Tr. 405). Davis testified that Jones specifically asked him not to place employees under work restrictions (Tr. 445-46). Specifically, Davis stated:

Well, in the course of reporting patient status to Mr. Jones and in conversations we had when he would call me, Mr. Jones was, many times, critical of the fact that restrictions had been placed on the patients' work activity. Some of them were less than friendly. They were generally critical if any restriction had been placed. He, on more than one occasion, asked me to write "No restrictions." (Tr. 405).

In December 1998, Saw terminated its contract with BayCoast (Tr. 405).[2] Robert Whitmore testified that OSHA's programmed inspections are based on employer LWDII rates (Tr. 364). The OSHA Occupational Injury and Illness Data Collection form, which is mailed to employers, including Saw, informs those employers that the injury and illness data collected from them "will be used to focus OSHA activities" including inspections (Tr. 365; Exh. C-274, p. 2, C-275, p. 2). The formula OSHA uses for targeting inspections changes frequently. For instance, in 1998, OSHA targeted employers for inspection if the employer's 1996 LWDII rate exceeded the average for its Standard Industrial Classification (SIC) code (Tr. 341). In 1999 OSHA used data from employers' 1997 OSHA form 196 injury summaries (Tr. 336). Employers with LWDIIs exceeding 16 were placed on OSHA's primary targeting list, while employees with LWDIIs between 10 and 16 were placed on the secondary list (Tr. 337-339). In 2000, inspections were planned for employers reporting an LWDII of over 14 on their 1998 injury summaries (Tr. 330-31, 339). Employers reporting an LWDII between 8 and 14 were placed on a secondary list, and might or might not be inspected, depending on OSHA's workload (Tr. 331). Whitmore testified that in any given year, employers would not know what injury rate would trigger an inspection two years down the line (Tr. 340-41). However, employers with extremely low LWDII rates were "not even on [OSHA's] radar screen" (Tr. 364). According to Whitmore,

OSHA has not targeted an employer with an LWDII of less than 4.6 since its inception (Tr. 433-44).

*12 SIC Codes. During 1995 and 1996, while Charles Kutan kept the OSHA logs, Saw listed its SIC code as 3317 Steel Pipe and Tubes (Tr. 269-70, 308; Exh. C-276, C-300). The description of the 3317 industrial code is: "Establishments primarily engaged in the production of welded or seamless steel pipe and tubes and heavy riveted steel pipe from purchased materials." Examples of the manufactured product include "pipe seamless steel-mfpm" (Exh. C-300). Charles Kutan testified that Saw was, and is, in the business of producing welded line pipe (Tr. 272). Kutan could not recall specifically telling Gary Jones Saw's SIC code, but stated that he would have shown him the paperwork, including the OSHA summaries on which the 3317 SIC code was set forth (Tr. 271, 293). Mr. Whitmore testified that during his formal interview with Dilip Bhargava, Saw's then president and CEO (Tr. 172), Bhargava told him that he believed that Saw fell under the 3317 industrial code (Tr. 391).

Whitmore testified that, in 1998, OSHA attempted to conduct a programmed inspection of Saw's facility based on Saw's 1996 LWDII rate, which at 9.55 exceeded the industry average of 5.6 for SIC 3317 (Tr. 308-09, 317-318). Gary Jones turned the OSHA inspectors away, claiming that Saw's SIC code was not 3317, but 3443 (Tr. 202-204, 232, 315, 318).[3] The description for SIC code 3443 Fabricated Plate Work (Boiler Shops) is: "Establishments primarily engaged in manufacturing power and marine boilers, pressure and non-pressure tanks, processing and storage vessels, heat exchangers, elements and similar products, by the process of cutting, forming and joining metal plates, shapes, bars, sheet, pipe mill products and tubing to custom or standard design, for factory or field assembly." (Exh. C-301). Chuck Kutan testified that Saw never manufactured boilers, pressure or non-pressure tanks, heat exchangers and/or elements (Tr. 272-273). Saw claims the 3443 classification was made in good faith, based on a bullet on the bottom of a second page listing possible sample products which includes "Pipe, large diameter: metal plate-made by plate fabricators." Though Saw does fabricate large diameter pipe from steel plates, Saw does not fabricate metal plate. All its plate is purchased (Tr. 110). Whitmore testified that no reasonable person could have believed that 3443 was the applicable SIC for Saw (Tr. 349-50).

Temporary employees. Saw admits that injuries suffered by temporary employees were not recorded on its OSHA 200 Logs or figured into its LWDII rates. Robert Murphey testified that he consulted the Blue Book (Tr. 132), which states in relevant part:

If [a temporary laborer is] subject to the supervision of the using firm, the temporary help supply service contractor is acting merely as a personnel department for the using firm, and the using firm must keep the records for the personnel supplied by the service. If the temporary workers remain subject primarily to the supervision of the supply service, the records must be kept by the service. In short, the records should usually be kept by the firm responsible for the day-to-day direction of the employee's activities.

***13** (Exh. C-278, p. 24). Murphey, however, testified that he believed injuries to temporary workers were recorded by Labor Ready, the personnel agency that supplied Saw's temporary laborers, because the Labor Ready representative, Gary Young, once asked him for a blank 200 log (Tr. 97-98, 110, 122).

Murphey also testified that he read the instructions for filling out the OSHA injury and illness data collection form in 1998 and 1999 (Tr. 135). The form asks employers to "Tell us about your employees and the hours they worked." The form directs the employer to estimate its average number of employees by "[adding] together the number of employees your establishment paid in every pay period during 1998 [and 1999]. Include all employees: full time, part time, temporary, seasonal, salaried and hourly," and divide the answer by the number of pay periods (Exh. C-274, C-275, p. 3). The form goes on to ask how many hours employees actually worked during the relevant years and directs the employer to "[I]include hours worked by salaried, hourly, part-time and seasonal workers, as well as hours workers subject to day to day supervision by your establishment" (e.g. temporary help service workers) (Exh. C-274, C-275, p. 3). Murphey stated that he either "missed that part," or "just didn't get [it] down." (Tr. 136).

Saw argues that Murphey was not the only one who misunderstood the reporting requirements. Labor Ready, the staffing agency that provided laborers for Jindal Steel kept its own OSHA 200 log for laborers supervised by Jindal (Jindal, Exh. C, pp. 231-32, 562). In addition, after reviewing Chapter IV of OSHA's Blue Book, Gary Jones advised Jindal's safety manager, Craig Wetherington, not to record injuries to temporary laborers. Though Wetherington told Jones that he believed injuries suffered by temporary employees were recordable, Jones insisted that such injuries should be recorded by the staffing agencies that provided the laborers (Jindal, Exh. A, pp. 221-22; Jindal; Tr. 561-62, 588-90).

Discussion

The Commission has defined a willful violation as one "committed with intentional, knowing or voluntary disregard for the requirements of the Act or with plain indifference to employee safety." Valdak Corp., 17 BNA OSHC 1135, 1136, 1993-95 CCH OSHD ¶30,759, p. 42,740 (No. 93-239, 1995), aff'd, 73 F.3d 1466 (8th Cir. 1996). Moreover, a series of disparate violations may be found to be willful if there is evidence that such violations are part of a deliberate pattern, practice, or course of conduct. See, Kaspar Wire Works, Inc. (Kaspar), 18 BNA OSHC 2178, 2000 BNA OSHC ¶32,134 (No. 90-2775, 2000), appeal filed, No. 00-1392 (D.C. Cir. Sept. 26, 2000).

***14** The Secretary maintains that Saw engaged in a deliberate pattern of underreporting in order to avoid being targeted for OSHA inspection. Saw maintains that the evidence in the record is insufficient to demonstrate Saw's motives, arguing that neither Gary Jones nor Robert Murphey knew or could have predicted the criteria OSHA would be using to target specific industries in any given year. However, any safety and health officer who knows that employers' LWDII are used to target inspections, would also know that an employer's chances of being inspected are reduced if it produces an extremely low LWDII. In any event, the Commission has held that the Secretary need not show that the employer had an evil or malicious motive to show willfulness. In other words, Complainant needn't prove why Saw engaged in the demonstrated practice of under-reporting. "The state of mind required for a willful violation need be only knowing, voluntary, or intentional." Id. at 2183-84. Complainant need only show that the demonstrated practice of under-reporting was deliberate. During the relevant time period, Saw failed to record 63.7% of all reportable cases on its OSHA 200. Of those cases that were reported, 40.6% were erroneously listed as involving no lost or restricted work days.[4] The sheer number of unreported injuries in this case cannot be explained by simple negligence. Those numbers, when viewed in light of the testimonial evidence, discussed more fully below, are sufficient to establish that the comprehensive under-reporting of injuries and lost workdays in this case was knowing and intentional and, therefore, willful.

Charles Kutan was hired in 1995 to act as Saw's human resources and safety director. Though Kutan had no prior training in safety and health, and though he had dual responsibility for human relations and safety, Kutan managed to correctly determine Saw's SIC code, and to record injuries as they occurred. A month after Gary Jones took over as human resources director in 1997, he was able to hire a safety director, first Ronnie Johnson and then Robert Murphey, with whom he could effectively split the duties that Kutan previously handled alone. Jones, Johnson and Murphey all had considerable safety and health experience. They inherited a pre-existing reporting system under which plant managers and supervisors were to report injuries and accidents to the safety director. Additional procedures, which should have increased reporting accuracy, were instituted. An on-site clinic was established, where a Licensed Vocational Nurse (LVN) could evaluate injuries, and provide limited treatment. The on-site nurse kept a sign-in log, which the safety manager could use to double check his OSHA log. When injured employees were seen by a physician, Return to Work (RTW) forms detailing employees' work restrictions were faxed to Saw from the off-site clinic. Despite the increase in the number and expertise of safety personnel, however, the timeliness and accuracy in recording injuries plummeted after Jones took over the management of Saw's safety department. Neither Jones nor Saw provided any explanation for the drop in the LWDII on Saw's 1997 OSHA 196 summary, which was 2.7, down from 9.55 the year before. Reporting accuracy did not improve when Robert Murphey took over responsibility for the logs in May 1998. Though Murphey claimed to have made every attempt to follow up on accidents and incidents at the Saw facility in order to ascertain whether all recordable injuries had been reported, and maintained that he used available accident reports and the sign-in logs from the on-site infirmary to ensure the accuracy of the OSHA 200 logs, injuries and lost workdays continued to be under-reported. LWDII rates of 3.06 and 4.18 were submitted for 1998 and 1999, respectively.

*15 Given the reporting system available to Murphey, this judge cannot credit Saw's contention that he did not possess actual knowledge, either of the occurrence, or of the extent, of cited employee injuries. First, Saw points out that Murphey "relied heavily" on accident reports submitted from first-line supervisors, reports which Saw maintains were infrequently filled out. Saw further argues that there was not enough information on the accident forms returned from the off-site clinic for Murphey to make a determination of recordability. Finally, Saw argues that the nurses' log, on which Murphey also relied, was incomplete and failed to provide Murphey with notice of cited injuries. Saw's arguments fail because, though there were lapses in reporting, there were so many redundancies in the way injuries were reported at Saw, it would have been impossible for almost two thirds of reportable injuries to go unrecognized by the safety manager. For instance, Saw's justifications do not explain Murphey's failure to record:

Item 22. On September 28, 1998, Myron Ferguson suffered a laceration to his thumb. His supervisor filled out an incident report (Exh. C-60). Nurse Stipes saw Ferguson on-site, and after dressing the wound she referred him to BayCoast emergency. In her clinical notes Stipes noted "considerable" bleeding (Exh. C-61). The clinic log clearly reflects that Mr. Ferguson was sent to the emergency room (Exh. C-248, p. 4). Ferguson's wound was sutured, and his treatment was listed on the RTW form generated by BayCoast for Saw (Exh. C-62, p. 6). In this case, there was an incident report, a log entry indicating a referral to the emergency room and an RTW form stating that sutures, recordable medical treatment had been provided (Exh. C-278, p. 43). It is difficult to conceive how Murphey's failure to report this injury could have been anything other than intentional.

Item 40. Alan Edgar suffered a probable second degree burn on March 22, 1999. Edgar reported the injury to Murphey on March 24; Murphey treated Edgar's burn for two days before taking him to the on-site clinic on March 26, after which Saw's nurse, Helen Stipe, treated the burn with repeated applications of a prescription antiseptic ointment (Exh. C-118). The log from the on-site clinic lists three visits by Alan Edgar; each listing contains a description of his injury and the medical treatment received (Exh. C-249, p. 5).

Saw is correct in stating that there was no accident report filed in this case. However, Murphey had actual knowledge of this injury, having treated the victim himself and taken him to the clinic nurse. He had access to the log, which details multiple repeat visits to the clinic for antiseptic dressings. The Blue Book clearly states that the "[a]pplication of ANTISEPTICS during second or subsequent visit to medical personnel" is recordable (Exh. C-278, p. 43). Saw's explanations do not account for Murphey's failure to record this item.

*16 Item 42. On April 2, 1999, Jose Garza suffered first and second degree burns to his shoulder. Bob Murphey treated the injury before sending Garza to the on-site clinic. Garza also visited the OccuCare clinic, where an accident report was generated, indicating that the burns were diagnosed as third degree. Ms. Stipe continued to administer repeated applications of antiseptics until April 8 (Exh. C-121, C-122). Five separate visits to the on-site clinic are listed in the clinic log; the record of each visit includes a description of the injury and the treatment administered (Exh. C-249, p. 6).

As in item 40, Murphey was shown to have actual knowledge of the unrecorded injury. Although the Blue Book clearly states that the "[a]pplication of ANTISEPTICS during second or subsequent visit to medical personnel" is recordable (Exh. C-278, p. 43), Murphey inexplicably failed to record this injury.

Item 51. On August 9, 1999, Reyes Garcia suffered a severe contusion to his right shin. Robert Murphey applied an ice pack to the site for twenty minutes before sending Garcia to the on-site clinic, where an additional ice pack was applied (Exh. C-142). After returning to work three days later, Garcia returned to the clinic no less than eight times for repeated treatments with ice and heat packs (Exh. C-142, C-249, p. 10). Each of the eight entries on the clinic log refers to both the shin injury and the treatment provided (Exh. C-249, p. 10).

Again, Murphey had first-hand knowledge of the occurrence of this injury, and the nurse's log clearly details the course of Mr. Garcia's treatment. The Blue Book clearly states that the "application of hot or cold COMPRESS(ES) during second or subsequent visit to medical personnel" is recordable medical treatment (Exh. C-278, p. 43). Yet Murphey failed to record this item.

The evidence establishes by a preponderance of the evidence that Murphey had actual knowledge of each of the clearly recordable injuries discussed above. This judge can only conclude that his failure to record these injuries was intentional. The Secretary made no attempt to prove that each of the cited items demonstrates such patent indicia of willfulness, and this judge declines to examine each instance for such evidence. Because the Secretary has alleged a single pattern of under-reporting, this sampling of clearly recordable injuries, of which Murphey had first hand knowledge, and yet failed to record is material evidence that Saw engaged in a deliberate practice of under-reporting.

Saw's intentional failure to record injuries sustained by temporary laborers is also evidence of an intentional practice of under-recording. Saw argues that its practice reflected its employees' good faith misinterpretation of the OSHA guidelines referring to temporary workers, and so cannot, as a matter of law, be found willful. However, though it is true that a finding of willfulness is not justified where the employer had a good faith opinion that the violative conditions conformed to the requirements of the cited standard, the Commission has held that the test of good faith for these purposes is an objective one. The employer's belief concerning a factual matter, or as here, concerning the interpretation of a standard, must have been reasonable under the circumstances. Calang Corp., 14 BNA OSHC 1789, 1987-90 CCH OSHD ¶29,080 (No. 85-319, 1990). In this case it is abundantly clear that the temporary labor contractor with whom Saw dealt acted merely as a conduit to provide labor for Saw. Saw was responsible for the day-to-day direction of the employees' activities. According to the plain language of the Blue Book, Saw was the firm responsible for keeping the required OSHA injury and illness records. The fact that the temporary agency also recorded injuries sustained by its temporary workers did not affect Saw's duty under the standard. For Jones, who is an attorney, and who worked in the employee relations and compliance department at Brown & Root before joining Saw Pipes, to have interpreted the Blue Book in any other way is not only unreasonable, but incredible. Moreover, the record establishes that Jindal's safety manager, Craig Wetherington, was not only knowledgeable in safety and health matters but specifically told Jones that the supervising employer had a duty to record such injuries. Jones not only ignored Wetherington, but instructed him to continue under-reporting Jindal's injuries or risk losing his job. Wetherington's testimony describes Jones' dealings with Jindal's safety personnel, but is strong evidence of state of mind, and of his bad faith in dealing with OSHA. Finally, Robert Murphey's claim to have "missed" those portions of the OSHA forms that clearly instructed him to include temporary workers in his calculations is no more believable than Jones' misinterpretation of OSHA regulations and is not credited.

*17 Lastly this judge finds that Jones' inexplicable change of Saw's SIC code to the code for boiler shops constitutes additional evidence of bad faith. After reading the relevant SIC definitions, this judge agrees that no reasonable person, and certainly not a professional with Jones' experience in safety and health, could have believed that 3443 was the applicable SIC for Saw. Even if Saw is correct in arguing that Jones could not have known that changing the SIC code would affect the likelihood of an OSHA inspection, the fact remains that Jones actually turned an OSHA inspector away in 1998 on grounds that OSHA had the wrong SIC code.

This judge credits the testimony of complainant's witnesses, Craig Wetherington and Carl Davis, both of whom concluded, based on their dealings with Gary Jones, that Saw was pursuing a course of conduct intended to reduce the number of injuries reported to OSHA. Those witnesses have no stake in the outcome of this matter, and this judge finds their testimony compelling. When viewed together with the percentage of unrecorded injuries occurring during the relevant time periods, and the clear recordability of a number of those injuries, this judge can only conclude that Saw's failure to report and under-reporting of injuries was part of a pattern of deliberate, that is, willful conduct.

PENALTY

While the Secretary has often exercised her authority to group related violations and propose a single penalty for a number of related violations, she chose not to do so in this case. Rather the violations were assessed individually, at $8000 apiece, resulting in an aggregate penalty of $536,000. While it is clear that the Secretary may propose multiple penalties for separate violations of the record keeping standard, Commission review of the proposed penalty is de novo, and the judge has discretion to assess a single penalty if deemed appropriate. See, Pepperidge Farm, Inc., 17, BNA OSHC 1993, 1997 CCH OSHD ¶31,301; citing, Miniature Nut and Screw Corp., 17 BNA OSHC 1557, 1996 CCH OSHD ¶30,986 (No. 93-2535, 1996). Complainant's departure from grouping similar violations results in an aggregate penalty that is disproportionate to the violation and overstates its gravity. The record demonstrates that Complainant did not base the gravity of Saw's under-reporting on the severity of the actual unreported injuries, which consist mainly of minor burns, contusions and foreign bodies in the eye. Rather Complainant argues that the gravity of the violation was high because Saw attempted to avoid inspections by falsely reducing its LWDII. However, the record does not clearly establish that Saw would actually have been inspected had it reported all injuries. A review of the safety violations, which, unlike record keeping violations, pose actual safety hazards to affected employees, does not indicate that this employer was attempting to cover up serious safety hazards by deferring OSHA inspections. In fact, for safety violations found at Saw's facility during the July 2000 inspection, OSHA proposed penalties of only $82,000, less than 16 percent of the penalties assessed for the record keeping violations.

In addition, at the hearing Complainant advanced the theory that Saw engaged in a single pattern and practice of willful non-reporting. Complainant argues that having generally established such a pattern, it follows that all unrecorded and under-recorded violations were willful. Complainant made no attempt to prove that Saw willfully failed to record each individual injury. The evidentiary path chosen by Complainant was considerably less onerous than showing willfulness in each and every instance. While choosing to make a single showing of a pattern and practice of under-reporting, Complainant nevertheless seeks to have each instance of under-reporting assessed individually, as though she had proved that Saw possessed a willful state of mind in each instance.

Taking these factors into account, I find that a more appropriate penalty would be reached by grouping all record keeping instances into a single violation with a single penalty.

While the individual injuries Saw failed to record were not serious, and though Saw did not appear to be shielding a large number of safety violations by under-reporting, the gravity of Respondent's under-reporting was nonetheless high. In this case, Saw crossed the line from simple non-reporting to attempting to influence work restrictions ordered by Dr. Davis. Work restrictions not only make an injury reportable, but, more fundamentally, are physician's orders that embody medical treatment, containing orders necessary for the patient's recovery. To urge a physician to limit or refrain from issuing work restrictions interferes with the patient's medical treatment and intrudes on the doctor/patient relationship. Saw, through Gary Jones, attempted to influence medical treatment to shield itself from inspections. I do not find it relevant that Dr. Davis did not yield to Saw's pressure, or that specific employees were not identified. I find that the gravity of this violation is, therefore, high and the statutory maximum penalty of $70,000 is appropriate.

ORDER

1. Willful citation 1, items 1 through 16, alleging violation of § 1904.2(a), are AFFIRMED.
2. Citation 1, item 17, alleging violation of § 1904.2(a), is VACATED.
3. Willful citation 1, items 18 through 20, alleging violation of § 1904.2(a), are AFFIRMED.
4. Citation 1, item 21, alleging violation of § 1904.2(a), is VACATED.
5. Willful citation 1, item 22, alleging violation of § 1904.2(a), is AFFIRMED.
6. Citation 1, item 23, alleging violation of § 1904.2(a), is VACATED.
7. Willful citation 1, items 24 through 26, alleging violation of § 1904.2(a), are AFFIRMED.
8. Citation 1, item 27, alleging violation of § 1904.2(a), is VACATED.
9. Willful citation 1, item 28, alleging violation of § 1904.2(a), is AFFIRMED.
10. Citation 1, item 29, alleging violation of § 1904.2(a), is VACATED.
11. Willful citation 1, items 30 through 33, alleging violation of § 1904.2(a), are AFFIRMED.

12. Citation 1, item 34, alleging violation of § 1904.2(a), is VACATED.
13. Willful citation 1, items 35 through 43, alleging violation of § 1904.2(a), are AFFIRMED.
14. Citation 1, item 44, alleging violation of § 1904.2(a), is VACATED.
15. Willful citation 1, items 45 through 62, alleging violation of § 1904.2(a), are AFFIRMED.
16. Citation 1, item 63, alleging violation of § 1904.2(a), is VACATED.
17. Willful citation 1, items 64 through 66, alleging violation of § 1904.2(a), are AFFIRMED.
18. Willful citation 1, item 67(1) through 67(7) alleging violation of § 1904.2(a), are AFFIRMED.
19. Citation 1, item 67(8), alleging violation of § 1904.2(a), is VACATED.
20. Willful citation 1, items 67(9) through 67(14), alleging violation of § 1904.2(a), are AFFIRMED.
21. Citation 1, items 67(15) and 67(16) alleging violation of § 1904.2(a), are VACATED.
22. The violations affirmed are combined for purposes of assessing a penalty. A single penalty of $70,000 is ASSESSED.

James H. Barkley
Judge, OSHR
Dated: May 20, 2002

FOOTNOTES

[1] This judge takes notice that, in a February 6, 1998, Compliance Letter, the Secretary interpreted medical treatment to include the use of casts, splints and/or orthopedic devices designed to immobilize a body part. Though this case was characterized as a restricted work activity case, it was also recordable as requiring medical treatment.
[2] Pursuant to the joint motion of the parties portions of the Jindal transcript were admitted and made part of this record by this judge's January 18, 2002 Order. Prior testimony from Jindal is labeled Exhs. A through J.
[3] At the hearing, this judge disallowed Respondent's attempts to impeach Dr. Davis' credibility by introducing evidence of a professional disciplinary action that involved neither Saw, nor any Saw employees. Such evidence was deemed irrelevant and prejudicial (Tr. 434-35, 442). In an effort to circumvent this judge's ruling, Saw has attached to its brief documents detailing actions taken by the Texas State Board of Medical Examiners in 1999 (Saw's post-hearing brief, Exh. D, E and F). Submission of those exhibits violates the spirit of this judge's ruling; Exhs. D, E and F are, therefore, stricken from the record and will be removed from Saw's brief.
[4] Robert Murphey claimed not to be familiar with SIC codes, but testified that when he was hired, and again in 1998 when Gary Jones turned the OSHA inspectors away, he was told Saw's SIC code was 3443 (Tr. 102-05).
[5] These percentages reflect all injury cases occurring between January 1998 and mid-2000, adjusted to conform to this judge's ruling on the contested items as set forth above.

Appendix F
OSHA: Employee Workplace Rights

U.S. Department of Labor
Elaine L. Chao, Secretary

Occupational Safety and Health Administration
John L. Henshaw, Assistant Secretary

OSHA 3021-08R, 2003

This informational booklet provides a general overview of a particular topic related to OSHA standards. It does not alter or determine compliance responsibilities in OSHA standards or the *Occupational Safety and Health Act of 1970*. Because interpretations and enforcement policy may change over time, you should consult current OSHA administrative interpretations and decisions by the Occupational Safety and Health Review Commission and the Courts for additional guidance on OSHA compliance requirements.

This publication is in the public domain and may be reproduced, fully or partially, without permission. Source credit is requested but not required.

CONTENTS

INTRODUCTION

The Occupational Safety and Health (OSH) Act of 1970 created the Occupational Safety and Health Administration (OSHA) within the Department of Labor and encouraged employers and employees to reduce workplace hazards and to implement safety and health programs. The Act gave employees many new rights and responsibilities. This booklet discusses these rights and responsibilities and encourages employees to work cooperatively with employers to promote safe and healthful workplaces that add value to everyone: businesses, workplaces, and workers' lives.

WORKER RIGHTS UNDER THE OSH ACT

The law encourages workers to be active players in their workplace's safety and health effort. It gives employees the right to

- Review copies of appropriate standards, rules, regulations, and requirements that the employer is required to have available at the workplace;
- Request information from the employer on safety and health hazards in the workplace, appropriate precautions to take, and procedures to follow if the employee is involved in an accident or is exposed to toxic substances;
- Gain access to relevant employee exposure and medical records;
- Request an OSHA inspection if they believe hazardous conditions or violations of standards exist in the workplace;
- Accompany an OSHA compliance officer during the inspection tour, or have an authorized employee representative do so;
- Respond to questions from the OSHA compliance officer;
- Observe any monitoring or measuring of hazardous materials and see the resulting records, as specified under the OSH Act and required by OSHA standards;
- Review or have an authorized representative review the employer's Log of Work-Related Occupational Injuries and Illnesses (OSHA 300) at a reasonable time and in a reasonable manner;
- Object to the timeframe set by OSHA for the employer to correct a violation by writing to the OSHA area director within 15 working days from the date the employer receives the citation;
- Submit a written request to the National Institute for Occupational Safety and Health for information on whether any substance in the workplace has potentially toxic effects in the concentration being used, and, if requested, have their names withheld from the employer;
- Be notified if the employer applies for a variance from an OSHA standard, and have an opportunity to testify at a variance hearing and appeal the final decision;
- Have their names withheld from their employer, by request to OSHA, if they sign and file a written complaint;
- Be advised of OSHA actions regarding a complaint, and request an informal review of any decision not to inspect the site or issue a citation;
- File a complaint if punished or discriminated against for acting as a "whistleblower" under the OSH Act or thirteen other federal statutes for which OSHA has jurisdiction, or for refusing to work when faced with imminent danger of death or serious injury and there is insufficient time for OSHA to inspect.

WORKER RIGHTS TO INFORMATION

Employer Responsibilities

Employers have a legal obligation to inform employees of OSHA safety and health standards that apply to their workplace. Upon request, the employer must make available copies of those standards and the OSH Act. The employer also must prominently display the official OSHA poster that describes rights and responsibilities under the OSH Act.

Protecting Employees Who Work with Hazardous Materials

Employers must establish a written, comprehensive hazard communication program to ensure that employees who work with or near hazardous materials are informed of the hazards and provided proper protection. A hazard communication program includes provisions for container labeling, material safety data sheets, and an employee training program. The program must include the following:

- A list of the hazardous chemicals in each workplace and material safety data sheets for each
- Methods the employer uses to inform employees of the hazards of non-routine tasks (for example, the cleaning of reactor vessels) and the hazards associated with chemicals in unlabeled pipes in their work areas
- A description of methods the employer at a multi-employer worksite will use to inform other employers at the site of the hazards to which their employees may be exposed

Employee Rights When an Employer Files a Variance

Some employers may not be able to comply fully with a new safety and health standard in the time provided due to shortages of personnel, materials, or equipment. In these situations, employers may apply to OSHA for a temporary variance from the standard. In other cases, employers may prefer to use methods or equipment that differ from those prescribed by OSHA, but which the employer believes are equal to or better than OSHA's requirements. In these cases, the employer may seek a permanent variance for the alternative approach.

The employer's application for a permanent or temporary variance must include certification of the following:

- The employer has informed workers of the variance application.
- The employee representative receives a copy of the variance application.
- The employer has posted a summary of the application wherever notices are normally posted in the workplace.

Employers also must inform employees that they have the right to request a hearing on the application. OSHA encourages employees, employers, and other interested groups to participate in the variance process.

Notices of variance applications are published in the *Federal Register,* inviting all interested parties to comment on the action.

WORKER RIGHTS TO ACCESS RECORDS AND TEST RESULTS

ACCESS TO EXPOSURE AND MEDICAL RECORDS

Employers must inform employees of the existence, location, and availability of their medical and exposure records when they begin employment and then at least annually. Employers also must provide these records to employees or their designated representatives upon request. Whenever an employer plans to stop doing business and there is no successor employer to receive and maintain these records, the employer must notify employees of their right of access to these records at least three months before closing the business.

RIGHT TO OBSERVE MONITORING PROCEDURES AND SEE TESTING RESULTS

OSHA standards require the employer to measure exposure to harmful substances. The employee (or employee representative) has the right to observe the testing and examine the records of the results. If the exposure levels are above the limit set by an OSHA standard, the employer must tell employees what will be done to reduce the exposure.

During an OSHA inspection, an OSHA industrial hygienist may conduct exposure tests if health hazards may be present in the workplace. The inspector may take samples to measure levels of dust, noise, fumes, or other hazardous materials.

OSHA will inform the employee or employee representative who participates in the inspection as to whether the employer is in compliance with OSHA standards. The inspector also will gather detailed information about the employer's efforts to control health hazards, including results of tests the employer may have conducted.

RIGHT TO REVIEW INJURY AND ILLNESS RECORDS

An employer with more than ten employees must maintain records of all work-related injuries and illnesses, and the employees or their representative have the right to review those records. Some industries with very low injury rates are exempt from these record keeping requirements.

WORKER RIGHTS TO PROMOTE WORKPLACE SAFETY

WORKING COOPERATIVELY TO REDUCE HAZARDS

OSHA encourages employers and employees to work together to reduce hazards. Employees should discuss safety and health problems with the employer, other workers, and, if a labor union exists, union representatives.

The OSHA area office can provide information on OSHA requirements. If the worksite is in a state with its own OSHA-approved occupational safety and health program, the state can provide similar information.

RIGHT TO REFUSE TO PERFORM UNSAFE WORK

Although nothing in the OSHA law specifically gives an employee the right to refuse to perform an unsafe or unhealthful job assignment, OSHA's regulations, which have been upheld by the U.S. Supreme Court, provide that an employee may refuse to work when faced with an imminent danger of death or serious injury. The conditions necessary to justify a work refusal are very stringent, however, and a work refusal should be taken only as a last resort. If time permits, the employee should report the unhealthful or unsafe condition to OSHA or another appropriate regulatory agency.

RECOURSE IF THE EMPLOYER DOES NOT CORRECT A HAZARD

An employee may file a complaint by phone, mail, e-mail, or fax with the nearest OSHA office and request an inspection if there are unsafe or unhealthful working conditions. When doing so, the employee may request that OSHA not reveal his or her name. If the OSHA area or state office determines that there are reasonable grounds for believing that a violation or danger exists, the office will investigate.

To file a complaint, call (800) 321-OSHA (6742); contact the nearest OSHA regional, areas, state plan, or consultation office; or file on online complaint at www. osha.gov. The teletypewriter (TTY) number is (877) 889-5627.

WORKER RIGHTS DURING THE INSPECTION PROCESS

RIGHT TO REPRESENTATION

The OSH Act gives employees or a workers' representative the right to accompany an OSHA compliance officer (also referred to as a compliance safety and health officer, CSHO, or inspector) during an inspection.

The labor union, if one exists, or the employees must choose the representative. Under no circumstances may the employer choose the workers' representative. If more than one union represents the employees, each union may choose a representative. Normally, union representatives will accompany the inspector in the areas of the facility where their members work. An OSHA inspector may conduct a comprehensive inspection of the entire workplace or a partial inspection limited to certain areas or aspects of the operation.

RIGHT TO HELP THE COMPLIANCE OFFICER

Workers have a right to talk privately to the compliance officer on a confidential basis, whether or not a workers' representative has been chosen. Workers are encouraged to

- point out hazards;
- describe accidents or illnesses that resulted from those hazards;
- discuss past worker complaints about hazards;

- inform the inspector if working conditions are not normal during the inspection.

RIGHTS TO INFORMATION FOLLOWING THE INSPECTION

At the end of the inspection, the OSHA inspector will meet with the employer and the employee representatives in a closing conference to discuss how any hazards that may have been found will be abated. If it is not practical to hold a joint conference, the compliance officer will hold separate conferences. OSHA will provide written summaries, on request.

HOW TO CHALLENGE THE ABATEMENT PERIOD

Whether or not the employer accepts OSHA's findings, the employee (or representative) has the right to contest the time OSHA allows for correcting a hazard. This contest must be filed in writing with the OSHA area director within 15 working days after the citation is issued. The Occupational Safety and Health Review Commission, an independent agency that is not part of the Department of Labor, will decide whether to change the abatement period.

Right to Information If No Inspection Is Conducted or No Citation Issued

The OSHA area director evaluates complaints from employees or their representatives and decides whether they are valid. If the area director decides not to inspect the workplace, he or she will send a certified letter to the complainant explaining the decision and the reasons for it. OSHA will inform complainants that they have the right to request further clarification of the decision from the OSHA area director. If still dissatisfied, they can appeal to the OSHA regional administrator for an informal review. Similarly, in the event that OSHA decides not to issue a citation after an inspection, employees have a right to further clarification from the area director and an informal review by the regional administrator.

WORKER RIGHTS TO PROTECTION FROM RETALIATION

RIGHT TO CONFIDENTIALITY

Employees who make a complaint to OSHA about safety and health hazards in their workplaces have a right to confidentiality. If the employee requests that his or her name not be used, OSHA will not tell the employer who filed the complaint or requested an inspection.

WHISTLE-BLOWER PROTECTIONS

Employees have a right to seek safety and health on the job without fear of punishment. That right is spelled out in Section 11(c) of the OSH Act. The law forbids the employer from punishing or discriminating against employees for exercising such rights as

- complaining to the employer, union, OSHA, or any other government agency about job safety and health hazards;
- participating in OSHA inspections, conferences, hearings, or other OSHA-related activities.

States administering their own occupational safety and health programs must have provisions at least as effective as those in the OSH Act to protect employees from discharge or discrimination. OSHA, however, retains its whistle-blower protection authority in all states regardless of the existence of an OSHA-approved state occupational safety and health program.

Workers who believe they have been punished for exercising safety and health rights must contact the nearest OSHA office within 30 days of the time they learn of the alleged discrimination. A representative of the employee's choosing can file the complaint for the worker. Following a complaint, OSHA will contact the complainant and conduct an in-depth interview to determine whether an investigation is necessary.

If the evidence shows that the employee has been punished for exercising safety and health rights, OSHA will ask the employer to restore that worker's job, earnings, and benefits. If the employer refuses, OSHA may take the employer to court. In such cases, a Department of Labor attorney will represent the employee to obtain this relief.

ADDITIONAL WHISTLE-BLOWER PROTECTIONS

Since passage of the OSH Act in 1970, Congress has expanded OSHA's whistle-blower protection authority to protect workers from discrimination under 13 additional federal statutes. The agency's investigators receive about 2000 complaints a year from employees who charge their employer with retaliation. Complaints must be reported to OSHA within set timeframes following the discriminatory action, as prescribed by each law.

These statutes, and the number of days employees have to file a complaint, are as follows:

- *Occupational Safety and Health Act of 1970* (30 days)
  - Provides discrimination protection for employees who exercise a variety of rights guaranteed under the Act, such as filing a safety and health complaint with OSHA and participating in an inspection.
- *Surface Transportation Assistance Act* (180 days)
  - Provides discrimination protections for truck drivers and other employees relating to the safety of commercial motor vehicles. Coverage includes all buses for hire and freight trucks with a gross vehicle weight greater than 10,001 pounds.
- *Asbestos Hazard Emergency Response Act* (90 days)
  - Provides discrimination protection for individuals who report violations of environmental laws relating to asbestos in elementary and secondary school systems, public or private.

- *International Safety Container Act (60 days)*
  - Provides discrimination protection for employees who report violations of the Act, which regulates shipping containers.
- *Energy Reorganization Act (180 days)*
  - Provides discrimination protection for employees of operators and sub-contractors of nuclear power plants licensed by the Nuclear Regulatory Commission and for employees of contractors working under contract with the Department of Energy.
- *Clean Air Act (30 days)*
  - Provides discrimination protection for employees who report violations of the Act, which provides for the development and enforcement of standards regarding air quality and air pollution.
- *Safe Drinking Water Act (30 days)*
  - Provides discrimination protection for employees who report violations of the Act, which requires that all drinking water systems in public buildings and new construction of all types be lead free.
- *Federal Water Pollution Control Act (30 days)*
  - Provides discrimination protection for employees who report hazard-ous pollution of waters that provide a natural habitant for living things. Also called the Clean Water Act.
- *Toxic Substances Control Act (30 days)*
  - Provides discrimination protection for employees who report violations of regulations involving the manufacture, distribution, and use of cer-tain toxic substances.
- *Solid Waste Disposal Act (30 days)*
  - Provides discrimination protection for employees who exercise certain rights under the Act, which provides assistance for the development of facilities for the recovery of energy and other resources from discarded materials and regulates hazardous waste management. Also called the Resource Conservation and Recovery Act.
- *Comprehensive Environmental Response, Compensation, and Liability Act (30 days)*
  - Provides discrimination protection for employees who exercise rights under the Act, which provides liability, compensation, cleanup, and emergency response for hazardous substances released into the environ-ment and for the cleanup of inactive hazardous waste disposal sites.
- *Wendell H. Ford Aviation Investment and Reform Act for the 21st Century (90 days)*
  - Provides discrimination protection for employees of air carriers, con-tractors, or subcontractors of air carriers who raise safety concerns.
- *Corporate and Criminal Fraud Accountability Act of 2002 (90 days)*
  - Provides discrimination protection for employees of publicly traded com-panies or brokerage firms or their contractors, subcontractors, or agents, who report violations of the Act, which covers mail, wire, bank, or secu-rities fraud or violations of laws related to fraud against stockholders.

- *Pipeline Safety Improvement Act of 2002 (180 days)*
  - Provides discrimination protection for employees who report violations of the federal law regarding pipeline safety and security or who refuse to violate such provisions.

OSHA's publication, "Protecting Whistleblowers" (OSHA 3164) provides additional information. It is available on the agency Web site (www.osha.gov). If you believe that you have been discriminated against, call (800) 321-OSHA (6742) to be connected to the nearest OSHA office to report your complaint.

WORKER RIGHTS IN STATE-PLAN STATES

States that assume responsibility for their own occupational safety and health programs must have provisions at least as effective as Federal OSHA's, including the protection of employee rights.

Any interested person or group, including employees, with a complaint concerning the operation or administration of a state plan may submit a complaint to the appropriate OSHA regional administrator. (See contact lists at the end of this booklet.) The OSHA regional administrator will investigate the complaints and inform the state and the complainant of these findings. When appropriate, OSHA will recommend corrective action.

WORKER RESPONSIBILITIES

Although OSHA does not cite employees for violations, the OSH Act requires that each employee "shall comply with all occupational safety and health standards and all rules, regulations, and orders issued under the Act" that are applicable. Each employee should

- read the OSHA poster at the jobsite;
- comply with all applicable OSHA standards;
- follow all lawful employer safety and health rules and regulations, and wear or use prescribed protective equipment while working;
- report hazardous conditions to the supervisor;
- report any job-related injury or illness to the employer, and seek treatment promptly;
- cooperate with the OSHA compliance officer conducting an inspection if he or she inquires about safety and health conditions in the workplace;
- exercise rights under the OSH Act in a responsible manner.

OSHA ASSISTANCE, SERVICES, AND PROGRAMS

OSHA can provide extensive help through a variety of programs, including assistance about safety and health programs, state plans, workplace consultations, voluntary protection programs, strategic partnerships, alliances, and training and education.

An overall commitment to workplace safety and health can add value to your business, your workplace, and your life.

Establishing a Safety and Health Management System

Working in a safe and healthful environment can stimulate innovation and creativity and result in increased performance and higher productivity. The key to a safe and healthful work environment is a comprehensive safety and health management system.

OSHA has electronic compliance assistance tools, or eTools, on its Web site that walk users through the steps required to develop a comprehensive safety and health program. The eTools are posted at www.osha.gov, and are based on guidelines that identify four general elements critical to a successful safety and health management system:

- Management leadership and employee involvement
- Worksite analysis
- Hazard prevention and control
- Safety and health training

State Programs

The Occupational Safety and Health Act of 1970 (OSH Act) encourages states to develop and operate their own job safety and health plans. OSHA approves and monitors these plans and funds up to 50 percent of each program's operating costs. State plans must provide standards and enforcement programs, as well as voluntary compliance activities, that are at least as effective as federal OSHA's.

Currently, 26 states and territories have their own plans. Twenty-three cover both private and public (state and local government) employees, and three states, Connecticut, New Jersey, and New York, cover only the public sector. For more information on state plans, visit www.osha.gov.

Consultation Assistance

Consultation assistance is available on request to employers who want help establishing and maintaining a safe and healthful workplace. Funded largely by OSHA, the service is provided at no cost to small employers and is delivered by state authorities through professional safety and health consultants.

Safety and Health Achievement Recognition Program

Under the consultation program, certain exemplary employers may request participation in OSHA's Safety and Health Achievement Recognition Program (SHARP). Eligibility for participation includes, but is not limited to, receiving a full-service, comprehensive consultation visit, correcting all identified hazards, and developing an effective safety and health management system.

Employers accepted into SHARP may receive an exemption from programmed inspections (not complaint or accident investigation inspections) for 1 year initially, or 2 years upon renewal. For more information about consultation assistance, visit www.osha.gov.

VOLUNTARY PROTECTION PROGRAMS

Voluntary Protection Programs (VPP) are designed to recognize outstanding achievements by companies that have developed and implemented effective safety and health management programs. There are three VPP programs: Star, Merit, and Demonstration. All are designed to

- recognize employers who have successfully developed and implemented effective and comprehensive safety and health management programs;
- encourage these employers to continuously improve their safety and health management programs;
- motivate other employers to achieve excellent safety and health results in the same outstanding way;
- establish a cooperative relationship between employers, employees, and OSHA.

VPP participation can bring many benefits to employers and employees, including fewer worker fatalities, injuries, and illnesses; lost-workday case rates generally 50 percent below industry averages; and lower workers' compensation and other injury- and illness-related costs. In addition, many VPP sites report improved employee motivation to work safely, leading to a better quality of life at work; positive community recognition and interaction; further improvement and revitalization of already-good safety and health programs; and a positive relationship with OSHA.

Additional information on VPP is available from OSHA regional offices listed at the end of this booklet. Also, see "Cooperative Programs" on OSHA's Web site.

COOPERATIVE PARTNERSHIPS

OSHA has learned firsthand that voluntary, cooperative partnerships with employers, employees, and unions can be a useful alternative to traditional enforcement and an effective way to reduce worker deaths, injuries, and illnesses. This is especially true when a partnership leads to the development and implementation of a comprehensive workplace safety and health management system.

ALLIANCE PROGRAM

Alliances enable organizations committed to workplace safety and health to collaborate with OSHA to prevent injuries and illnesses in the workplace. OSHA and its allies work together to reach out to, educate, and lead the nation's employers and their employees in improving and advancing workplace safety and health.

Alliances are open to all, including trade or professional organizations, businesses, labor organizations, educational institutions, and government agencies. In some cases, organizations may be building on existing relationships with OSHA through other cooperative programs.

There are few formal program requirements for Alliances, which are less structured than other cooperative agreements, and the agreements do not include an enforcement component. However, OSHA and the participating organizations must define, implement, and meet a set of short- and long-term goals that fall into three categories: training and education, outreach and communication, and promotion of the national dialogue on workplace safety and health.

STRATEGIC PARTNERSHIP PROGRAM

OSHA Strategic Partnerships are agreements among labor, management, and government to improve workplace safety and health. These partnerships encourage, assist, and recognize the efforts of the partners to eliminate serious workplace hazards and achieve a high level of worker safety and health. Whereas OSHA's Consultation Program and VPP entail one-on-one relationships between OSHA and individual worksites, most strategic partnerships build cooperative relationships with groups of employers and employees.

For more information about this program, contact your nearest OSHA office or visit OSHA's Web site.

Occupational Safety and Health Training

The OSHA Training Institute in Arlington Heights, Illinois, provides basic and advanced training and education in safety and health for federal and state compliance officers, state consultants, other federal agency personnel, and private-sector employers, employees, and their representatives.

In addition, 20 OSHA Training Institute Education Centers at 35 locations throughout the United States deliver courses on OSHA standards and occupational safety and health issues to thousands of students a year.

TRAINING GRANTS

OSHA awards grants to nonprofit organizations to provide safety and health training and education to employers and workers in the workplace. Grants often focus on high-risk activities or hazards or may help nonprofit organizations in training, education, and outreach. OSHA expects each grantee to develop a program that addresses a safety and health topic named by OSHA, recruit workers and employers for the training, and conduct the training. Grantees are also expected to follow up with students to find out how they applied the training in their workplaces.

For more information on training or grants, contact OSHA Office of Training and Education, 2020 Arlington Heights Rd., Arlington Heights, IL 60005; or call (847) 297-4810.

OTHER ASSISTANCE MATERIALS

OSHA has a variety of materials and tools on its Web site at www.osha.gov. These include eTools such as Expert Advisors and Electronic Compliance Assistance Tools, information on specific health and safety topics, regulations, directives, publications, videos, and other information for employers and employees.

OSHA also has an extensive publications program. For a list of items, visit OSHA's Web site at www.osha.gov or contact the OSHA Publications Office, U.S. Department of Labor, 200 Constitution Avenue, NW, N-3101, Washington, DC 20210. Telephone (202) 693-1888 or fax to (202) 693-2498.

In addition, OSHA's CD-ROM includes standards, interpretations, directives, and more. It is available for sale from the U.S. Government Printing Office. To order, write to the Superintendent of Documents, U.S. Government Printing Office, Washington, DC 20402, or phone (202) 512-1800.

To contact OSHA to report an emergency, file a complaint, or seek OSHA advice, assistance, or products, call (800) 321-OSHA (6742) or contact your nearest OSHA regional office listed at the end of this publication. The teletypewriter (TTY) number is (877) 889-5627.

Employees can also file a complaint online and get more information on OSHA federal and state programs by visiting OSHA's Web site at www.osha.gov.

OSHA REGIONAL OFFICES

Region I
(CT,* ME, MA, NH, RI, VT*)
Boston, MA 02203
(617) 565-9860

Region II
(NJ,* NY,* PR,* VI*)
201 Varick Street, Room 670
New York, NY 10014
(212) 337-2378

Region III
(DE, DC, MD,* PA,* VA,* WV)
The Curtis Center
170 S. Independence Mall West Suite 740 West
Philadelphia, PA 19106-3309
(215) 861-4900

Region IV
(AL, FL, GA, KY,* MS, NC,* SC,* TN*)
Atlanta Federal Center
61 Forsyth Street SW, Room 6T50
Atlanta, GA 30303
(404) 562-2300

Region V
(IL, IN,* MI,* MN,* OH, WI)
230 South Dearborn Street, Room 3244
Chicago, IL 60604
(312) 353-2220

Region VI
(AR, LA, NM,* OK, TX)
525 Griffin Street, Room 602
Dallas, TX 75202
(214) 767-4731 or 4736 x224

Region VII
(IA,* KS, MO, NE)
City Center Square
1100 Main Street, Suite 800
Kansas City, MO 64105
(816) 426-5861

Region VIII
(CO, MT, ND, SD, UT,* WY*)
1999 Broadway, Suite 1690
PO Box 46550
Denver, CO 80202-5716
(303) 844-1600

Region IX
(American Samoa, AZ,* CA,* HI, NV,* Northern Mariana Islands)
71 Stevenson Street, Room 420
San Francisco, CA 94105
(415) 975-4310

Region X
(AK,* ID, OR,* WA*)
1111 Third Avenue, Suite 715
Seattle, WA 98101-3212
(206) 553-5930

*These states and territories operate their own OSHA-approved job safety and health programs. (Connecticut, New Jersey, and New York plans cover public employees only.) States with approved programs must have a standard that is identical to, or at least as effective as, the federal standard.

Note: To get contact information for OSHA area offices, OSHA-approved state plans, and OSHA consultation projects, visit www.osha.gov or call (800) 321-OSHA (6742).

Appendix G
OSHA Field Inspection Reference Manual CPL 2.103, Section 8 — Chapter IV. Post-Inspection Procedures

CHAPTER IV: POST-INSPECTION PROCEDURES

A. ABATEMENT

1. **Period**. The abatement period shall be the shortest interval within which the employer can **reasonably** be expected to correct the violation. An abatement date shall be set forth in the citation as a specific date, not a number of days. When the abatement period is very short (i.e., 5 working days or less) and it is uncertain when the employer will receive the citation, the abatement date shall be set so as to allow for a mail delay and the agreed-upon abatement time. When abatement has been witnessed by the CSHO during the inspection, the abatement period shall be "Corrected During Inspection" on the citation.

2. **Reasonable Abatement Date**. The establishment of the shortest practicable abatement date requires the exercise of professional judgment on the part of the CSHO.

   NOTE: Abatement periods exceeding 30 calendar days should not normally be necessary, or safety violations. Situations may arise, however, especially for health violations, where extensive structural changes are necessary or where new equipment or parts cannot be delivered within 30 calendar days. When an initial abatement date is granted that is in excess of 30 calendar days, the reason, if not self-evident, shall be documented in the case file.

3. **Verification of Abatement**. The Area Director is responsible for determining if abatement has been accomplished. When abatement is not accomplished during the inspection or the employer does not notify the Area Director by letter of the abatement, verification shall be determined by telephone and documented in the case file.

NOTE: If the employer's abatement letter indicates that a condition has not been abated, but the date has passed, the Area Director shall contact the employer for an explanation. The Area Director shall explain Petition for Modification of Abatement (PMA) procedures to the employer, if applicable.

4. **Effect of Contest Upon Abatement Period**. In situations where an employer contests either (1) the period set for abatement or (2) the citation itself, the abatement period generally shall be considered not to have begun until there has been an affirmation of the citation and abatement period. In accordance with the Act, the abatement period begins when a final order of the Review Commission is issued, and this abatement period is not tolled while an appeal to the court is ongoing unless the employer has been granted a stay. In situations where there is an employee contest of the abatement date, the abatement requirements of the citation remain unchanged.

   a. Where an employer has contested only the proposed penalty, the abatement period continues to run unaffected by the contest.

   b. Where the employer does not contest, he must abide by the date set forth in the citation even if such date is within the 15-working-day notice of contest period. Therefore, when the abatement period designated in the citation is 15 working days or less and a notice of contest has not been filed, a follow-up inspection of the worksite may be conducted for purposes of determining whether abatement has been achieved within the time period set forth in the citation. A failure to abate notice may be issued on the basis of the CSHO's findings.

   c. Where the employer has filed a notice of contest to the initial citation within the contest period, the abatement period does not begin to run until the entry of a final Review Commission order. Under these circumstances, any follow-up inspection within the contest period shall be discontinued and a failure to abate notice shall not be issued.

NOTE: There is one exception to the above rule. If an early abatement date has been designated in the initial citation and it is the opinion of the CSHO and/or the Area Director that a situation classified as imminent danger is presented by the cited condition, appropriate imminent danger proceedings may be initiated notwithstanding the filing of a notice of contest by the employer.

   d. If an employer contests an abatement date in good faith, a Failure to Abate Notice shall not be issued for the item contested until a final order affirming a date is entered, the new abatement period, if any, has been completed, and the employer has still failed to abate.

5. **Long-Term Abatement Date for Implementation of Feasible Engineering Controls**. Long-term abatement is abatement which will be completed

more than one year from the citation issuance date. In situations where it is difficult to set a specific abatement date when the citation is originally issued (e.g., because of extensive redesign requirements consequent upon the employer's decision to implement feasible engineering controls and uncertainty as to when the job can be finished, the CSHO shall discuss the problem with the employer at the closing conference and, in appropriate cases, shall encourage the employer to seek an informal conference with the Area Director.

   a. **Final Abatement Date**. The CSHO and the Assistant Area Director shall make their best judgment as to a reasonable abatement date. A specific date for final abatement shall, in all cases, be included in the citation. The employer shall not be permitted to propose an abatement plan setting its own abatement dates. If necessary, an appropriate petition may be submitted later by the employer to the Area Director to modify the abatement date. (See D.2. of this chapter for PMAs.)

   b. **Employer Abatement Plan**. The employer is required to submit an abatement plan outlining the anticipated long-term abatement procedures.

NOTE: A statement agreeing to provide the affected Area Offices with written periodic progress reports shall be part of the long-term abatement plan.

6. **Feasible Administrative, Work Practice and Engineering Controls**. Where applicable, the CSHO shall discuss control methodology with the employer during the closing conference.

   a. **Definitions**.

     (1) **Engineering Controls**. Engineering controls consist of substitution, isolation, ventilation and equipment modification.

     (2) **Administrative Controls**. Any procedure which significantly limits daily exposure by control or manipulation of the work schedule or manner in which work is performed is considered a means of administrative control. The use of personal protective equipment is **not** considered a means of administrative control.

     (3) **Work Practice Controls**. Work practice controls are a type of administrative controls by which the employer modifies the manner in which the employee performs assigned work. Such modification may result in a reduction of exposure through such methods as changing work habits, improving sanitation and hygiene practices, or making other changes in the way the employee performs the job.

     (4) **Feasibility**. Abatement measures required to correct a citation item are feasible when they can be accomplished by the employer. The CSHO, following current directions and guidelines, shall inform the employer, where appropriate, that a determination will be made as to whether engineering or administrative controls are feasible.

(a) **Technical Feasibility**. Technical feasibility is the existence of technical know-how as to materials and methods available or adaptable to specific circumstances which can be applied to cited violations with a reasonable possibility that employee exposure to occupational hazards will be reduced.

(b) **Economic Feasibility**. Economic feasibility means that the employer is financially able to undertake the measures necessary to abate the citations received.

NOTE: If an employer's level of compliance lags significantly behind that of its industry, allegations of economic infeasibility will not be accepted.

b. Responsibilities
 (1) The CSHO shall document the underlying facts which give rise to an employer's claim of infeasibility.
  (a) When economic infeasibility is claimed the CSHO shall inform the employer that, although the cost of corrective measures to be taken will generally not be considered as a factor in the issuance of a citation, it may be considered during an informal conference or during settlement negotiations.
  (b) Serious issues of feasibility should be referred to the Area Director for determination.
 (2) The Area Director is responsible for making determinations that engineering or administrative controls are or are not feasible.
c. **Reducing Employee Exposure**. Whenever feasible, engineering, administrative, or work practice controls can be instituted even though they are not sufficient to eliminate the hazard (or to reduce exposure to or below the permissible exposure limit [PEL]). Nonetheless, they are required in conjunction with personal protective equipment to reduce exposure to the lowest practical level.

B. CITATIONS

1. Issuing Citations.
 a. **Sending Citations to the Employer**. Citations shall be sent by certified mail; hand delivery of citations to the employer or an appropriate agent of the employer may be substituted for certified mailing if it is believed that this method would be more effective. A signed receipt shall be obtained whenever possible; otherwise the circumstances of delivery shall be documented in the file.
 b. **Sending Citations to the Employee**. Citations shall be mailed to employee representatives no later than one day after the citation is sent to the employer. Citations shall also be mailed to any employee upon request.

c. **Follow-up Inspections**. If a follow-up inspection reveals a failure to abate, the time specified for abatement has passed, and no notice of contest has been filed, a Notification of Failure to Abate Alleged Violation (OSHA-2B) may be issued immediately without regard to the contest period of the initial citation.

2. Amending or Withdrawing Citation and Notification of Penalty in Part or In Its Entirety.

   a. **Citation Revision Justified**. Amendments to or withdrawal of a citation shall be made when information is presented to the Area Director which indicates a need for such revision under certain conditions which may include:

   (1) Administrative or technical error.
      (a) Citation of an incorrect standard.
      (b) Incorrect or incomplete description of the alleged violation.
   (2) Additional facts establish a valid affirmative defense.
   (3) Additional facts establish that there was no employee exposure to the hazard.
   (4) Additional facts establish a need for modification of the correction date, or the penalty, or reclassification of citation items.

   b. **Citation Revision Not Justified**. Amendments to or withdrawal of a citation shall not be made by the Area Director under certain conditions which include:

   (1) Valid notice of contest received.
   (2) The 15 working days for filing a notice of contest has expired and the citation has become a final order.
   (3) Employee representatives have not been given the opportunity to present their views unless the revision involves only an administrative or technical error.
   (4) Editorial and/or stylistic modifications.

   c. **Procedures for Amending or Withdrawing Citations**. The following procedures are to be followed in amending or withdrawing citations. The instructions contained in this section, with appropriate modification, are also applicable to the amendment of the Notification of Failure to Abate Alleged Violation, OSHA-2B Form:

   (1) Withdrawal of or modifications to the citation and notification of penalty, shall normally be accomplished by means of an informal settlement agreement (ISA). (See D.4.b. of this chapter for further information on ISA's).
   (2) Changes initiated by the Area Director without an informal conference are exceptions. In such cases the procedures given below shall be followed:
      (a) If proposed amendments to citation items change the classification of the items; e.g., serious to other-than-serious, the original citation items shall be withdrawn and new, appropriate citation items issued.

(b) The amended Citation and Notification of Penalty Form (OSHA-2) shall clearly indicate that:

    1 The employer is obligated under the Act to post the amendment to the citation along with the original citation until the amended violation has been corrected or for 3 working days, whichever is longer;

    2 The period of contest of the amended portions of the OSHA-2 will begin from the day following the date of receipt of the amended Citation and Notification of Penalty; and

    3 The contest period is not extended as to the unamended portions of the original citation.

(c) A copy of the original citation shall be attached to the amended Citation and Notification of Penalty Form when the amended form is forwarded to the employer.

(d) When circumstances warrant it, a citation may be withdrawn in its entirety by the Area Director. Justifying documentation shall be placed in the case file. If a citation is to be withdrawn, the following procedures apply:

    1 A letter withdrawing the Citation and Notification of Penalty shall be sent to the employer. The letter shall refer to the original citation and penalty, state that they are withdrawn and direct that the letter be posted by the employer for 3 working days in those locations where the original citation was posted.

    2 When applicable to the specific situation (e.g., an employee representative participated in the walkaround inspection, the inspection was in response to a complaint signed by an employee or an employee representative, or the withdrawal resulted from an informal conference or settlement agreement in which an employee representative exercised the right to participate), a copy of the letter shall also be sent to the employee or the employee representative as appropriate.

C.   PENALTIES

1. **General Policy**. The penalty structure provided under Section 17 of the Act is designed primarily to provide an incentive toward correcting violations voluntarily, not only to the offending employer but, more especially, to other employers who may be guilty of the same infractions of the standards or regulations.

    a. While penalties are not designed primarily as punishment for violations, the Congress has made clear its intent that penalty amounts should be sufficient to serve as an effective deterrent to violations.

b. Large proposed penalties, therefore, serve the public purpose intended under the Act; and criteria guiding approval of such penalties by the Assistant Secretary are based on meeting this public purpose. (See OSHA Instruction CPL 2.80.)

c. The penalty structure outlined in this section is designed as a general guideline. The Area Director may deviate from this guideline if warranted, to achieve the appropriate deterrent effect.

2. **Civil Penalties**
   a. **Statutory Authority.** Section 17 provides the Secretary with the statutory authority to propose civil penalties for violations of the Act.
      (1) Section 17(b) of the Act provides that any employer who has received a citation for an alleged violation of the Act which is determined to be of a serious nature shall be assessed a civil penalty of up to $7000 for each violation. (See OSHA Instruction CPL 2.51H, or the most current version, for congressional exemptions and limitations placed on penalties by the Appropriations Act.)
      (2) Section 17(c) provides that, when the violation is specifically determined not to be of a serious nature, a proposed civil penalty of up to $7000 may be assessed for each violation.
      (3) Section 17(i) provides that, when a violation of a posting requirement is cited, a civil penalty of up to $7000 shall be assessed.
   b. **Minimum Penalties.** The following guidelines apply:
      (1) The proposed penalty for any willful violation shall not be less than $5000. The $5000 penalty is a statutory minimum and not subject to administrative discretion. See C.2.m.(1)(a)1, below, for applicability to small employers.
      (2) When the adjusted proposed penalty for an other-than-serious violation (citation item) would amount to less than $100, no penalty shall be proposed for that violation.
      (3) When, however, there is a citation item for a posting violation, this minimum penalty amount does not apply with respect to that item since penalties for such items are mandatory under the Act.
      (4) When the adjusted proposed penalty for a serious violation (citation item) would amount to less than $100, a $100 penalty shall be proposed for that violation.
   c. **Penalty Factors.** Section 17(j) of the Act provides that penalties shall be assessed on the basis of four factors:
      (1) The gravity of the violation.
      (2) The size of the business.
      (3) The good faith of the employer.
      (4) The employer's history of previous violations.
   d. **Gravity of Violation.** The gravity of the violation is the primary consideration in determining penalty amounts. It shall be the basis for calculating the basic penalty for both serious and other violations. To

determine the gravity of a violation the following two assessments shall be made:

(1) The severity of the injury or illness which could result from the alleged violation.

(2) The probability that an injury or illness could occur as a result of the alleged violation.

e. **Severity Assessment**. The classification of the alleged violations as serious or other-than-serious, in accordance with the instructions in Chapter III, C.2., is based on the severity of the injury or illness that could result from the violation. This classification constitutes the first step in determining the gravity of the violation. A severity assessment shall be assigned to a hazard to be cited according to the most serious injury or illness which could reasonably be expected to result from an employee's exposure as follows:

(1) **High Severity**: Death from injury or illness; injuries involving permanent disability; or chronic, irreversible illnesses.

(2) **Medium Severity**: Injuries or temporary, reversible illnesses resulting in hospitalization or a variable but limited period of disability.

(3) **Low Severity**: Injuries or temporary, reversible illnesses not resulting in hospitalization and requiring only minor supportive treatment.

(4) **Minimal Severity**: Other-than-serious violations. Although such violations reflect conditions which have a direct and immediate relationship to the safety and health of employees, the injury or illness most likely to result would probably not cause death or serious physical harm.

f. **Probability Assessment**. The probability that an injury or illness will result from a hazard has no role in determining the classification of a violation but does affect the amount of the penalty to be proposed.

(1) **Categorization**. Probability shall be categorized either as greater or as lesser probability.

(a) Greater probability results when the likelihood that an injury or illness will occur is judged to be relatively high.

(b) Lesser probability results when the likelihood that an injury or illness will occur is judged to be relatively low.

(2) **Violations**. The following circumstances may normally be considered, as appropriate, when violations likely to result in injury or illness are involved:

(a) Number of workers exposed.

(b) Frequency of exposure or duration of employee overexposure to contaminants.

(c) Employee proximity to the hazardous conditions.

(d) Use of appropriate personal protective equipment (PPE).

(e) Medical surveillance program.

(f) Youth and inexperience of workers, especially those under 18 years old.

(g) Other pertinent working conditions.

(3) **Final Probability Assessment**. All of the factors outlined above shall be considered together in arriving at a final probability assessment. When strict adherence to the probability assessment procedures would result in an unreasonably high or low gravity, the probability may be adjusted as appropriate based on professional judgment. Such decisions shall be adequately documented in the case file.

g. **Gravity-Based Penalty**. The gravity-based penalty (GBP) is an unadjusted penalty and is calculated in accordance with the following procedures:

(1) The GBP for each violation shall be determined based on an appropriate and balanced professional judgment combining the severity assessment and the final probability assessment.

(2) For serious violations, the GBP shall be assigned on the basis of the following scale:

Severity	Probability	GBP	Gravity
High	Greater	$5000	High ($5000+)
Medium	Greater	$3500	—
Low	Greater	$2500	Moderate
High	Lesser	$2500	—
Medium	Lesser	$2000	—
Low	Lesser	$1500	Low

NOTE: The gravity of a violation is defined by the GBP. A **high gravity** violation is one with a GBP of $5000 or greater. A **moderate gravity** violation is one with GBP of $2000 to $3500. A **low gravity** violation is one with a GBP of $1500.

(3) The highest gravity classification (high severity and greater probability) shall normally be reserved for the most serious violative conditions, such as those situations involving danger of death or extremely serious injury or illness. If the Area Director determines that it is appropriate to achieve the necessary deterrent effect, a GBP of $7000 may be proposed. The reasons for this determination shall be documented in the case file.

(4) For other-than-serious safety and health violations, there is no severity assessment.

(5) The Area Director may authorize a penalty between $1000 and $7000 for an **other-than-serious** violation when it is determined to be appropriate to achieve the necessary deterrent effect. The reasons for such a determination shall be documented in the case file.

Probability	GBP
Greater	$1000–$7000
Lesser	$0

(6) A GBP may be assigned in some cases without using the severity and the probability assessment procedures outlined in this section when these procedures cannot appropriately be used.

(7) The Penalty Table (Table IV-1) may be used for determining appropriate adjusted penalties for serious and other-than-serious violations.

h. **Gravity Calculations for Combined or Grouped Violations**. Combined or grouped violations will normally be considered as one violation and shall be assessed one GBP. The following procedures apply to the calculation of penalties for combined and grouped violations:

(1) The severity and the probability assessments for combined violations shall be based on the instance with the highest gravity. It is not necessary to complete the penalty calculations for each instance or subitem of a combined or grouped violation if it is clear which instance will have the highest gravity.

(2) For grouped violations, the following special guidelines shall be adhered to:

(a) **Severity Assessment**. There are two considerations to be kept in mind in calculating the severity of grouped violations:

1   The severity assigned to the grouped violation shall be no less than the severity of the most serious reasonably predictable injury or illness that could result from the violation of any single item.

2   If a more serious injury or illness is reasonably predictable from the grouped items than from any single violation item, the more serious injury or illness shall serve as the basis for the calculation of the severity factor of the grouped violation.

(b) **Probability Assessment**. There are two considerations to be kept in mind in calculating the probability of grouped violations:

1   The probability assigned to the grouped violation shall be no less than the probability of the item which is most likely to result in an injury or illness.

2   If the overall probability of injury or illness is greater with the grouped violation than with any single violation item, the greater probability of injury or illness shall serve as the basis for the calculation of the probability assessment of the grouped violation.

(3) In egregious cases an additional factor of up to the number of violation instances may be applied. Such cases shall be handled in

accordance with OSHA Instruction CPL 2.80. Penalties calculated with this additional factor shall not be proposed without the concurrence of the Assistant Secretary. (See also C.2.k.(2)(c)**4** of this chapter.)

i. **Penalty Adjustment Factors**. The GBP may be reduced by as much as 95 percent depending upon the employer's "good faith," "size of business," and "history of previous violations." Up to 60-percent reduction is permitted for size; up to 25-percent reduction for good faith, and 10-percent for history.

(1) Since these adjustment factors are based on the general character of a business and its safety and health performance, the factors generally shall be calculated only once for each employer. After the classification and probability ratings have been determined for each violation, the adjustment factors shall be applied subject to the limitations indicated in the following paragraphs.

(2) Penalties assessed for violations that are classified as high severity and greater probability shall be adjusted only for size and history.

(3) Penalties assessed for violations that are classified as repeated shall be adjusted only for size.

(4) Penalties assessed for regulatory violations, which are classified as willful, shall be adjusted for size. Penalties assessed for serious violations, which are classified as willful, shall be adjusted for size and history.

NOTE: If one violation is classified as willful, no reduction for good faith can be applied to any of the violations found during the same inspection. The employer cannot be willfully in violation of the Act and at the same time, be acting in good faith.

(5) The rate of penalty reduction for size of business, employer's good faith and employer's history of previous violations shall be calculated on the basis of the criteria described in the following paragraphs:

(a) **Size**. A maximum penalty reduction of 60 percent is permitted for small businesses. "Size of business" shall be measured on the basis of the maximum number of employees of an employer at all workplaces at any one time during the previous 12 months.

1 The rates of reduction to be applied are as follows:

2 When a small business (1-25 employees) has one or more serious violations of high gravity or a number of serious violations of moderate gravity, indicating a lack of concern for employee safety and health, the CSHO may recommend that only a partial reduction in penalty shall be permitted for size of business.

(b) **Good Faith**. A penalty reduction of up to 25 percent, based on the CSHO's professional judgment, is permitted in recognition of an employer's "good faith."

  1 The 25% credit for "good faith" normally requires a written safety and health program. In exceptional cases, the compliance officer may recommend the full 25% for a smaller employer (1-25 employees) who has implemented an efficient safety and health program, but has not reduced it to writing.

     a Provides for appropriate management commitment and employee involvement; worksite analysis for the purpose of hazard identification; hazard prevention and control measures; and safety and health training.

     NOTE: One example of a framework for such a program is given in OSHA's voluntary "Safety and Health Program Management Guidelines" (*Federal Register*, Vol. 54, No. 16, January 26, 1989, pp. 3904-3916, or later revisions as published).

     b Has deficiencies that are only incidental.

  2 A reduction of 15 percent shall normally be given if the employer has a documentable and effective safety and health program, but with more than only incidental deficiencies.

  3 No reduction shall be given to an employer who has no safety and health program or where a willful violation is found.

  4 Only these percentages (15% or 25%) may be used to reduce penalties due to the employer's good faith. No intermediate percentages shall be used.

  5 Where young workers (i.e., less than 18 years old) are employed, the CSHO's evaluation must consider whether the employer's safety and health program appropriately addresses the particular needs of such workers with regard to the types of work they perform and the hazards to which they are exposed.

     c **History**. A reduction of 10 percent shall be given to employers who have not been cited by OSHA for any serious, willful, or repeated violations in the past three years.

     d **Total**. The total reduction will normally be the sum of the reductions for each adjustment factor.

j. **Effect on Penalties If Employer Immediately Corrects or Initiates Corrective Action**. Appropriate penalties will be proposed with respect to an alleged violation even though, after being informed of such alleged violation by the CSHO, the employer immediately corrects or initiates steps to correct the hazard.

k. **Failure to Abate**. A Notification of Failure to Abate an Alleged Violation (OSHA-2B) shall be issued in cases where violations have not been corrected as required.

(1) **Failure to Abate**. Failure to abate penalties shall be applied when an employer has not corrected a previously cited violation which had become a final order of the Commission. Citation items become final order of the Review Commission when the abatement date for that item passes, if the employer has not filed a notice of contest prior to that abatement date. See D.5. of this chapter for guidance on determining final dates of settlements and Review Commission orders.

(2) **Calculation of Additional Penalties**. A GBP for unabated violations is to be calculated for failure to abate a serious or other-than-serious violation on the basis of the facts noted upon reinspection. This recalculated GBP, however, shall not be less than that proposed for the item when originally cited, except as provided in C.2.k.(4), below.

(a) In those instances where no penalty was initially proposed, an appropriate penalty shall be determined after consulting with the Assistant Area Director. In no case shall the unadjusted penalty be less than $1000 per day.

(b) Only the adjustment factor for size — based upon the circumstances noted during the reinspection — shall then be applied to arrive at the daily proposed penalty.

(c) The daily proposed penalty shall be multiplied by the number of calendar days that the violation has continued unabated, except as provided below:

1   The number of days unabated shall be counted from the day following the abatement date specified in the citation or the final order. It will include all calendar days between that date and the date of reinspection, excluding the date of reinspection.

2   Normally the maximum total proposed penalty for failure to abate a particular violation shall not exceed 30 times the amount of the daily proposed penalty.

3   At the discretion of the Area Director, a lesser penalty may be proposed with the reasons for doing so (e.g., achievement of an appropriate deterrent effect) documented in the case file.

4   If a penalty in excess of the normal maximum amount of 30 times the amount of the daily proposed penalty is deemed appropriate by the Area Director, the case shall be treated under the violation-by-violation (egregious) penalty procedures established in OSHA Instruction CPL 2.80.

(3) **Partial Abatement**.
   (a) When the citation has been partially abated, the Area Director may authorize a reduction of 25 percent to 75 percent to the amount of the proposed penalty calculated as outlined in C.2.k.(2), above.
   (b) When a violation consists of a number of instances and the follow-up inspection reveals that only some instances of the violation have been corrected, the additional daily proposed penalty shall take into consideration the extent that the violation has been abated.

   EXAMPLE: Where 3 out of 5 instances have been corrected, the daily proposed penalty (calculated as outlined in C.2.k.(2), above, without regard to any partial abatement) may be reduced by 60 percent.

(4) **Good Faith Effort to Abate**. When the CSHO believes, and so documents in the case file, that the employer has made a good faith effort to correct the violation and had good reason to believe that it was fully abated, the Area Director may reduce or eliminate the daily proposed penalty that would otherwise be justified.

l. **Repeated Violations**. Section 17(a) of the Act provides that an employer who repeatedly violates the Act may be assessed a civil penalty of not more than $70,000 for each violation.

(1) **Gravity-Based Penalty Factors**. Each violation shall be classified as serious or other-than-serious. A GBP shall then be calculated for repeated violations based on facts noted during the current inspection. Only the adjustment factor for size, appropriate to the facts at the time of the reinspection, shall be applied.

(2) **Penalty Increase Factors**. The amount of the increased penalty to be assessed for a repeated violation shall be determined by the size of the employer.
   (a) **Smaller Employers**. For employers with 250 or fewer employees, the GBP shall be doubled for the first repeated violation and quintupled if the violation has been cited twice before. If the Area Director determines that it is appropriate to achieve the necessary deterrent effect, the GBP may be multiplied by **10**.
   (b) **Larger Employers**. For employers with more than 250 employees, the GBP shall be multiplied by **5** for the first repeated violation and multiplied by **10** for the second repeated violation.

(3) **Other-Than-Serious, No Initial Penalty**. For a repeated other-than-serious violation that otherwise would have no initial penalty, a GBP penalty of $200 shall be assessed for the first repeated violation, $500 if the violation has been cited twice before, and $1000 for a third repetition.

NOTE: This penalty will not have the penalty increase factors applied as discussed under C.2.l.(2).

(4) **Regulatory Violations.** For repeated instances of regulatory violations, the initial penalty shall be doubled for the first repeated violation and quintupled if the violation has been cited twice before. If the Area Director determines that it is appropriate to achieve the necessary deterrent effect, the initial penalty may be multiplied by **10**.

NOTE: See Chapter III, C.2.f., for additional guidance on citing repeated violations.

m. **Willful Violations.** Section 17(a) of the Act provides that an employer who willfully violates the Act may be assessed a civil penalty of not more than $70,000 but not less than $5000 for each violation.

(1) **Gravity-Based Penalty Factors.** Each willful violation shall be classified as serious or other-than-serious.

(a) **Serious Violations.** For willful serious violations, a gravity of **high**, **medium moderate**, or **low** shall be assigned based on the GBP of the underlying serious violation, as described at C.2.g.(2).

1   The adjustment factor for size shall be applied as shown in the following chart:

Employees	Percent Reduction
10 or less	80
11-20	60
21-30	50
31-40	40
41-50	30
51-100	20
101-250	10
251 or more	0

2   The adjustment factor for history shall be applied as described at C.2.i.(5)(c); i.e., a reduction of 10 percent shall be given to employers who have not been cited by OSHA for any serious, willful, or repeated violations in the past 3 years. There shall be no adjustment for good faith.

3   The proposed penalty shall then be determined from the table below:

Penalties to Be Proposed									
Total percentage reduction for size and/or history	0%	10%	20%	30%	40%	50%	60%	70%	80%	90%
High gravity	$70,000	$63,000	$56,000	$49,000	$42,000	$35,000	$28,000	$21,000	$14,000	$7000
Moderate gravity	$55,000	$49,000	$44,000	$38,000	$33,000	$27,500	$22,000	$16,500	$11,000	$5500
Low gravity	$40,000	$36,000	$32,000	$28,000	$24,000	$20,000	$16,000	$12,000	$8000	$5000

    4 In no case shall the proposed penalty be less than $25,000 $5,000.

  (b) **Other-Than-Serious Violations**. For willful other-than-serious violations, the minimum willful penalty of $5000 shall be assessed.

 (2) **Regulatory Violations**. In the case of regulatory violations (see C.2.n., below) that are determined to be willful, the unadjusted initial penalty shall be multiplied by **10**. In no event shall the penalty, after adjustment for size, be less than $5000.

n. **Violation of 29 CFR Parts 1903 and 1904 Regulatory Requirements**. Except as provided in the Appropriations Act, Section 17 of the Act provides that an employer who violates any of the posting requirements shall be assessed a civil penalty of up to $7000 for each violation and may be assessed a like penalty for record keeping violations.

 (1) **General Application**. Unadjusted penalties for regulatory violations, including posting requirements, shall have the adjustment factors for size and history applied (excluding willful violations, see C.2.m.(2), above).

 (2) **Posting Requirements**. Penalties for violation of posting requirements shall be proposed as follows:

  (a) **OSHA Notice (Poster)**. If the employer has not displayed (posted) the notice furnished by the Occupational Safety and Health Administration as prescribed in 29 CFR 1903.2 (a), an other-than-serious citation shall normally be issued. The unadjusted penalty for this alleged violation shall be $1000 provided that the employer has received a copy of the poster or had knowledge of the requirement.

  (b) **Annual Summary**. If an employer fails to post the summary portion of the OSHA-200 Form during the month of February as required by 29 CFR 1904.5(d)(1), and/or fails to complete the summary prior to February 1, as required by 29 CFR 1904.5(b), even if there have been no injuries, an other-than-serious citation shall be issued. The unadjusted penalty for this violation shall be $1000.

  (c) **Citation**. If an employer received a citation that has not been posted as prescribed in 29 CFR 1903.16, an other-than-serious citation shall normally be issued. The unadjusted penalty shall be $3000.

(3) **Reporting and Record Keeping Requirements**. Section 17(c) of the Act provides that violations of the recordkeeping and reporting requirements may be assessed civil penalties of up to $7000 for each violation.

(a) **OSHA-200 Form**. If the employer does not maintain the Log and Summary of Occupational Injuries and Illnesses, OSHA-200 Form, as prescribed in 29 CFR Part 1904, an other-than-serious citation shall be issued. There shall be an unadjusted penalty of $1000 for each year the form was not maintained, for each of the preceding 3 years.

1   When no recordable injuries or illnesses have occurred at a workplace during the current calendar year, the OSHA 200 need not be completed until the end of the calendar year for certification of the summary.

2   An OSHA-200 with significant deficiencies shall be considered as not maintained.

(b) **OSHA-101 Forms**. If the employer does not maintain the Supplementary Record, OSHA 101 Form (or equivalent), as prescribed in 29 CFR Part 1904, an other-than-serious citation shall be issued. There shall be an unadjusted penalty of $1000 for each OSHA-101 Form not maintained.

1   A penalty of $1000 for each OSHA-101 Form not maintained at all up to a maximum of $7000.

2   A penalty of $1000 for each OSHA-101 Form inaccurately maintained up to a maximum of $3000.

3   Minor inaccuracies shall be cited, but with no penalties.

4   If large numbers of violations or other circumstances indicate that the violations are willful, then other penalties including, violation-by-violation, may be applied.

(c) **Reporting**. Employers are required to report either orally or in writing to the nearest Area Office within 8 hours, any occurrence of an employment accident which is fatal to one or more employees or which results in the hospitalization of three or more employees.

1   An other-than-serious citation shall be issued for failure to report such an occurrence. The unadjusted penalty shall be $5000.

2   If the Area Director determines that it is appropriate to achieve the necessary deterrent effect, an unadjusted penalty of $7000 may be assessed.

3   If the Area Director becomes aware of an incident required to be reported under 29 CFR 1904.8 through some means other than an employer report, prior to the elapse of the 8-hour reporting period and an inspection of the incident is made, a citable violation for failure to report does not exist.

(4) **Grouping**. Violations of the posting and record keeping requirements which involve the same document (e.g., summary portion of the OSHA-200 Form was neither posted nor maintained) shall be grouped as an other-than-serious violation for penalty purposes. The unadjusted penalty for the grouped violations would then take on the highest dollar value of the individual items, which will normally be $1000.

(5) **Access to Records**.

    (a) **29 CFR Part 1904**. If the employer fails upon request to provide records required in 1904.2 for inspection and copying by any employee, former employee, or authorized representative of employees, a citation for violation of 29 CFR 1904.7(b)(1) shall normally be issued. The unadjusted penalty shall be $1000 for each form not made available.

        1 Thus, if the OSHA-200 for the 3 preceding years is not made available, the unadjusted penalty would be $3000.

        2 If the employer is to be cited for failure to maintain these records, no citation of 1904.7 shall be issued.

    (b) **29 CFR 1910.20**. If the employer is cited for failing to provide records as required under 29 CFR 1910.20 for inspection and copying by any employee, former employee, or authorized representative of employees, an unadjusted penalty of $1000 shall be proposed for each record; i.e., either medical record or exposure record, on an individual employee basis. A maximum $7000 may be assessed for such violations. This policy does not preclude the use of violation-by-violation penalties where appropriate. (See OSHA Instruction CPL 2.80.)

    EXAMPLE: If all the necessary evidence is established where an authorized employee representative requested exposure and medical records for 3 employees and the request was denied by the employer, a citation would be issued for 6 instances of violation of 29 CFR 1910.20, with an unadjusted penalty of $6000.

(6) **Notification Requirements**. When an employer has received advance notice of an inspection and fails to notify the authorized employee representative as required by 29 CFR 1903.6, an other-than-serious citation shall be issued. The violation shall have an unadjusted penalty of $2000.

TABLE IV-1

Percent Reduction	Penalty (in Dollars)							
0	1000	1500	2000	2500	3000	3500	5000	7000
10	900	1350	1800	2250	2700	3150	4500	6300
15	850	1275	1700	2125	2550	2975	4250[a]	5950[a]
20	800	1200	1600	2000	2400	2800	4000	5600
25	750	1125	1500	1875	2250	2625	3750	5250
30	700	1050	1400	1750	2100	2450	3500	4900
35	650	975	1300	1625	1950	2275	3250[a]	4550[a]
40	600	900	1200	1500	1800	2100	3000	4200
45	550	825	1100	1375	1650	1925	2750[a]	3850[a]
50	500	750	1000	1250	1500	1750	2500	3500
55	450	675	900	1125	1350	1575	2250[a]	3150[a]
60	400	600	800	1000	1200	1400	2000	2800
65	350	525	700	875	1050	1225	1750[a]	2450[a]
70	300	450	600	750	900	1050	1500	2100
75	250	375	500	625	750	875	1250[a]	1750[a]
85	150	225	300	375	450	525	750[a]	1050[a]
95	100[b]	100[b]	100	125	150	175	250[a]	350[a]

[a] Starred figures represent penalty amounts that would not normally be proposed for high gravity serious violations because no adjustment for good faith is made in such cases. They may occasionally be applicable for other-than-serious violations where the Area Director has determined a high unadjusted penalty amount to be warranted.

[b] Administratively, OSHA will not issue a penalty less than $100 for a serious violation.

3. Criminal Penalties.

    a.   The Act and the U.S. Code provide for criminal penalties in the following cases:

        (1) Willful violation of an OSHA standard, rule, or order causing the death of an employee. (Section 17(e).)

        (2) Giving unauthorized advance notice. (Section 17(f).)

        (3) Giving false information. (Section 17(g).)

        (4) Killing, assaulting or hampering the work of a CSHO. (Section 17(h)(2).)

    b.   Criminal penalties are imposed by the courts after trials and not by the Occupational Safety and Health Administration or the Occupational Safety and Health Review Commission.

D. POST-CITATION PROCESSES.

1. Informal Conferences.

    a.   **General**. Pursuant to 29 CFR 1903.19, the employer, any affected employee or the employee representative may request an informal conference. When an informal conference is conducted, it shall be conducted within the 15 working day contest period. If the employer's intent to contest is not clear, the Area Director shall contact the employer for clarification.

b. **Procedures**. Whenever an informal conference is requested by the employer, an affected employee or the employee representative, both parties shall be afforded the opportunity to participate fully. If either party chooses not to participate in the informal conference, a reasonable attempt shall be made to contact that party to solicit their input prior to signing an informal settlement agreement if the adjustments involves more than the penalty. If the requesting party objects to the attendance of the other party, separate informal conferences may be held. During the conduct of a joint informal conference, separate or private discussions shall be permitted if either party so requests. Informal conferences may be held by any means practical.

   (1) The employer shall be requested to complete and post the form found at the end of the informal conference letter until after the informal conference has been held.

   (2) Documentation of the Area Director's actions notifying the parties of the informal conference shall be placed in the case file.

c. **Participation by OSHA Officials**. The inspecting CSHOs and their Assistant Area Directors shall be notified of an upcoming informal conference and, if practicable, given the opportunity to participate in the informal conference (unless, in the case of the CSHO, the Area Director anticipates that only a penalty adjustment will result).

   (1) At the discretion of the Area Director, one or more additional OSHA employees (in addition to the Area Director) may be present at the informal conference. In cases in which proposed penalties total $100,000 or more, a second OSHA staff member shall attend the informal conference.

   (2) The Area Director shall ensure that notes are made indicating the basis for any decisions taken at or as a result of the informal conference. It is appropriate to tape record the informal conference and to use the tape recording in lieu of written notes.

d. **Conduct of the Informal Conference**. The Area Director shall conduct the informal conference in accordance with the following guidelines:

   (1) **Opening Remarks**. The opening remarks shall include discussions of the following:

      (a) Purpose of the informal conference.

      (b) Rights of participants.

      (c) Contest rights and time restraints.

      (d) Limitations, if any.

      (e) Settlements of cases.

      (f) Other relevant information.

(g) If the Area Director states any views on the legal merits of the employer's contentions, it should be made clear that those views are personal opinions only.

(2) **Closing**. At the conclusion of the discussion the main issues and potential courses of action shall be summarized. A copy of the summary, together with any other relevant notes or tapes of the discussion made by the Area Director, shall be placed in the case file.

e. **Decisions**. At the end of the informal conference, the Area Director shall make a decision as to what action is appropriate in the light of facts brought up during the conference.

(1) Changes to citations, penalties or abatement dates normally shall be made by means of an informal settlement agreement in accordance with current OSHA procedures; the reasons for such changes shall be documented in the case file. For more detail on settlement agreements, see D.4.b., below.

(2) Employers shall be informed that they are required by 29 CFR 1903.19 to post copies of all amendments to the citation resulting from informal conferences. Employee representatives must also be provided with copies of such documents. This regulation covers amended citations, citation withdrawals and settlement agreements.

f. **Failure to Abate**. If the informal conference involves an alleged failure to abate, the Area Director shall set a new abatement date in the informal settlement agreement, documenting for the case file the time that has passed since the original citation, the steps that the employer has taken to inform the exposed employees of their risk and to protect them from the hazard, and the measures that will have to be taken to correct the condition.

2. **Petitions for Modification of Abatement Date (PMA)**. Title 29 CFR 1903.14a governs the disposition of PMAs. If the employer requests additional abatement time after the 15-working-day contest period has passed, the following procedures for PMAs are to be observed:

a. **Filing Date**. A PMA must be filed in writing with the Area Director who issued the citation no later than the close of the next working day following the date on which abatement was originally required.

(1) If a PMA is submitted orally, the employer shall be informed that OSHA cannot accept an oral PMA and that a written petition must be mailed by the end of the next working day after the abatement date. If there is not sufficient time to file a written petition, the employer shall be informed of the requirements below for late filing of the petition.

(2) A late petition may be accepted only if accompanied by the employer's statement of exceptional circumstances explaining the delay.

   b. **Failure to Meet All Requirements**. If the employer's letter does not meet all the requirements of 1903.14a(b)(1)-(5), the employer shall be contacted within 10 working days and notified of the missing elements. A reasonable amount of time for the employer to respond shall be specified during this contact with the employer.

     (1) If no response is received or if the information returned is still insufficient, a second attempt (by telephone or in writing) shall be made. The employer shall be informed of the consequences of a failure to respond adequately; namely, that the PMA will not be granted and the employer may, consequently, be found in failure to abate.

     (2) If the employer responds satisfactorily by telephone and the Area Director determines that the requirements for a PMA have been met, appropriate documentation shall be placed in the case file.

   c. **Delayed Decisions**. Although OSHA policy is to handle PMAs as expeditiously as possible, there are cases where the Area Director's decision on the PMA is delayed because of deficiencies in the PMA itself, a decision to conduct a monitoring inspection and/or the need for Regional Office or National Office involvement. Requests for additional time (e.g., 45 days) for the Area Director to formulate a position shall be sent to the Review Commission through the Regional Solicitor. A letter conveying this request shall be sent at the same time to the employer and the employee representatives.

   d. **Area Office Position on the PMA**. After 15 working days following the PMA posting, the Area Director shall determine the Area Office position, agreeing with or objecting to the request. This shall be done within 10 working days following the 15 working days (if additional time has not been requested from the Review Commission; in the absence of a timely objection, the PMA is automatically granted even if not explicitly approved). The following action shall be taken:

     (1) If the PMA requests an abatement date which is two years or less from the issuance date of the citation, the Area Director has the authority to approve or object to the petition.

     (2) Any PMA requesting an abatement date which is more than two years from the issuance date of the citation requires the approval of the Regional Administrator as well as the Area Director.

     (3) If the PMA is approved, the Area Director shall notify the employer and the employee representatives by letter.

     (4) If supporting evidence justifies it (e.g., employer has taken no meaningful abatement action at all or has otherwise exhibited bad faith), the Area Director or the Regional Administrator, as appropriate and after consultation with the Regional Solicitor, shall object to the PMA. In such a case, all relevant documentation shall be sent to the Review Commission in accordance with 29 CFR 1903.14a(d). Both the employer and the employee representatives shall be notified of this action by letter, with return receipt requested.

(a) The letters of notification of the objection shall be mailed on the same date that the agency objection to the PMA is sent to the Review Commission.

(b) When appropriate, after consultation with the Regional Solicitor, a failure to abate notification may be issued in conjunction with the objection to the PMA.

e. **Employee Objections**. Affected employees or their representatives may file an objection in writing to an employer's PMA with the Area Director within 10 working days of the date of posting of the PMA by the employer or its service upon an authorized employee representative.

(1) Failure to file such a written objection with the 10-working-day period constitutes a waiver of any further right to object to the PMA.

(2) If an employee or an employee representative objects to the extension of the abatement date, all relevant documentation shall be sent to the Review Commission.

(a) Confirmation of this action shall be mailed (return receipt requested) to the objecting party as soon as it is accomplished.

(b) Notification of the employee objection shall be mailed (return receipt requested) to the employer on the same day that the case file is forwarded to the Commission.

3. **Services Available to Employers**. Employers requesting abatement assistance shall be informed that OSHA is willing to work with them even after citations have been issued.

4. **Settlement of Cases By Area Directors**.

a. **General**. Area Directors are granted settlement authority, using the following policy guidelines to negotiate settlement agreements.

(1) Except for egregious cases, or cases which affect other jurisdictions, Area Directors are authorized to enter into Informal Settlement Agreements with an employer before the employer files a written notice of contest.

NOTE: After the employer has filed a written notice of contest, the Area Director may proceed toward a Formal Settlement Agreement with the concurrence of the Regional Solicitor in cases where a settlement appears probable without the need for active participation by an attorney.

(2) Area Directors are authorized to change abatement dates, to reclassify violations (e.g., willful to serious, serious to other-than-serious), and to modify or withdraw a penalty, a citation or a citation item if the employer presents evidence during the informal conference which convinces the Area Director that the changes are justified.

(a) If an employer, having been cited as willfully or repeatedly violating the Act, decides to correct all violations, but wishes to purge himself or herself of the adverse public perception attached to a willful or repeated violation classification and is willing to pay all or almost all of the penalty and is willing to make significant additional concessions, then a Section 17 designation may be applicable. Decisions to make a Section 17 designation shall be based on whether the employer is willing to make significant concessions.

NOTE: Significant concessions may include the company entering into a corporate-wide settlement agreement subject to OSHA Instruction CPL 2.90, providing employee training of a specified type and frequency, hiring a qualified safety and health consultant and implementing the recommendations, effecting a comprehensive safety and health program, reporting new construction jobs or other worksites to OSHA, or waiving warrants for specified inspections/periods.

(b) A Section 17 designation also may be considered if the employer has advanced substantial reasons why the original classification is questionable but is willing to pay the penalty as proposed.

NOTE: Where the original classification clearly was excessive, Section 17 is not appropriate. Instead, the citation shall be amended to the appropriate classification.

(3) The Area Director has authority to actively negotiate the amount of penalty reduction, depending on the circumstances of the case and what improvements in employee safety and health can be obtained in return.

(4) Employers shall be informed that they are required by 29 CFR 1903.19 to post copies of all amendments or changes resulting from informal conferences. Employee representatives must also be provided with copies of such documents. This regulation covers amended citations, citation withdrawals and settlement agreements.

b. **Pre-Contest Settlement (Informal Settlement Agreement)**. Pre-contest settlements generally will occur during, or immediately following, the informal conference and prior to the completion of the 15-working-day contest period.

(1) If a settlement is reached during the informal conference, an Informal Settlement Agreement shall be prepared and the employer representative shall be invited to sign it. The Informal Settlement Agreement shall be effective upon signature by both the Area Director and the employer representative so long as the contest period has

not expired. Both shall date the document as of the day of actual signature.

(a) If the employer representative requests more time to consider the agreement and if there is sufficient time remaining of the 15-working-day period, the Area Director shall sign and date the agreement and provide the signed original for the employer to study while considering whether to sign it. A letter explaining the conditions under which the agreement will become effective shall be given (or mailed by certified mail, return receipt requested) to the employer and a record kept in the case file.

(b) The Area Director shall sign and date the agreement and provide the original (in person, or by certified mail, return receipt requested) to the employer if any other circumstances warrant such action; the agreement may also be sent to the employer for signature, and returned to the Area Director, via facsimile if circumstances warrant.

   1 If the signed agreement is provided to the employer, a copy shall be kept in the case file and the employer informed in writing that no changes are to be made to the original by the employer without explicit prior authorization for such changes from the Area Director.

   2 In every case the Area Director shall give formal notice in writing to the employer that the citation will become final and unreviewable at the end of the contest period unless the employer either signs the agreement or files a written notice of contest.

   3 If the employer representative wishes to make any changes to the text of the agreement, the Area Director must agree to and authorize the proposed changes prior to the expiration of the contest period.

      a If the changes proposed by the employer are acceptable to the Area Director, they shall be authorized and the exact language to be written into the agreement shall be worked out mutually. The employer shall be instructed to incorporate the agreed-upon language into the agreement, sign it and return it to the Area Office as soon as practicable by telefacsimile, if possible.

      b Annotations incorporating the exact language of any changes authorized by the Area Director shall be made to the retained copy of the agreement, and a dated record of the authorization shall be signed by the Area Director and placed in the case file.

   4 Upon receipt of the Informal Settlement Agreement signed by the employer, the Area Director shall ensure that any modified text of the agreement is in accordance with the notations made in the case file.

      a If so, the citation record shall be updated in IMIS in accordance with current procedures.

      b If not, and if the variations substantially change the terms of the agreement, the agreement signed by the employer shall be considered as a notice of intent to contest and handled accordingly. The employer shall be so informed as soon as possible.

    5 A reasonable time shall be allowed for return of the agreement from the employer.

      a After that time, if the agreement has still not been received, the Area Director shall presume that the employer is not going to sign the agreement; and the citation shall be treated as a final order until such time as the agreement is received, properly signed prior to the expiration of the contest period.

      b The employer shall be required to certify that the informal settlement agreement was signed prior to the expiration of the contest period.

  (2) If the Area Director's settlement efforts are unsuccessful and the employer contests the citation, the Area Director shall state the terms of the final settlement offer in the case file.

c. **Procedures for Preparing the Informal Settlement Agreement**. The Informal Settlement Agreement shall be prepared and processed in accordance with current OSHA policies and practices. For guidance for determining final dates of settlements and Review Commission orders see D.5., below.

d. **Post-Contest Settlement (Formal Settlement Agreement)**. Post-contest settlements will generally occur before the complaint is filed with the Review Commission.

  (1) Following the filing of a notice of contest, the Area Director shall, unless other procedures have been agreed upon, notify the Regional Solicitor when it appears that negotiations with the employer may produce a settlement. This shall normally be done at the time when the notice of contest transmittal memorandum is sent to the Regional Solicitor.

  (2) If a settlement is later requested by the employer with the Area Director, the Area Director shall communicate the terms of the settlement to the Regional Solicitor who will then draft the settlement agreement.

e. **Corporate-Wide Settlement Agreements**. Corporate-wide Settlement Agreements (CSAs) may be entered into under special circumstances to obtain formal recognition by the employer of cited hazards and formal acceptance of the obligation to seek out and abate those hazards throughout all workplaces under its control. Guidelines, policies and procedures for entering into CSA negotiations are found in OSHA Instruction CPL 2.90.

5. **Guidance for Determining Final Dates of Settlements and Review Commission Orders**.

   a. **Citation/Notice of Penalty Not Contested**. The Citation/Notice of Penalty and abatement date becomes a final order of the Commission on the date the 15-working-day contest period expires.

   b. **Citation/Notice of Penalty Resolved by Informal Settlement Agreement (ISA)**. The ISA becomes final, with penalties due and payable, 15 working days after the date of the last signature.

   NOTE: A later due date for payment of penalties may be set by the terms of the ISA.

   NOTE: The Review Commission does NOT review the ISA.

   c. **Citation/Notice of Penalty Resolved by Formal Settlement Agreement (FSA)**. The Citation/Notice of Penalty becomes final 30 days after docketing of the Administrative Law Judge's (ALJ's) **Order** "approving" the parties' stipulation and settlement agreement, assuming there is no direction for review. The Commission's **Notice of Docketing** specifies the date upon which the decision becomes a final order. If the FSA is "approved" by a Commission's **Order**, it will become final after 60 days.

   NOTE: A later due date for payment of penalties may be set by the terms of the FSA.

   NOTE: Settlement is permitted and encouraged by the Commission at any stage of the proceedings. (See 29 CFR 2200.100(a).)

   d. **Citation/Notice of Penalty Resolved by an ALJ Decision**. The ALJ decision/report becomes a final order of the Commission 30 days after docketing unless the Commission directs a review of the case. The Commission's **Notice of Docketing** specifies the date upon which the decision becomes a final order.

   e. **ALJ Decision is Reviewed by Commission**. According to Section 11 of the OSH Act, the Commission decision becomes final 60 days after the **Notice of Commission Decision** if no appeal has been filed with the U.S. Court of Appeals. The **Notice of Commission Decision** specifies the date the Commission decision is issued.

   f. **Commission Decision Reviewed by the U.S. Court of Appeals**. The U.S. Court of Appeals' decision becomes final 90 days after the entry of the judgment, if no appeal has been filed with the U.S. Supreme Court.

E. REVIEW COMMISSION.

1. **Transmittal of Notice of Contest and Other Documents to Commission.**

   a. **Notice of Contest.** In accordance with the Occupational Safety and Health Review Commission (OSHRC) revised Rules of Procedure (51 F.R. 32020, No. 173, September 8, 1986), the original notice of contest, together with copies of all relevant documents (all contested Citations and Notifications of Penalty and Notifications of Failure to Abate Alleged Violation, and proposed additional penalty) shall be transmitted by the Area Director to the OSHRC postmarked prior to the expiration of 15 working days after receipt of the notice of contest (29 CFR 2200.33). The Regional Solicitor shall be consulted in questionable cases.

   (1) The envelope that contained the notice of contest shall be retained in the case file with the postmark intact.

   (2) Where the Area Director is certain that the notice of contest was not mailed; i.e., postmarked, within the 15-working-day period allowed for contest, the notice of contest shall be returned to the employer who shall be advised of the statutory time limitation. The employer shall be informed that OSHRC has no jurisdiction to hear the case because the notice of contest was not filed within the 15 working days allowed and, therefore, that the notice of contest will not be forwarded to the OSHRC. A copy of all untimely notices of contest shall be retained in the case file.

   (3) If the notice of contest is submitted to the Area Director after the 15-working-day period, but the notice contests only the reasonableness of the abatement period, it shall be treated as a Petition for Modification of Abatement and handled in accordance with the instructions in D.2. of this chapter.

   (4) If written communication is received from an employer containing objection, criticism or other adverse comment as to a citation or proposed penalty, which does not clearly appear to be a notice of contest, the Area Director shall contact the employer as soon as possible to clarify the intent of the communication. Such clarification must be obtained within 15 working days after receipt of the communication so that if, in fact, it is a notice of contest, the file may be forwarded to the Review Commission within the allowed time. The Area Director shall make a memorandum for the case file regarding the substance of this communication.

   (5) If the Area Director determines that the employer intends the document to be a notice of contest, it shall be transmitted to the OSHRC in accordance with E.1.a., above. If the employer did not intend the document to be a notice of contest, it shall be

retained in the case file with the memorandum of the contact with the employer. If no contact can be made with the employer, communications of the kind referred to in E.1. a.(4), above, shall be timely transmitted to the OSHRC.

(6) If the Area Director's contact with the employer reveals a desire for an informal conference, the employer shall be informed that an informal conference does not stay the running of the 15-working-day period for contest.

b. **Documents to Executive Secretary**. The following documents are to be transmitted within the 15-working-day time limit to the Executive Secretary, Occupational Safety and Health Review Commission, 1825 K Street, N.W., Washington, D.C. 20006:

NOTE: In order to give the Regional Solicitor the maximum amount of time to prepare the information needed in filing a complaint with the Review Commission, the notice of contest and other documents shall not be forwarded to the Review Commission until the final day of the 15-working-day period.

(1) **All Notices of Contest**. The originals are to be transmitted to the Commission and a copy of each retained in the case file.

(2) **All Contested Citations and Notices of Proposed Penalty or Notice of Failure to Abate Issued in the Case**. The signed copy of each of these documents shall be taken from the case file and sent to the Commission after a copy of each is made and placed in the case file.

(3) **Certification Form**. The certification form shall be used for all contested cases and a copy retained in the case file. It is essential that the original of the certification form, properly executed, be transmitted to the Commission.

(a) When listing the Region number in the heading of the form, do not use Roman numerals. Use 1, 2, 3, 4, 5, 6, 7, 8, 9, or 10. Insert "C" in the CSHO Job Title block if a safety CSHO or "I," if a health CSHO.

(b) Item 3 on the certification form shall be filled in by inserting only the word "employer" or "employee" in the space provided. This holds true even when the notice of contest is filed by an attorney for the party contesting the action. An item "4" shall be added where other documents, such as additional notices of contest, are sent to the Commission.

(c) Use a date stamp with the correct date for each item in the document list under the column headed "Date."

(d) Be sure to have the name and address of the Regional Solicitor or attorney who will handle the case inserted in the box containing the printed words "FOR THE SECRETARY OF LABOR." The Commission notifies this person of the hearing date and other official actions on the case. If this box is not filled in by the Area Director, delay in receipt of such notifications by the appropriate Regional Solicitor or attorney could result.

(4) **Documents Sent to OSHRC**. In most cases, the envelope sent to the OSHRC Executive Secretary will contain only four documents — the certification form, the employer's letter contesting OSHA's action, and a copy of the Citation and Notification of Penalty Form (OSHA-2) or of the Notice of Failure to Abate Form (OSHA-2B).

c. **Petitions for Modification of Abatement Dates (PMAs)**.

(1) In accordance with the OSHRC Rules of Procedure the Secretary or duly authorized agent shall have the authority to approve petitions for modification of abatement filed pursuant to 29 CFR 2200.37(b) and (c).

(2) The purpose of this transfer of responsibility is to facilitate the handling and to expedite the processing of PMAs to which neither the Secretary nor any other affected party objects. The Area Director who issued the citation is the authorized agent of the Secretary and shall receive, process, approve, disapprove or otherwise administer the petitions in accordance with 29 CFR 2200.37 and 2200.38, 29 CFR 1903.14a, and D.2. of this chapter. In general, the Area Director shall:

(a) Ensure that the formal requirements of 2200.37(b) and (c) and 1903.14a are met.

(b) Approve or disapprove uncontested PMA's within 15 working days from the date the petition was posted where all affected employees could have notice of the petition.

(c) Forward to the Review Commission within 10 working days after the 15-working-day approval period all petitions objected to by the Area Director or affected employees.

(d) File a response setting forth the reasons for opposing granting of the PMA within 10 working days after receipt of the docketing by the Commission.

2. **Transmittal of File to Regional Solicitor**.

a. **Notification of the Regional Solicitor**. Under the Commission's Rules of Procedure the Secretary of Labor is required to file a complaint with the Commission within 20 calendar days after the Secretary's receipt of a notice of contest.

b. **Subpoena**. The Commission's rules provide that any person served with a subpoena, whether merely to testify in any Commission hearing or to produce records and testify in such hearing, shall, within 5 days after the serving of the subpoena, move to revoke the subpoena if the person does not intend to comply with the subpoena. These time limitations must be complied with, and expeditious handling of any subpoena served on OSHA employees is necessary. In addition, OSHA personnel may be subpoenaed to participate in nonthird-party OSHA actions. In both types of cases, the Solicitor will move to revoke the subpoena on OSHA personnel. Therefore, when any such subpoena is served on OSHA personnel, the Regional Solicitor shall immediately be notified by telephone.

3. **Communications with Commission Employees**. There shall be no ex parte communication, with respect to the merits of any case not concluded, between the Commission, including any member, officer, employee, or agent of the Commission who is employed in the decisional process, and any of the parties or interveners. Thus, CSHOs, Area Directors, Regional Administrators, or other field personnel shall refrain from any direct or indirect communication relevant to the merits of the case with Administrative Law Judges or any members or employees of the Commission. All inquiries and communications shall be handled through the Regional Solicitor.

4. **Dealings with Parties While Proceedings Are Pending before the Commission**.

   a. Clearance with Regional Solicitor. After the notice of contest is filed and the case is within the jurisdiction of the Commission, there shall be no investigations of or conferences with the employer without clearance from the appropriate Regional Solicitor. Such requests shall be referred promptly to the Regional Solicitor for a determination of the advisability, scope and timing of any investigation, and the advisability of and participation in any conference. To the maximum extent possible, there shall be consultation with the Solicitor on questions of this nature so as to insure no procedural or legal improprieties.

   b. **Inquiries**. Once a notice of contest has been filed, all inquiries relating to the general subject matter of the Citation and Notification of Penalty raised by any of the parties of the proceedings, including the employer and affected employees or authorized employee representative, shall be referred promptly to the Regional Solicitor. Similarly, all other inquiries, such as from prospective witnesses, insurance carriers, other Government agencies, attorneys, etc., shall be referred to the Regional Solicitor.

Appendix H
Workplace Violence: Hazard Awareness

The following references may help increase awareness of violence in the workplace.

- *Fatal occupational injuries by event or exposure and major private industry sector.* U.S. Department of Labor (DOL), Bureau of Labor Statistics (BLS), (2005), 48 KB PDF, 5 pages. Includes the category "Assaults and Violent Acts" and reports 792 such acts in 2005.
- *Workplace Violence Awareness and Prevention.* OSHA and The Long Island Coalition for Workplace Violence Awareness and Prevention, (1996, February). Includes facts and figures about workplace violence, elements of a workplace violence prevention program, and a sample program.
- *Cal/OSHA Guidelines for Workplace Security.* State of California, (1995, March 30). Characterizes establishments, profiles, and motives of the agent or assailant, and identifies preventive measures by type. In California, the majority (60 percent) of workplace homicides involved a person entering a small late-night retail establishment. Nonfatal Type II events involving assaults to service providers, especially to health-care providers, may represent the most prevalent category of workplace violence resulting in physical injury.
- *Occupational Violence.* National Institute for Occupational Safety and Health (NIOSH) Safety and Health Topic. Reports that an average of 1.7 million people were victims of violent crime while working or on duty in the United States each year from 1993 through 1999 according to the Bureau of Justice Statistics (BJS). Includes NIOSH publications as well as other U.S. government occupational violence links including a psychological first aid manual for mental health providers.
- *Violence on the Job.* NIOSH Publication No. 2004-100d, (2004). Discusses practical measures for identifying risk factors for violence at work and taking strategic action to keep employees safe. Available as streaming video, Flash media, PDF transcript, or as a CD-ROM.
- *Violence: Occupational Hazards in Hospitals.* U.S. Department of Health and Human Services (DHHS), NIOSH Publication No. 2002-101 (2002, April). Also available as a 105 KB PDF, fifteen pages. Increases employee and employer awareness of the risk factors for violence in hospitals and provides strategies for reducing exposure to these factors.
- *Stress at Work.* DHHS, NIOSH Publication No. 99-101 (1999). Highlights knowledge about the causes of stress at work and outlines steps that can be taken to prevent job stress. Defines job stress as the harmful physical and emotional responses that occur when the requirements of the job do not

match the capabilities, resources, or needs of the worker. Job stress can lead to poor health and even injury. Explores a combination of organizational change and stress management as the most useful approach for preventing stress at work.

- *Violence in the Workplace.* NIOSH Fact Sheet (1997, June). Provides basic information on workplace violence, including risk factors and prevention strategies. Homicide is reported as the second leading cause of death on the job, second only to motor vehicle crashes, and homicide is the leading cause of workplace death among females.

- *Violence in the Workplace — Risk Factors and Prevention Strategies.* DHHS, NIOSH Current Intelligence Bulletin 57 (1996, July). Reviews what is known about fatal and nonfatal violence in the workplace to determine the focus needed for prevention and research efforts. Reports that each week in the United States, an average of twenty workers are murdered and 18,000 are assaulted while at work. These staggering figures should not be an accepted cost of doing business in our society — nor should death or injury be an inevitable result of one's chosen occupation.

- *Preventing Homicide in the Workplace.* DHHS, NIOSH Publication No. 93-109 (1995, May). Reports workplaces with the highest rates of occupational homicide were taxicab establishments, liquor stores, gas stations, detective/protective services, justice/public order establishments (including courts, police protection establishments, legal counsel and prosecution establishments, correctional institutions, and fire protection establishments), grocery stores, jewelry stores, hotels/motels, and eating/drinking places. Taxicab establishments had the highest rate of occupational homicide — nearly forty times the national average and more than three times the rate of liquor stores, which had the next highest rate.

- *Homicide in U.S. Workplaces: A Strategy for Prevention and Research.* DHHS, NIOSH Publication No. 92-103 (1992, September), 435 KB PDF, thirteen pages. Serves as a foundation for the development of a national strategy for use in prioritizing research and targeting interventions to prevent work-related homicides. Recommends study of various environmental approaches such as improved lighting, locked drop-safes, work areas openly visible to the public, and increased staffing. Behavioral strategies such as training in conflict resolution and non-violent response should also be examined.

- *Violence in the Workplace: Oregon, 1991–1995.* Oregon Department of Consumer and Business Services (1996, December), 66 KB PDF, eleven pages. Provides a study of Workers' Compensation Claims Caused by Violent Acts, 1991 to 1995. In 1988, 32 percent of the violent claims came from the state hospital system (80 percent of the violent claims from state government). Because of the problems at the hospital, the staff was increased and given additional training, and the resident population was reduced. As a result, the number of violent claims at state hospitals dropped from 202 in 1988 to 61 in 1991; it fell further to 39 in 1995. The patients discharged were placed in residential care programs run by private-sector companies.

There was no increase in violent claims from these companies during the 1988–1991 period.

- *Workplace Violence: Issues in Response.* Federal Bureau of Investigation (FBI) (2004, March 1). Also available as a 6.2 MB PDF, eighty pages. Shares expertise of representatives from law enforcement, private industry, government, law, labor, professional organizations, victim services, the military, academia, mental health, as well as the FBI on this important issue. This monograph resulted from a June 2002 symposium hosted by the FBI's National Center for the Analysis of Violent Crime entitled "Violence in the Workplace."
- *Violence in the Workplace 1993–1999.* U.S. Department of Justice (DOJ), Bureau of Justice Statistics (BJS) (2001). Presents data for 1993 through 1999 from the National Crime Victimization Survey estimating the extent of workplace crime in the United States. Workplace violence accounted for 18 percent of all violent crime during the seven-year period. Of the occupations examined, police officers experienced such crimes at the highest rate (260.8 per 1000 police officers), whereas college or university professors and teachers had the lowest rate (1.6 per 1000 teachers). Government employees had violent victimization rates (28.6 per 1000 government workers) that were higher than those people who work for private companies (9.9 per 1000 workers) or self-employed people (7.4 per 1000).
- Sygnatur, Eric F. and Guy A. Toscano. "Work-Related Homicides: The Facts." *Compensation and Working Conditions* 3.8 (2000, Spring), 76 KB PDF, six pages. Provides information on work-related homicides, including information about the perpetrators, demographics of the decedents, and other relevant facts about these events, such as the time of the incident, the location, and the type of establishment in which the homicide occurred. Contrary to popular belief, the majority of these incidents are not crimes of passion committed by disgruntled coworkers and spouses, but rather result from robberies.
- *Most Workplace Violence on Women Hidden, Says Center Report.* University of Albany (UA), Center for Women in Government. Summarizes and comments on a report addressing workplace violence, emphasizing data specific to women. Two-thirds of the nonfatal attacks on women are committed by patients or residents in institutional settings. Husbands, boyfriends, and ex-partners commit 15 percent of all workplace homicides against women. Women are more likely to suffer serious injury from workplace violence than men. Women who are victims of violent workplace crimes are twice as likely as men to know their attackers.
- *Training Resources.* University of Minnesota (UM), Minnesota Center Against Violence and Abuse. Provides training resources specific to workplace violence, including U.S. Office of Personnel Management guide, the OSHA guidelines, and a prevention guide from the State of Mississippi.
- *Workplace Violence: Can You Close the Door On It?* American Nurses Association (ANA) (1994). Heightens awareness regarding workplace violence and recommends nine steps to prevent it.

Appendix I
The Fair Labor Standards Act (FLSA)

OVERVIEW

The Fair Labor Standards Act (FLSA), which prescribes standards for the basic minimum wage and overtime pay, affects most private and public employment. It requires employers to pay covered employees who are not otherwise exempt at least the federal minimum wage and overtime pay of one-and-one-half-times the regular rate of pay. For non-agricultural operations, it restricts the hours that children under age sixteen can work and forbids the employment of children under age eighteen in certain jobs deemed too dangerous. For agricultural operations, it prohibits the employment of children under age sixteen during school hours and in certain jobs deemed too dangerous. The Act is administered by the Employment Standards Administration's Wage and Hour Division within the U.S. Department of Labor.

COMPLIANCE ASSISTANCE MATERIALS

Basic Information

Employment Law Guide — Minimum Wage and Overtime Pay — Describes the minimum wage and overtime pay requirements.

General Information on the FLSA — The Fair Labor Standards Act (FLSA) establishes minimum wage, overtime pay, record keeping, and child labor standards affecting full-time and part-time workers in the private sector and in federal, state, and local governments.

The Handy Reference Guide to the FLSA — The FLSA establishes minimum wage, overtime pay, record keeping, and youth employment standards affecting full-time and part-time workers in the private sector and in federal, state, and local governments.

Minimum Wage Rates for Each State

Questions and Answers About the Minimum Wage — The federal minimum wage for covered nonexempt employees is $5.85 per hour effective July 24, 2007; $6.55 per hour effective July 24, 2008; and $7.25 per hour effective July 24, 2009. The federal minimum wage provisions are contained in the FLSA. Many states also have minimum wage laws.

YouthRules! Web Site — Contains information for teens who have a job or are looking for a job and for their parents, teachers, and employers.

U.S. Department of Labor's (DOL) FairPay Overtime Initiative

Child Labor Requirements in Nonagricultural Occupations

Child Labor Requirements in Agricultural Occupations

Frequently Asked Questions Regarding Child Labor

General Information from the Employment Standards Administration, Wage and Hour Division (PDF)

Filing a complaint — DOL's Wage and Hour Division manages complaints regarding violations of the various laws and regulations it administers. To file a complaint concerning one of these laws, contact your nearest Wage and Hour Division office or call the Department's Toll-Free Wage and Hour Help Line at 1-866-4-US-WAGE.

FACT SHEETS

Topical Index — Alphabetical listing of Wage and Hour Division fact sheets.

E-TOOLS

Fair Labor Standards Act (FLSA) Coverage and Employment Status Advisor — Helps employers and employees understand and determine coverage under the FLSA.

Fair Labor Standards Act (FLSA) Hours Worked Advisor — Helps employers and employees determine which work-related activities are considered "hours worked" and thus hours for which employees must be paid.

Fair Labor Standards Act (FLSA) Overtime Security Advisor — Helps employees and employers determine whether a particular employee is exempt from the FLSA's minimum wage and overtime pay requirements.

Fair Labor Standards Act (FLSA) Section 14(c) Advisor — Helps employers, employees, and their family members understand FLSA Section 14(c), which authorizes employers, after receiving a certificate from DOL, to pay less than the federal minimum wage to workers who have disabilities for the work being performed.

Fair Labor Standards Act (FLSA) Child Labor Rules Advisor — Helps young workers and their employers, parents, and educators understand the FLSA's child labor provisions, which dictate the hours youth can work and the jobs they may and may not perform.

Poster Advisor — Helps employers determine which federal DOL posters they are required to display and print required posters free of charge.

POSTERS

Fair Labor Standards Act (FLSA) Minimum Wage Poster — Describes the requirement that every employer of employees subject to the FLSA's minimum wage and overtime provisions must post a notice explaining the Act. (Español) (Chinese)

RECORD KEEPING

Every employer covered by the Fair Labor Standards Act (FLSA) must keep certain records for each covered, nonexempt worker. There is no required form for the records, but the records must include accurate information about the employee and data about the hours worked and the wages earned. The following is a listing of the basic records that an employer must maintain:

Employee's full name, as used for social security purposes, and on the same record, the employee's identifying symbol or number if such is used in place of name on any time, work, or payroll records

Address, including zip code

Birth date, if younger than nineteen

Sex and occupation

Time and day of week when employee's workweek begins; hours worked each day and total hours worked each workweek

Basis on which employee's wages are paid

Regular hourly pay rate

Total daily or weekly straight-time earnings

Total overtime earnings for the workweek

All additions to or deductions from the employee's wages

Total wages paid each pay period

Date of payment and the pay period covered by the payment

APPLICABLE LAWS AND REGULATIONS

The Fair Labor Standards Act of 1938 (PDF) — Establishes minimum wages, overtime pay, record keeping, and child labor standards for private-sector and government workers.

29 CFR Parts 510 to 794

RELATED TOPICS AND LINKS

Employment Law Guide — Prevailing Wages in Construction Contracts — Requires contractors and subcontractors performing federally financed construction projects to pay their laborers and mechanics at rates not less than the prevailing wage rates and fringe benefits for corresponding classes of laborers and mechanics employed on similar projects in the area. **See also Compliance Assistance by Law — The Davis-Bacon and Related Acts.**

Employment Law Guide — Prevailing Wages in Service Contracts — Describes the coverage and basic requirements of the McNamara-O'Hara Service Contract Act (SCA). **See also Compliance Assistance by Law — The McNamara-O'Hara Service Contract Act.**

Employment Law Guide — Wages in Supply and Equipment Contracts — Describes the coverage and basic requirements of the Walsh-Healey Public Contracts Act (PCA) for contractors with contracts in excess of $10,000 for the manufacturing or furnishing of materials, supplies, articles, or equipment to the U.S. government or the District of Columbia. Covered contractors must pay employees on contracts the federal minimum wage of $5.85 per hour effective July 24, 2007; $6.55 per hour effective July 24, 2008; and $7.25 per hour effective July 24, 2009. **See also Compliance Assistance by Law — The Walsh-Healey Public Contracts Act.**

Employment Law Guide — Hours and Safety Standards Act in Construction Contracts — Describes the Contract Work Hours and Safety Standards Act (CWHSSA) which requires contractors and subcontractors with covered contracts to pay laborers and mechanics employed in the performance of the contracts one-and-one-half times their basic rate of pay for all hours worked over forty in a workweek. **See also Compliance Assistance by Law — The Contract Work Hours and Safety Standards Act (CWHSSA).**

Employment Law Guide — "Kickbacks" in Federally Funded Construction — Prohibits a contractor or subcontractor from in any way inducing an employee to give up any part of the compensation to which he or she is entitled under his or her contract and requires contractors and subcontractors on certain federally funded construction contracts to submit weekly statements of compliance. **See also Compliance Assistance by Law — The Copeland "Anti-Kickback" Act.**

YouthRules! Web Site — Contains information for teens who have a job or are looking for a job and for their parents, teachers, and employers.

See also Compliance Assistance by Topic — Wages and Hours Worked.

DOL CONTACTS

Employment Standards Administration (ESA), Wage and Hour Division, 200 Constitution Avenue, NW, Room S-3502, Washington, DC 20210; Tel: 1-866-4USWAGE (1-866-487-9243), TTY: 1-877-889-5627.

For questions on other DOL laws, please call DOL's toll-free help line at 1-866-4-USA-DOL. Live assistance is available in English and Spanish, Monday through Friday from 8:00 a.m. to 8:00 p.m. Eastern Time. Additional service is available in more than 140 languages through a translation service. Tel: 1-866-4-USA-DOL; TTY: 1-877-889-5627.

Appendix J
CPL 02-00-124 — CPL 2-0.124 — Multi-Employer Citation Policy

Record Type:	Instruction
Directive Number:	CPL 02-00-124
Old Directive Number:	CPL 2-0.124
Title:	Multi-Employer Citation Policy.
Information Date:	12/10/1999

U.S. DEPARTMENT OF LABOR   Occupational Safety and Health Administration

Directive Number:Cpl 2-0.124	**Effective Date**: December 10, 1999
Subject: Multi-Employer Citation Policy	

Abstract

Purpose:	To clarify the Agency's multi-employer citation policy
Scope:	OSHA-wide
References:	OSHA Instruction CPL 2.103 (the FIRM)
Suspensions:	Chapter III, Paragraph C. 6. of the FIRM is suspended and replaced by this directive
State Impact:	This Instruction describes a Federal Program Change. Notification of State intent is required, but adoption is not.
Action Offices:	National, Regional, and Area Offices
Originating Office:	Directorate of Compliance Programs
Contact:	Carl Sall, (202) 693-2345 Directorate of Construction N3468 FPB 200 Constitution Ave., NW Washington, DC 20210.

By and Under the Authority of
R. Davis Layne
Deputy Assistant Secretary, OSHA.

TABLE OF CONTENTS

I.   *Purpose.* This Directive clarifies the Agency's multi-employer citation policy and suspends Chapter III. C. 6. of OSHA's Field Inspection Reference Manual (FIRM).

II. *Scope.* OSHA-Wide

III. *Suspension.* Chapter III. Paragraph C. 6. of the FIRM (CPL 2.103) is suspended and replaced by this Directive.

IV. *References.* OSHA Instructions:

     CPL 02-00.103; OSHA Field Inspection Reference Manual (FIRM), September 26, 1994.

     ADM 08-0.1C, OSHA Electronic Directive System, December 19, 1997.

V.   *Action Information*

     A. *Responsible Office.* Directorate of Construction.

     B. *Action Offices.* National, Regional and Area Offices

     C. *Information Offices.* State Plan Offices, Consultation Project Offices

     Federal Program Change. This Directive describes a Federal Program Change for which State adoption is not required. However, the States shall respond via the two-way memorandum to the Regional Office as soon as the State's intent regarding the multi-employer citation policy is known, but no later than 60 calendar days after the date of transmittal from the Directorate of Federal-State Operations.

Force and Effect of Revised Policy. The revised policy provided in this Directive is in full force and effect from the date of its issuance. It is an official Agency policy to be implemented OSHA-wide.

Changes in Web Version of FIRM. A note will be included at appropriate places in the FIRM as it appears on the Web indicating the suspension of Chapter III paragraph 6. C. and its replacement by this Directive, and a hypertext link will be provided connecting viewers with this Directive.

Background. OSHA's Field Inspection Reference Manual (FIRM) of September 26, 1994 (CPL 2.103), states at Chapter III, paragraph 6. C., the Agency's citation policy for multi-employer worksites. The Agency has determined that this policy needs clarification. This directive describes the revised policy.

A. *Continuation of Basic Policy.* This revision continues OSHA's existing policy for issuing citations on multi-employer worksites. However, it gives clearer and more detailed guidance than did the earlier description of the policy in the FIRM, including new examples explaining when citations should and should not be issued to exposing, creating, correcting, and controlling employers. These examples, which address common situations and provide general policy guidance, are not intended to be exclusive. In all cases, the decision on whether to issue citations should be based on all of the relevant facts revealed by the inspection or investigation.

B. *No Changes in Employer Duties.* This revision neither imposes new duties on employers nor detracts from their existing duties under the OSH Act. Those duties continue to arise from the employers' statutory duty to comply with OSHA standards and their duty to exercise reasonable diligence to determine whether violations of those standards exist.

Multi-Employer Worksite Policy. The following is the multi-employer citation policy:

A. *Multi-Employer Worksites.* On multi-employer worksites (in all industry sectors), more than one employer may be citable for a hazardous condition that violates an OSHA standard. A two-step process must be followed in determining whether more than one employer is to be cited.

1. *Step One.* The first step is to determine whether the employer is a creating, exposing, correcting, or controlling employer. The definitions in paragraphs (B) through (E) below explain and give examples of each. Remember that an employer may have multiple roles (see paragraph H). Once you determine the role of the employer, go to Step Two to determine if a citation is appropriate (Note that only exposing employers can be cited for General Duty Clause violations).

2. *Step Two.* If the employer falls into one of these categories, it has obligations with respect to OSHA requirements. Step Two is to determine if the employer's actions were sufficient to meet those obligations. The extent of the actions required of employers varies based on which category applies. Note that the extent of the measures that a controlling employer must take to satisfy its duty to exercise reasonable care to prevent and detect violations is less than what is required of an employer with respect to protecting its own employees.

B. The Creating Employer

1. *Step 1: Definition:* The employer that caused a hazardous condition that violates an OSHA standard.

2. *Step 2: Actions Taken:* Employers must not create violative conditions. An employer that does so is citable even if the only employees exposed are those of other employers at the site.

   a. ***Example 1:*** Employer Host operates a factory. It contracts with Company S to service machinery. Host fails to cover drums of a chemical despite S's repeated requests that it do so. This results in airborne levels of the chemical that exceed the Permissible Exposure Limit.

   Analysis: Step 1: Host is a creating employer because it caused employees of S to be exposed to the air contaminant above the PEL. **Step 2**: Host failed to implement measures to prevent the accumulation of the air contaminant. It could have met its OSHA obligation by implementing the simple engineering control of covering the drums. Having failed to implement a feasible engineering control to meet the PEL, Host is citable for the hazard.

   b. ***Example 2:*** Employer M hoists materials onto Floor 8, damaging perimeter guardrails. Neither its own employees nor employees of other employers are exposed to the hazard. It takes effective steps to keep all employees, including those of other employers, away from the unprotected edge and informs the controlling employer of the problem. Employer M lacks authority to fix the guardrails itself.

   Analysis: Step 1: Employer M is a creating employer because it caused a hazardous condition by damaging the guardrails. **Step 2:** While it lacked the authority to fix the guardrails, it took immediate and effective steps to keep all employees away from the hazard and notified the controlling employer of the hazard. Employer M is not citable since it took effective measures to prevent employee exposure to the fall hazard.

C. *The Exposing Employer*

  1. *Step 1: Definition:* An employer whose own employees are exposed to the hazard. See Chapter III, section (C)(1)(b) for a discussion of what constitutes exposure.

  2. *Step 2: Actions Taken:* If the exposing employer created the violation, it is citable for the violation as a creating employer. If the violation was created by another employer, the exposing employer is citable if it (1) knew of the hazardous condition or failed to exercise reasonable diligence to discover the condition, and (2) failed to take steps consistent with its authority to protect is employees. If the exposing employer has authority to correct the hazard, it must do so. If the exposing employer lacks the authority to correct the hazard, it is citable if it fails to do each of the following: (1) ask the creating and/or controlling employer to correct the hazard; (2) inform its employees of the hazard; and (3) take reasonable alternative protective measures. In extreme circumstances (e.g., imminent danger situations), the exposing employer is citable for failing to remove its employees from the job to avoid the hazard.

  a. ***Example 3:*** Employer Sub S is responsible for inspecting and cleaning a work area in Plant P around a large, permanent hole at the end of each day. An OSHA standard requires guardrails. There are no guardrails around the hole and Sub S employees do not use personal fall protection, although it would be feasible to do so. Sub S has no authority to install guardrails. However, it did ask Employer P, which operates the plant, to install them. P refused to install guardrails.

  Analysis: **Step 1**: Sub S is an exposing employer because its employees are exposed to the fall hazard. **Step 2**: While Sub S has no authority to install guardrails, it is required to comply with OSHA requirements to the extent feasible. It must take steps to protect its employees and ask the employer that controls the hazard - Employer P - to correct it. Although Sub S asked for guardrails, since the hazard was not corrected, Sub S was responsible for taking reasonable alternative protective steps, such as providing personal fall protection. Because that was not done, Sub S is citable for the violation.

  b. ***Example 4:*** Unprotected rebar on either side of an access ramp presents an impalement hazard. Sub E, an electrical subcontractor, does not have the authority to cover the rebar. However, several times Sub E asked the general contractor, Employer GC, to cover the rebar. In the meantime, Sub E instructed its employees to use a different access route that avoided most of the uncovered rebar and required them to keep as far from the rebar as possible.

Analysis: **Step 1**: Since Sub E employees were still exposed to some unprotected rebar, Sub E is an exposing employer. **Step 2**: Sub E made a good faith effort to get the general contractor to correct the hazard and took feasible measures within its control to protect its employees. Sub E is not citable for the rebar hazard.

D. *The Correcting Employer*
  1. *Step 1: Definition:* An employer who is engaged in a common undertaking, on the same worksite, as the exposing employer and is responsible for correcting a hazard. This usually occurs where an employer is given the responsibility of installing and/or maintaining particular safety/health equipment or devices.
  2. *Step 2: Actions Taken:* The correcting employer must exercise reasonable care in preventing and discovering violations and meet its obligations of correcting the hazard.
      a. ***Example 5:*** Employer C, a carpentry contractor, is hired to erect and maintain guardrails throughout a large, 15-story project. Work is proceeding on all floors. C inspects all floors in the morning and again in the afternoon each day. It also inspects areas where material is delivered to the perimeter once the material vendor is finished delivering material to that area. Other subcontractors are required to report damaged/missing guardrails to the general contractor, who forwards those reports to C. C repairs damaged guardrails immediately after finding them and immediately after they are reported. On this project few instances of damaged guardrails have occurred other than where material has been delivered. Shortly after the afternoon inspection of Floor 6, workers moving equipment accidentally damage a guardrail in one area. No one tells C of the damage and C has not seen it. An OSHA inspection occurs at the beginning of the next day, prior to the morning inspection of Floor 6. None of C's own employees are exposed to the hazard, but other employees are exposed.

      Analysis: Step 1: C is a correcting employer since it is responsible for erecting and maintaining fall protection equipment. **Step 2:** The steps C implemented to discover and correct damaged guardrails were reasonable in light of the amount of activity and size of the project. It exercised reasonable care in preventing and discovering violations; it is not citable for the damaged guardrail since it could not reasonably have known of the violation.

E. *The Controlling Employer*
  1. *Step 1: Definition:* An employer who has general supervisory authority over the worksite, including the power to correct safety

and health violations itself or require others to correct them. Control can be established by contract or, in the absence of explicit contractual provisions, by the exercise of control in practice. Descriptions and examples of different kinds of controlling employers are given below.

2. *Step 2: Actions Taken:* A controlling employer must exercise reasonable care to prevent and detect violations on the site. The extent of the measures that a controlling employer must implement to satisfy this duty of reasonable care is less than what is required of an employer with respect to protecting its own employees. This means that the controlling employer is not normally required to inspect for hazards as frequently or to have the same level of knowledge of the applicable standards or of trade expertise as the employer it has hired.

3. *Factors Relating to Reasonable Care Standard.* Factors that affect how frequently and closely a controlling employer must inspect to meet its standard of reasonable care include the following:
   a. The scale of the project.
   b. The nature and pace of the work, including the frequency with which the number or types of hazards change as the work progresses.
   c. How much the controlling employer knows both about the safety history and safety practices of the employer it controls and about that employer's level of expertise.
   d. More frequent inspections are normally needed if the controlling employer knows that the other employer has a history of non-compliance. Greater inspection frequency may also be needed, especially at the beginning of the project, if the controlling employer had never before worked with this other employer and does not know its compliance history.
   e. Less frequent inspections may be appropriate where the controlling employer sees strong indications that the other employer has implemented effective safety and health efforts. The most important indicator of an effective safety and health effort by the other employer is a consistently high level of compliance. Other indicators include the use of an effective, graduated system of enforcement for non-compliance with safety and health requirements coupled with regular jobsite safety meetings and safety training.

4. *Evaluating Reasonable Care.* In evaluating whether a controlling employer has exercised reasonable care in preventing and discovering violations, consider questions such as whether the controlling employer:
   a. Conducted periodic inspections of appropriate frequency (frequency should be based on the factors listed in G.3.);

    b. Implemented an effective system for promptly correcting hazards; and

    c. Enforces the other employer's compliance with safety and health requirements with an effective, graduated system of enforcement and follow-up inspections.

5. *Types of Controlling Employers*

    a. *Control Established by Contract.* In this case, **the Employer Has a Specific Contract Right to Control Safety**: To be a controlling employer, the employer must itself be able to prevent or correct a violation or to require another employer to prevent or correct the violation. One source of this ability is explicit contract authority. This can take the form of a specific contract right to require another employer to adhere to safety and health requirements and to correct violations the controlling employer discovers.

    (1) ***Example 6:*** Employer GH contracts with Employer S to do sandblasting at GH's plant. Some of the work is regularly scheduled maintenance and so is general industry work; other parts of the project involve new work and are considered construction. Respiratory protection is required. Further, the contract explicitly requires S to comply with safety and health requirements. Under the contract GH has the right to take various actions against S for failing to meet contract requirements, including the right to have non-compliance corrected by using other workers and back-charging for that work. S is one of two employers under contract with GH at the worksite, where a total of five employees work. All work is done within an existing building. The number and types of hazards involved in S's work do not significantly change as the work progresses. Further, GH has worked with S over the course of several years. S provides periodic and other safety and health training and uses a graduated system of enforcement of safety and health rules. S has consistently had a high level of compliance at its previous jobs and at this site. GH monitors S by a combination of weekly inspections, telephone discussions and a weekly review of S's own inspection reports. GH has a system of graduated enforcement that it has applied to S for the few safety and health violations that had been committed by S in the past few years. Further, due to respirator equipment problems S violates respiratory protection requirements two days before GH's next scheduled inspection of S. The next day there is an OSHA inspection. There is no notation of the equipment problems in S's inspection reports to GH and S made no mention of it in its telephone discussions.

Analysis: Step 1: GH is a controlling employer because it has general supervisory authority over the worksite, including contractual authority to correct safety and health violations. **Step 2:** GH has taken reasonable steps to try to make sure that S meets safety and health requirements. Its inspection frequency is appropriate in light of the low number of workers at the site, lack of significant changes in the nature of the work and types of hazards involved, GH's knowledge of S's history of compliance and its effective safety and health efforts on this job. GH has exercised reasonable care and is not citable for this condition.

(2) *Example 7:* Employer GC contracts with Employer P to do painting work. GC has the same contract authority over P as Employer GH had in Example 6. GC has never before worked with P. GC conducts inspections that are sufficiently frequent in light of the factors listed above in (G)(3). Further, during a number of its inspections, GC finds that P has violated fall protection requirements. It points the violations out to P during each inspection but takes no further actions.

Analysis: Step 1: GC is a controlling employer since it has general supervisory authority over the site, including a contractual right of control over P. **Step 2:** GC took adequate steps to meet its obligation to discover violations. However, it failed to take reasonable steps to require P to correct hazards since it lacked a graduated system of enforcement. A citation to GC for the fall protection violations is appropriate.

(3) *Example 8:* Employer GC contracts with Sub E, an electrical subcontractor. GC has full contract authority over Sub E, as in Example 6. Sub E installs an electric panel box exposed to the weather and implements an assured equipment grounding conductor program, as required under the contract. It fails to connect a grounding wire inside the box to one of the outlets. This incomplete ground is not apparent from a visual inspection. Further, GC inspects the site with a frequency appropriate for the site in light of the factors discussed above in (G)(3). It saw the panel box but did not test the outlets to determine if they were all grounded because Sub E represents that it is doing all of the required tests on all receptacles. GC knows that Sub E has implemented an effective safety and health program. From previous experience it also knows Sub E is familiar with the applicable safety requirements and is technically competent. GC had asked Sub E if the electrical equipment is OK for use and was assured that it is.

Analysis: Step 1: GC is a controlling employer since it has general supervisory authority over the site, including a contractual right of control over Sub E. **Step 2:** GC exercised reasonable care. It had determined that Sub E had technical expertise, safety knowledge and had implemented safe work practices. It conducted inspections with appropriate frequency. It also made some basic inquiries into the safety of the electrical equipment. Under these circumstances GC was not obligated to test the outlets itself to determine if they were all grounded. It is not citable for the grounding violation.

b. *Control Established by a Combination of Other Contract Rights:* Where there is no explicit contract provision granting the right to control safety, or where the contract says the employer does *not* have such a right, an employer may still be a controlling employer. The ability of an employer to control safety in this circumstance can result from a combination of contractual rights that, together, give it broad responsibility at the site involving almost all aspects of the job. Its responsibility is broad enough so that its contractual authority necessarily involves safety. The authority to resolve disputes between subcontractors, set schedules and determine construction sequencing are particularly significant because they are likely to affect safety. (Note that citations should only be issued in this type of case after consulting with the Regional Solicitor's office).

(1) *Example 9:* Construction manager M is contractually obligated to: set schedules and construction sequencing, require subcontractors to meet contract specifications, negotiate with trades, resolve disputes between subcontractors, direct work and make purchasing decisions, which affect safety. However, the contract states that M does not have a right to require compliance with safety and health requirements. Further, Subcontractor S asks M to alter the schedule so that S would not have to start work until Subcontractor G has completed installing guardrails. M is contractually responsible for deciding whether to approve S's request.

Analysis: Step 1: Even though its contract states that M does not have authority over safety, the combination of rights actually given in the contract provides broad responsibility over the site and results in the ability of M to direct actions that necessarily affect safety. For example, M's contractual obligation to determine whether to approve S's request to alter the schedule has direct safety implications. M's decision relates directly to whether S's employees will be protected from a fall hazard. M is a controlling employer. **Step 2:** In this example, if M refused to alter the schedule, it would be citable for the fall hazard violation.

(2) ***Example 10:*** Employer ML's contractual authority is limited to reporting on subcontractors' contract compliance to owner/developer O and making contract payments. Although it reports on the extent to which the subcontractors are complying with safety and health infractions to O, ML does not exercise any control over safety at the site.

Analysis: **Step 1:** ML is not a controlling employer because these contractual rights are insufficient to confer control over the subcontractors and ML did not exercise control over safety. Reporting safety and health infractions to another entity does not, by itself (or in combination with these very limited contract rights), constitute an exercise of control over safety. **Step 2:** Since it is not a controlling employer it had no duty under the OSH Act to exercise reasonable care with respect to enforcing the subcontractors' compliance with safety; there is therefore no need to go to Step 2.

c. *Architects and Engineers:* Architects, engineers, and other entities are controlling employers only if the breadth of their involvement in a construction project is sufficient to bring them within the parameters discussed above.

(1) ***Example 11:*** Architect A contracts with owner O to prepare contract drawings and specifications, inspect the work, report to O on contract compliance, and to certify completion of work. A has no authority or means to enforce compliance, no authority to approve/reject work and does not exercise any other authority at the site, although it does call the general contractor's attention to observed hazards noted during its inspections.

Analysis: Step 1: A's responsibilities are very limited in light of the numerous other administrative responsibilities necessary to complete the project. It is little more than a supplier of architectural services and conduit of information to O. Its responsibilities are insufficient to confer control over the subcontractors and it did not exercise control over safety. The responsibilities it does have are insufficient to make it a controlling employer. Merely pointing out safety violations did not make it a controlling employer. NOTE: In a circumstance such as this it is likely that broad control over the project rests with another entity. **Step 2:** Since A is not a controlling employer it had no duty under the OSH Act to exercise reasonable care with respect to enforcing the subcontractors' compliance with safety; there is therefore no need to go to Step 2.

(2) ***Example 12:*** Engineering firm E has the same contract authority and functions as in Example 9.

Analysis: **Step 1:** Under the facts in Example 9, E would be considered a controlling employer. **Step 2:** The same type of analysis described in Example 9 for Step 2 would apply here to determine if E should be cited.

d. *Control Without Explicit Contractual Authority.* Even where an employer has no explicit contract rights with respect to safety, an employer can still be a controlling employer if, in actual practice, it exercises broad control over subcontractors at the site (see Example 9). NOTE: Citations should only be issued in this type of case after consulting with the Regional Solicitor's office.

(1) *Example 13:* Construction manager MM does not have explicit contractual authority to require subcontractors to comply with safety requirements, nor does it explicitly have broad contractual authority at the site. However, it exercises control over most aspects of the subcontractors' work anyway, including aspects that relate to safety.

Analysis: Step 1: MM would be considered a controlling employer since it exercises control over most aspects of the subcontractor's work, including safety aspects. **Step 2:** The same type of analysis on reasonable care described in the examples in (G)(5)(a) would apply to determine if a citation should be issued to this type of controlling employer.

F. *Multiple Roles*

1. *A creating, correcting or controlling employer* will often also be an exposing employer. Consider whether the employer is an exposing employer before evaluating its status with respect to these other roles.

2. *Exposing, creating and controlling employers* can also be correcting employers if they are authorized to correct the hazard.

Index